GW00726115

MATHEMATICS FOR TELECOMMUNICATIONS AND ELECTRICAL ENGINEERING

VOLUME I

GENERAL TECHNICAL SERIES

General Editor:
AIR COMMODORE J. R. MORGAN, O.B.E., (Retd.), B.Sc. (Eng.), M.I.Mech.E., F.R.Ae.S., R.A.F.
Formerly Director of Studies, Royal Air Force Technical College and Deputy Director, Educational Services, Air Ministry.

RADIO
J. D. Tucker, A.M.Brit.I.R.E. and
D. F. Wilkinson, B.Sc. (Eng.), M.I.E.E.
(*In three volumes*)

MATHEMATICS FOR TELECOMMUNICATIONS AND ELECTRICAL ENGINEERING
W. H. Grinsted, O.B.E., M.I.E.E., F.C.G.I. and
D. F. Spooner, B.Sc. (Lond.)
(*In two volumes*)

ELECTRICAL INSTALLATION TECHNOLOGY AND PRACTICE
J. O. Paddock, M.A.S.E., H.N.C., C.G.L.I. Full Tech. Cert.
R. A. W. Galvin, B.Sc., A.M.A.S.E.E., Grad.I.E.E., C.G.Cert. in Elec. Eng. Practice

ENGINEERING SCIENCE AND CALCULATIONS
W. E. Fisher, O.B.E., D.Sc., M.I.Mech.E.
(*In three volumes*)

WORKSHOP PROCESSES FOR MECHANICAL ENGINEERING TECHNICIANS
R. T. Pritchard, M.I.Prod.E., Full Tech. Cert. C.G.L.I.
(*In two volumes*)

THE PRINCIPLES OF TELECOMMUNICATIONS ENGINEERING
H. R. Harbottle, O.B.E., B.Sc. (Eng.), D.F.H., M.I.E.E. and
B. L. G. Hanman, B.Sc. (Eng.)

ELEMENTARY TELECOMMUNICATION PRACTICE
J. R. G. Smith, C.Eng., M.I.E.E.

SIMPLIFIED CALCULUS
F. L. Westwater, O.B.E., R.N. (Retd), M.A. (Edin.), M.A. (Cantab.), M.I.E.E.

ENGINEERING DRAWING AND MATERIALS
FOR MECHANICAL ENGINEERING TECHNICIANS
H. Ord, A.R.Ae.S., F.R.Econ.S.
(*In four volumes*)

ENGINEERING DRAWING FOR G.1 AND G.2 COURSES
H. Ord, A.R.Ae.S., F.R.Econ.S.

MATHEMATICS FOR TELECOMMUNICATIONS

AND ELECTRICAL ENGINEERING

VOLUME I

BY

D. F. SPOONER
B.Sc. (Lond.)

Senior Lecturer in Mathematics, Woolwich Polytechnic
Chief Examiner in Mathematics for Telecommunications
to the City and Guilds of London Institute

AND

W. H. GRINSTED
O.B.E., M.I.E.E., F.C.G.I.

Formerly Member of the City and Guilds Institute's Advisory
Committee on Telecommunications Engineering

THE ENGLISH UNIVERSITIES PRESS LTD
ST. PAUL'S HOUSE WARWICK LANE
LONDON · E.C.4

First impression 1952
Second impression 1959
Third impression 1963
Second edition 1965
Second impression 1969

SBN 340 04792 5

Printed in Great Britain for The English Universities Press, Limited, by Richard Clay (The Chaucer Press), Ltd, Bungay, Suffolk.

AUTHORS' PREFACE

THESE two volumes are intended for home study and correspondence course students as well as for class work in technical schools and colleges. They cover the ground of the Mathematics syllabus in the City and Guilds Telecommunications Technicians' Course (New Scheme, No. 49). The present volume deals with the work of the first two years of this course. It also covers all the mathematics required by Electrical and Telecommunication students about to enter the O stage of the Ordinary National Courses. Volume II covers the work of the third and fourth years of the Telecommunication Technicians' Mathematics syllabus and also completes the normal Ordinary National Certificate Syllabus.

The two volumes are also appropriate text-books for the City and Guilds Electrical Technicians' Course (No. 57); Volume I covers fully the requirements of the first two years of this course; Volume II is more than adequate for the remaining two years.

Thus together they provide class texts for both Ordinary National Certificate and City and Guilds courses for electrical engineers.

It is hoped also that these books may provide stimulating subject matter for work in the upper forms of secondary schools with a technological bias, and also be of value to those telecommunication and other engineers who desire to improve their mathematical equipment by private study, or pursue mathematics for its own sake.

A working knowledge of elementary arithmetic including decimals is assumed, also some acquaintance with the meaning of algebraic terms and symbols. On the technical side the student is expected to be learning the principles, construction and operation of the telecommunication equipment referred to in the examples and exercises.

The main purposes of the authors are briefly:

(*a*) to interest students in mathematics for its own sake, as well as for its relevance to technology;

(*b*) to help them to realise how a mathematical statement will describe and summarise the whole range of a physical relation, affording the same advantage over statements of individual facts that a master key has over individual keys;

(*c*) to give them confidence in handling mathematical tools, and in discovering for themselves their usefulness;

(*d*) to develop a sound sense of degrees of accuracy, and of the uses and limitations of approximations.

There are four important hurdles which every student must learn to surmount:

1. The realisation that his problem *can* be tackled mathematically.
2. The expression of his problem in mathematical language.
3. The mathematical operations needed.
4. The interpretation of the mathematical results in terms of the original practical problem.

Some books on mathematics are concerned mainly with the third hurdle, whereas many students never survive the first two! At the risk of trespassing in the fields of Physics and Telecommunications this present series endeavours to assist the student over all four hurdles.

D. F. S.
W. H. G.

PREFACE TO THE SECOND EDITION

SINCE this volume was first published, variations in syllabus, both in National Certificate and City and Guilds Technicians courses, have involved minor changes with each fresh impression.

In this second edition two major additions are made, which I believe will enhance the value of the book to the reader. The chapter on trigonometry has been enlarged to include the study of the scalene triangle, involving the Sine and Cosine Rules, which now appear in the Mathematics A syllabus of the City and Guilds Telecommunications course.

By courtesy of the City and Guilds of London Institute, past papers in Practical Mathematics and Mathematics A, the first and second year examinations of the Telecommunications Technicians course, are now included as an appendix to this edition.

D. F. S.

CONTENTS

		PAGE
	THE USE OF S.I. UNITS	ix
	INTRODUCTION TO VOLUME I	xiii
CHAP.		
1.	FACTORS, INDICES, AND THE SHORTHAND OF ALGEBRA	1
	Simple formulæ—Meaning and use of indices—Numbers in standard form.	
	EXERCISE 1(*a*)	4
	EXERCISE 1(*b*)	20
2.	THE LANGUAGE OF GRAPHS	23
	EXERCISE 2(*a*)	31
	EXERCISE 2(*b*)	42
3.	LOGARITHMS AND THE SLIDE-RULE	45
	EXERCISE 3(*a*)	53
	EXERCISE 3(*b*)	57
	EXERCISE 3(*c*)	67
4.	THE DISCOVERY AND MANIPULATION OF FORMULÆ	70
	Use of logarithms in evaluating formulæ—Simple algebraic factors.	
	EXERCISE 4(*a*)	91
	EXERCISE 4(*b*)	103
	REVISION PAPERS A	107
5.	EQUATIONS	112
	Simple equations in one unknown—Fractions and brackets—Simple problems leading to linear equations—Simultaneous linear equations—Problems.	
	EXERCISE 5(*a*)	122
	EXERCISE 5(*b*)	132
6.	THE GRAPHS OF SOME MATHEMATICAL FUNCTIONS	135
	Graphs of the simpler mathematical functions—Linear and quadratic functions—Simple applications—The inverse proportion function $y = a/x$.	
	EXERCISE 6(*a*)	145
	EXERCISE 6(*b*)	161
	REVISION PAPERS B	165

CHAP. PAGE

7. "TRIGEOMETRY"—THE SPATIAL ASPECT OF MATHEMATICS 171

Lines and angles—Simple properties of triangles and polygons—Congruence and similarity—Pythagoras—Circle properties, including tangents—Methods of measuring areas.

EXERCISE 7(a) 189
EXERCISE 7(b) 206
EXERCISE 7(c) 234

8. AN INTRODUCTION TO TRIGONOMETRY 240

Sine, cosine and tangent—Examples involving right-angled triangles—Angles greater than 90°—The quadrant rules—Analogy with production of A.C. voltages—Area of any triangle—Sine and Cosine Rules.

EXERCISE 8(a) 253
EXERCISE 8(b) 271
EXERCISE 8(c) 274e

9. AN INTRODUCTION TO MECHANICS 275

Parallelogram of vectors—Triangle of forces—Moments—Problems involving poled lines, levers—Hooke's Law and elasticity—Stress, strain, and strength (simple treatment only).

EXERCISE 9 310
REVISION PAPERS C 316

ANSWERS TO THE EXERCISES 322

APPENDIX—PAST EXAMINATION PAPERS . . . 336

Practical Mathematics (1st Year C. & G.) 1962, 1963, 1964—Mathematics A (2nd Year C. & G.) 1962, 1963, 1964—Answers to the above.

LOGARITHM AND TRIGONOMETRIC TABLES . . . 353

INDEX 364

THE USE OF SI UNITS

THE name Système International d'Unités (International System of Units) with its internationally recognised abbreviation SI is given to the metric system now accepted as standard in the United Kingdom for scientific and technological measurements. For full details the reader is referred to the British Standards Institution booklet *The Use of SI Units* published in January 1969.

For the electrical engineer this is substantially the MKS system based on the metre, kilogramme and second as measures of length, mass (not weight!) and time. No change is involved in electrical measurements. An innovation in some fields is the adoption of the Newton (N) as the unit of force, defined as the force required to give a mass of one kilogramme an acceleration of one metre per second squared. The practical (gravitational) units, the pound weight (lb. wt., more recently lbf.) and the gramme weight, are no longer acceptable; although doubtless they will persist in everyday use for some time to come.

The content of the present volume is very little affected by the adoption of SI units, concerned as it is with mathematical operations, in which the units used are of little importance. A summary of the recommended units, and some useful relations between SI units and others in common use, is given below.

Table of SI Units

Quantity	*SI Unit*	*Symbol*	*Equivalents in Common Use*
time	second	s	
length	metre	m	39·37 in., 3·28 ft.
mass	kilogramme	kg	2·20 lb.
force	newton	N	0·2248 lbf., 0·1019 kgf. or 101·9 gm. wt.
pressure, stress	newton per sq. metre	N/m^2	145·0 lbf./in.^{2}
energy, work	joule	J	0·7374 ft.lbf., 0·2388 calories

Quantity	*SI Unit*	*Symbol*	*Equivalents in Common Use*
power	watt	W	$1{\cdot}341 \times 10^{-3}$ h.p.
current	ampere	A	
charge (quantity of electricity)	coulomb	C	
electric potential	volt	V	
capacitance	farad	F	
inductance	henry	H	
resistance, reactance, impedance	ohm	Ω	
conductance, susceptance, admittance	mho	$1/\Omega$	
frequency	hertz	Hz	preferred name for cycle per second (c/s)

In all the above multiples and sub-multiples commonly used are:

Factor	10^{6}	10^{3}	10^{-2}	10^{-3}	10^{-6}	10^{-12}
Prefix	mega-	kilo-	centi-	milli-	micro-	pico-
Symbol	M	k	C	m	u	P

INTRODUCTION TO VOLUME I

THE rapid technological developments of the last half-century have brought fresh challenges to the mathematical scientist as well as to the engineer. Each fresh advance in technical skill and inventiveness has been accompanied by corresponding refinements in the mathematical skills which interpret their results and suggest new pathways of research. The harnessing of electrical and magnetic energy, for instance, has become increasingly fruitful as the laws governing their behaviour have been crystallised into mathematical formulæ, which are sometimes of a startling simplicity. We have a beautiful example of such a formula in Ohm's Law, expressed succinctly as

$$V = IR$$

connecting the voltage V volts applied across a resistance R ohms with the current I amperes flowing through it.

In the words of a distinguished American engineer, Professor W. L. Everitt : *

> "Much of the advance of the electrical industry has been due to the simplicity of Ohm's Law and the wide range over which it operates. It is probably because the fundamental law is linear, and therefore fairly simple, that mathematical analysis has been relied on more, and carried further and produced greater results in electrical than in any other branch of engineering."

As an engineer, the student is not so much interested in abstract mathematical expressions as in their application to physical relations. The simplicity of many of the fundamental laws and of the mathematical formulæ corresponding to them enables him to make an early start in their application. It is appropriate, therefore, that the early chapters of this book should deal with the meaning of simple algebraic

* *Communication Engineering*, W. L. Everitt. (McGraw-Hill, 1937.)

expressions and practical formulæ (Chapter 1), with the tools for their evaluation, such as logarithms and the slide rule (Chapter 3) and with their practical manipulation (Chapter 4).

The construction and interpretation of graphs of practical data receive early treatment in Chapter 2, while in Chapter 5 simple equations are introduced. In Chapter 6 the relation between the graphical and the algebraic forms of the same mathematical statement is studied in simple cases.

Chapter 7 introduces the student informally to the rudiments of Geometry, and in the concluding chapters are laid the foundations of Trigonometry (including the idea of a "sine wave") and of Mechanics, together with an elementary approach to the strength of materials (common ground for any branch of engineering).

Throughout the book examples and exercises are chosen wherever possible for their application in some practical field (not confined exclusively to telecommunications) and "drill" examples have been reduced to a minimum. It is hoped that the student will perform the experiments suggested, and invent others for himself. Three groups of graded revision papers are provided, dealing with the work covered in Chapters 1–4, 1–6, and 1–9 respectively. Harder sections in the text are marked with an asterisk, and may be "skipped" at a first reading.

The authors wish to acknowledge their indebtedness to Mr. W. E. Sawdy, B.Sc., for preparing the diagrams for publication, and to Mrs. A. L. Spooner for typing the manuscript.

CHAPTER 1

FACTORS, INDICES, AND THE SHORTHAND OF ALGEBRA

1.1. Prime Factors

In Arithmetic it is often useful to express numbers in their prime factors:

e.g., $$42 = 2 \times 21 = 2 \times 3 \times 7$$

where 2, 3, and 7 are the prime-number factors of 42; *i.e.*, they cannot be broken down into simpler factors.

In some numbers prime factors are repeated:

e.g., $$\begin{aligned} 96 &= 2 \times 48 \\ &= 2 \times 2 \times 24 \\ &= 2 \times 2 \times 2 \times 12 \\ &= 2 \times 2 \times 2 \times 2 \times 6 \\ &= 2 \times 2 \times 2 \times 2 \times 2 \times 3 \end{aligned}$$

We usually use a shorthand form of this last statement, and write

$$96 = 2^5 \times 3$$

where the small " index " 5 simply " indicates " the number of times the factor 2 occurs. The expression 2^5 is called the " fifth power of two ", or just " two to the fifth ".

Similarly, the student may check for himself that

$$648 = 2^3 \times 3^4,$$

and $1008 = 2^4 \times 3^2 \times 7$ (2^3 is called " two cubed " and 3^2 is known as " three squared ").

1.2. Powers of 10

As our number system is based on TEN, we frequently have to deal with the powers of TEN. For example, in expressing 300,000 in its most convenient factors:

$$300{,}000 = 3 \times 10 \times 10 \times 10 \times 10 \times 10 = 3 \times 10^5$$

We shall be handling powers of ten very extensively in Chapter 3 (see also para. 1.13).

1.3. Powers in Algebra

We use the same index notation when we use letters instead of numbers, *e.g.*, $a \times a \times a$ we write as a^3 (this could represent the volume in cubic feet of a cube of side a ft.).

Similarly, $x \times x \times x \times y \times y = x^3 \times y^2$, which is usually written x^3y^2 for short.

This generalised arithmetic is the beginning of algebra.

1.4. H.C.F. and L.C.M.

The Highest Common Factor (H.C.F.) of a set of numbers is important, and is conveniently found from their prime factors, *e.g.*, the H.C.F. of 648 and 1008 is the largest number which divides exactly into both 648 and 1008 (*i.e.*, is a factor of both).

But $648 = 2^3 \times 3^4$ and $1008 = 2^4 \times 3^2 \times 7$, so that the H.C.F. will contain the third power of 2 and the second power of 3. There is no factor 7 in 648, so this prime number will not occur in the H.C.F.

Hence the H.C.F. of 648 and 1008

$$= 2^3 \times 3^2 = 8 \times 9 = 72$$

Note that this H.C.F. goes exactly nine times into 648, and 14 times into 1008.

The Least Common Multiple (L.C.M.) of the same two numbers is the smallest number into which 648 and 1008 will each divide exactly.

In this case the L.C.M. must contain the factor 2 four times, the factor 3 four times, and the factor 7 once in order to "contain" 648 and 1008 exactly.

Hence the L.C.M. of 648 and 1008

$$= 2^4 \times 3^4 \times 7$$
$$= 1008 \times 3^2 \quad (\text{since } 1008 = 2^4 \times 3^2 \times 7)$$
$$= 9072$$

Note that 648 goes exactly 2×7, *i.e.*, 14 times, and 1008 exactly 3^2, *i.e.*, 9 times into this L.C.M.

In subtracting or adding fractions we need to express them

with the same denominator. The best (lowest) common denominator is simply the L.C.M. of the denominators of the separate fractions.

e.g.,
$$\frac{23}{648} - \frac{11}{1008} = \frac{23 \times 14}{9072} - \frac{11 \times 9}{9072}$$
$$= \frac{322 - 99}{9072} = \frac{223}{9072}$$

Note: There are other arithmetical ways of obtaining H.C.F.s and L.C.M.s, but the above processes, using prime factors, are immediately applicable to algebraic expressions too.

For example, for the expressions a^3bc and a^2b^4d:

$$\text{H.C.F.} = a^2b, \quad \text{L.C.M.} = a^3b^4cd.$$

1.5. Some Practical Considerations

(*a*) When small numbers only are involved, it is often unnecessary to put each into its prime factors in order to calculate their H.C.F. or L.C.M. The H.C.F. is then clear by inspection, *e.g.*, the H.C.F. of 24 and 32 is clearly 8. The L.C.M. is best found by taking the largest number concerned and finding by trial the smallest number by which this must be multiplied to include each of the other numbers as a complete factor, *e.g.*, consider the numbers 24, 16, 18,

24 is the largest number
48 (2 × 24) contains 16 but not 18
72 (3 × 24) contains 18 but not 16
144 (6 × 24) contains 16 and 18

and is clearly the smallest number which contains 24, 16, 18, *i.e.*, the L.C.M. of 24, 16, 18 is 144.

The above working should be done mentally; but you are advised at this stage to check the result by putting each number in prime factors.

(*b*) In many calculations it is worth while to leave intermediate pieces of working in factors (not necessarily prime factors) rather than to work each out separately. This often saves labour and tends to produce more accurate work.

Example 1(*a*). *A drum carries 7 miles of wire, the wire itself weighing 3 cwt. How many feet of wire are there to the pound?*

$$3 \text{ cwt.} = 3 \times 112 \text{ lb.}$$
$$7 \text{ miles} = 7 \times 1760 \times 3 \text{ ft.}$$

∴ It follows that 1 lb. of wire has a length

$$\frac{\overset{110}{\overset{440}{\cancel{7} \times \cancel{1760} \times \cancel{3}}}}{\underset{\underset{4}{16}}{\cancel{3} \times \cancel{112}}} \text{ ft.}$$

$$= 110 \text{ ft.}$$

i.e., 1 lb. of this wire measures 110 ft.

Note: If we multiply out completely at each stage, we obtain $\frac{36960}{336}$ ft. to be simplified. Clearly much more formidable!

(*c*) The following rules of divisibility help in finding numerical factors:

If the last digit is divisible by 2 the number has a factor 2.
If the last 2 digits are divisible by 4 the number has a factor 4.
If the last digit is 5 or 0 the number has a factor 5 or 10.
If the sum of the digits is divisible by 3 or 9, then the number itself is divisible by 3 or 9.

EXERCISE 1(*a*)

Express in prime factors the following:

1. 231. 2. 243. 3. 216. 4. 294. 5. 1683. 6. 3250.
7. 40 gallons expressed in pints. 8. 1 mile reduced to inches.
9. 6 sq. miles in acres. 10. $2\frac{1}{4}$ cu. yd. in cu. in.

Express each of the following factorised numbers as a single number:

11. $2^4 \times 3^5$. 12. $3 \times 5^2 \times 7^3$. 13. $2^{11} \times 11^2$.

Work out the L.C.M. and the H.C.F. of each of the following sets of numbers:

14. 72, 42. 15. 24, 36, 32. 16. 128, 144.
17. 56, 63, 84. 18. 231, 294. 19. 24, 33, 36, 72.
20. 96, 112, 192.

Evaluate the following:

21. $\frac{4}{21} + \frac{3}{28}$. 22. $\frac{19}{24} - \frac{5}{56}$. 23. $6\frac{5}{12} + 3\frac{7}{16}$.
24. $3\frac{11}{18} + 4\frac{32}{45}$. 25. $4\frac{5}{14} - 2\frac{7}{10}$. 26. $3\frac{7}{12} - 5\frac{11}{15} + 4\frac{5}{18}$.

Arrange the following sets of fractions in descending order of magnitude; in each case express with a common denominator first:

27. $\frac{5}{12}, \frac{6}{15}, \frac{7}{18}, \frac{9}{20}$. 28. $\frac{17}{36}, \frac{15}{32}, \frac{19}{42}$. 29. $\frac{22}{63}, \frac{19}{56}, \frac{18}{49}$.

(Check by putting as decimal fractions.)

30. Convert 2 tons to pounds, stating your result in prime factors. If a certain kind of steel joist is delivered in lengths each weighing 35 lb., how many lengths will there be in a delivery of 2 tons?

31. A sheet of material 10 ft. long and 3 ft. 9 in. wide is to be cut into squares without leaving any waste. What is the largest size these squares may be?

32. The details of a selector rack include pieces of angle-iron of the same cross-section whose lengths are $16\frac{1}{2}$, $27\frac{1}{2}$, and 33 in. respectively. What is the shortest length in which the angle-iron may be ordered if it is to be cut up into pieces of any one of these three sizes without any waste?

33. Two gear-wheels have 16 and 36 teeth respectively. If they start turning with two particular teeth in mesh, how few revolutions will each wheel perform before the same two teeth are in mesh again?

34. When three resistances of 8, 12, and 18 ohms are connected in parallel their combined resistance is $\dfrac{1}{\frac{1}{8}+\frac{1}{12}+\frac{1}{18}}$ ohms. Evaluate this resistance.

(The denominator is the total "conductance" in mhos—it is better to work this out first.)

35. Three lighthouses are seen to flash simultaneously. If one flashes every 4 sec., another every 6 sec., and the third every 10 sec., prove that they all three flash together at exactly 1-min. intervals.

36. Four motor-cycles circle a track in 1 min. 24 sec., 1 min. 30 sec., 1 min. 36 sec., and 1 min. 40 sec. respectively. If they were to start level and maintain these times, how long after will they again be all four abreast at the starting-point?

1.6. Simple Formulæ—The Shorthand of Algebra

Algebra begins as the attempt to generalise statements in arithmetic, and express these generalisations in a concise "shorthand" statement.

For example, we know from experience that the area of a rectangular sheet of metal 5 ft. long and 3 ft. wide is 5×3, *i.e.*, 15 sq. ft. If pressed for an explanation, we would point out that the sheet may be divided into 3 strips, each containing 5 unit squares, making 15 unit squares in all. (See Fig. 1.)

It is but a short step from this particular observation to a generalised statement that the area of a rectangular sheet in

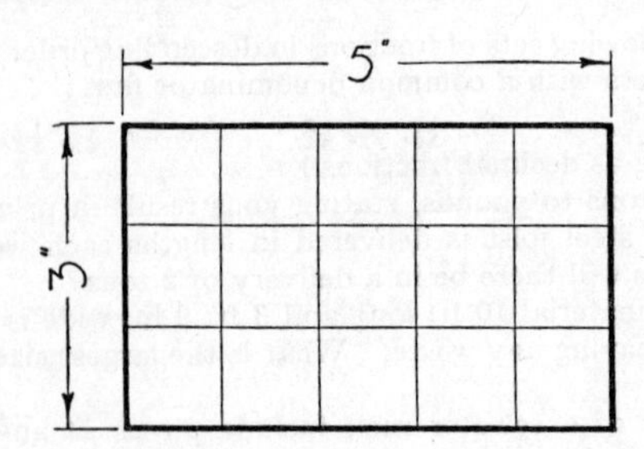

FIG. 1.

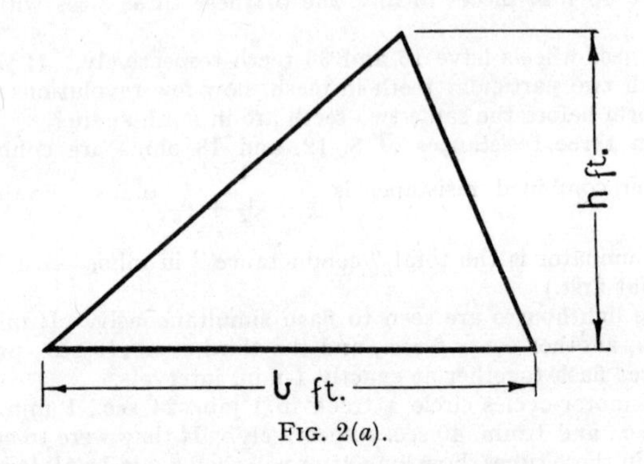

FIG. 2(*a*).

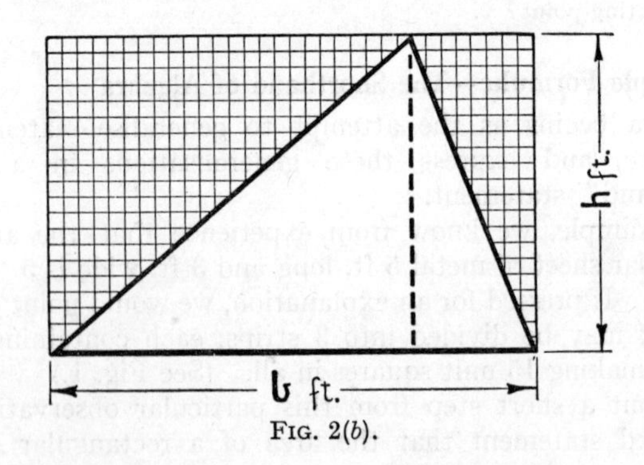

FIG. 2(*b*).

square feet is obtained by multiplying the number of feet in its length by the number of feet in its breadth. In terser but less exact phraseology: " area of rectangle equals length times breadth ". This is clumsy jargon—the invention of Algebra reduces this to a concise " formula ": $A = l \times b$, where A sq. ft. is the area of rectangle l ft. by b ft.

It is customary to omit the multiplication sign when letters are multiplied, and write $A = lb$ as the final shorthand statement or formula. Equally well we could write: $A = bl$.

A similar rule for finding the area of a triangle is " half the base times the perpendicular height " (*i.e.*, half the area of a rectangle with sides equal to the base and height of the triangle).

Referring to Figs. 2(*a*) and (*b*), a formula for the area, A sq. ft., of the triangle is thus:

$$A = \frac{bh}{2}$$

Note in Fig. 2(*b*) that the dotted " height " of the triangle divides it into two small triangles, each equal to half the area of a small rectangle. The whole triangle is thus equal to half the whole rectangle, whose area is bh sq. ft.

1.7. Historical Note

It is interesting to note that the earliest use of a formula arose from the problems of land measurement in the Babylonian and Egyptian empires of antiquity. Elsewhere nomad peoples only needed to count, for their wealth was measured by head of cattle, sheep, or camels. But in the civilisations in the Euphrates basin and the Nile Delta arose a new kind of wealth, the ownership of land. The measurement of triangles and rectangles for such civilisations, in which rich irrigated land was precious, became a practical necessity. The temple priests of Egypt were the first practical geometers,* and were occupied with the practical problems of mensuration and surveying rather than with general theorems. The earliest reference to their work is contained in a papyrus in the British Museum written by Ahmes about 1700 B.C.

* " Geo-metry " means earth measurement.

1.8. Other Mensuration Formulæ

The Circle

The fundamental measurements for a circle are its diameter (d in.), its radius (r in.), and its circumference (c in.) (Fig. 3).

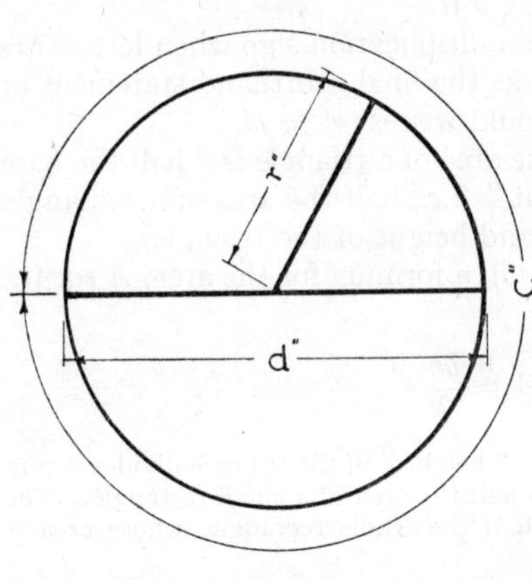

Fig. 3.

In any circle the circumference is approximately $3\frac{1}{7}$ times the diameter (more accurately 3·14159 times). In other words, the ratio of the circumference to the diameter is about $3\frac{1}{7}$. This ratio is generally denoted by π (the Greek letter " pi "). The value of π as determined by measurement is, of course, only approximate. Industrious mathematicians have found ways of calculating it to as many places of decimals as may be wished, but it never " comes out ". In this it resembles, for instance, the square root of two. Such numbers are called " incommensurable ".

In the language of the formula, we may write

$$c = \pi d \quad \text{or} \quad \pi = \frac{c}{d}$$

or, as the diameter is exactly twice the radius

$$c = 2\pi r$$

Area of the Circle

The formula for the area of a circle can be demonstrated as follows:

Draw a large circle on paper (*e.g.*, 4 in. in radius) and divide it into a number of equal parts or " sectors " (Fig. 4(*a*)). Cut out these sectors, and paste them " sardine-fashion " on to a sheet of card, thus forming a crude rectangle (Fig. 4(*b*)).

The width of this " rectangle " is very nearly the radius of the

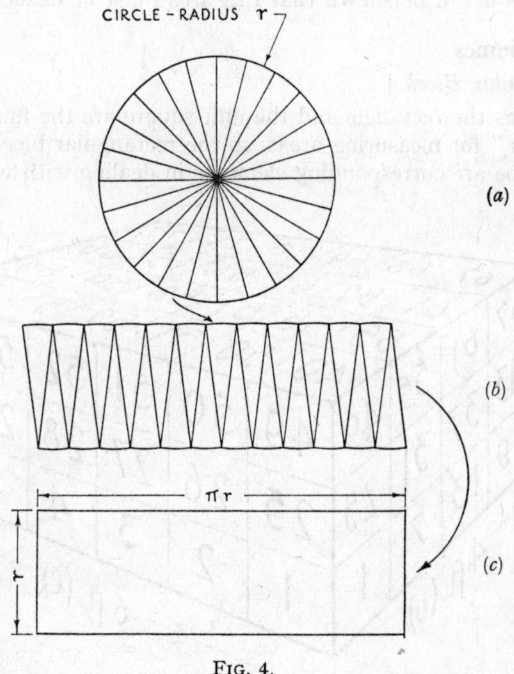

FIG. 4.

circle (r in.), and its length very nearly half the circle's circumference (πr in.).

The larger the number of sectors used, the closer the resulting figure will approach the true rectangle of Fig. 4(*c*). We therefore assume that the area (A sq. in.) of the circle is exactly equal to the area of this true rectangle.

i.e., $$A = \pi r \times r$$

But we write $r \times r$ as r^2

$$\therefore \quad A = \pi r^2$$

which is the familiar rule for calculating the area of a circle.

(The above is not a formal proof, but by the methods of the Calculus it can be shown that this area must be exactly πr^2.)

1.9. Volumes

Rectangular Block

Just as the rectangle and the unit square are the fundamental " bricks " for measuring areas, so the rectangular block and the unit cube are corresponding elements in dealing with volumes.

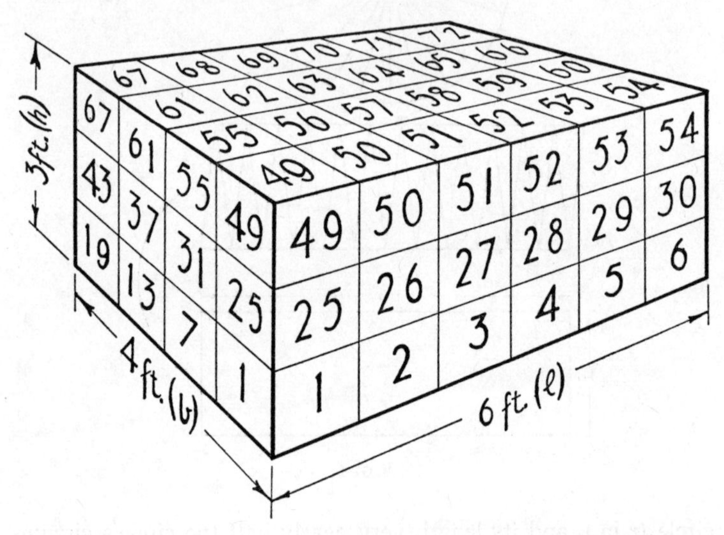

Fig. 5.

Consider a rectangular block 6 ft. long, 4 ft. broad, and 3 ft. high (Fig. 5). Imagine it sliced into 3 horizontal layers each 1 ft. thick, and each slab divided into unit cubes.

The bottom layer clearly contains 6×4, *i.e.*, 24 unit cubes, numbered in 4 rows of 6 from 1 to 24; similarly, the middle and top rows are consecutively numbered 25 to 48 and 49 to 72 respectively.

The total volume is clearly $6 \times 4 \times 3$ cu. ft. (3 layers each containing 4 rows, 6 cubes to each row).

Generalising, the volume, V cu. ft., of a rectangular block whose dimensions are l ft. by b ft. by h ft. is given by the formula

$$V = lbh \text{ (" Length} \times \text{Breadth} \times \text{Height ")}$$

But note that bh sq. ft. is the cross-sectional area A sq. ft. of the block. So we see that the volume can equally well be regarded as " length $\times$ area of cross-section " or $V = lA$. Again, as lb sq. ft. is the base area B sq. ft., the volume can also be expressed as " base area $\times$ height ", *i.e.*, $V = Bh$.

The Prism and the Cylinder

We see here a little of the potency of a formula. By a slight shift of emphasis, we can use a formula to interpret facts in

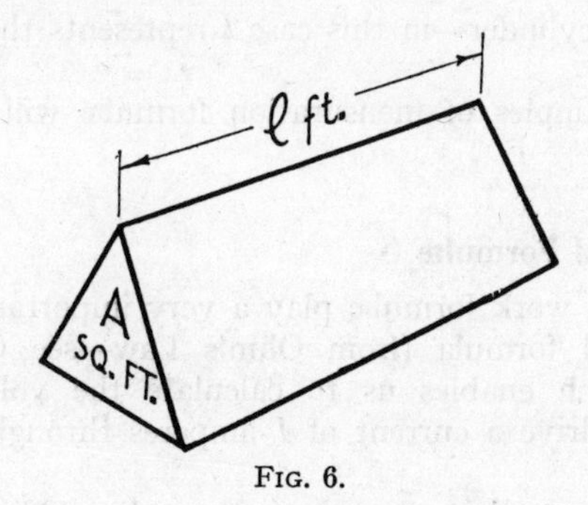

FIG. 6.

different ways and suggest a different viewpoint.

In this instance, examination of a simple formula gives us the key to finding the volume of a prism, *i.e.*, a body with uniform cross-section. This cross-section may be any shape—a triangle for example (Fig. 6) or a circle (Fig. 7), which gives us the ordinary cylinder.

In all such figures, the volume V cu. ft. will be $V = lA$, where l ft. is the length, and A sq. ft. the cross-sectional area. There are in fact l slices of unit thickness, each of which contains A cu. ft.

Thus for the cylinder $A = \pi r^2$ (para. 1.8) and hence $V = l \times \pi r^2$, or we usually write $V = \pi r^2 l$, for the order of multiplication cannot affect the result. Remember too that a flat

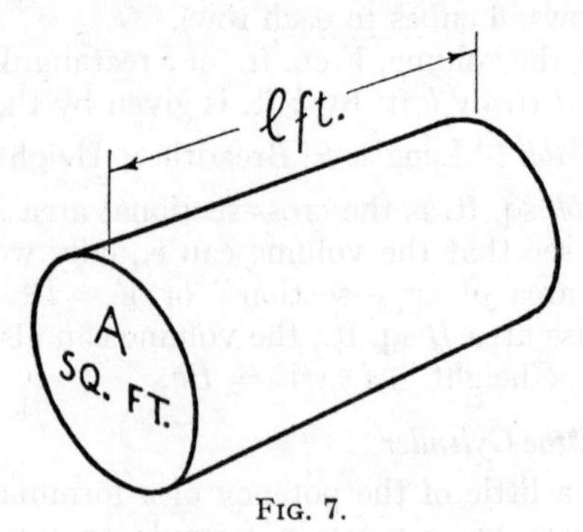

FIG. 7.

disc is also a cylinder—in this case l represents the *thickness* of the disc.

Further examples of mensuration formulæ will be found in Chapter 4.

1.10. Electrical Formulæ

In electrical work formulæ play a very important part. The most essential formula (from Ohm's Law—see Chapter 4) is $V = IR$, which enables us to calculate the voltage V volts necessary to drive a current of I amperes through a resistance R ohms.

For example, to drive 3 amperes through a 400-ohm resistor a supply voltage of 3×400, *i.e.*, 1200 volts would be required.

Many communication engineers refer colloquially to "IR drops" around a circuit when they are considering drops in voltage (a particularly useful idea in estimating bias voltages in valve circuits).

The implications of Ohm's Law will be treated more fully in Chapter 4.

1.11. The Index Laws

At the beginning of this chapter we introduced the ideas of indices and powers. Before we can use these ideas in working formulæ we need to know the rules for their use—how they behave when multiplied or divided, and so on. We will begin with multiplication.

Multiplication of Powers

E.g., to multiply 10^3 by 10^4.

In full $10^3 \times 10^4 = 10 \times 10 \times 10 \quad \times 10 \times 10 \times 10 \times 10$

$$= 10^7$$

We started with 3 factors (each 10) multiplied together and then multiplied by 4 more factors, making now 7 factors in all. We note that multiplying the powers involves adding the indices $(3 + 4 = 7)$.

This can clearly be generalised as

$$10^p \times 10^q = 10^{p+q}$$

where p and q are indices.

Division of Powers

Similarly, if we divide 10^7 by 10^4 we clearly obtain 10^3,

for $$\frac{10^7}{10^4} = \frac{10 \times 10 \times 10 \times 10 \times 10 \times 10 \times 10}{10 \times 10 \times 10 \times 10}$$

$$= 10 \times 10 \times 10$$

$$= 10^3$$

Thus dividing one power by another power involves subtracting the indices $(7 - 4 = 3)$.

In general, $$\frac{10^a}{10^b} = 10^{a-b}$$

where a and b are indices.

Raising to a Power

Note that $8^2 = 8 \times 8 = 64$

But $8 = 2 \times 2 \times 2$ and $64 = 8 \times 8$

$$= 2^3 \qquad\qquad = 2 \times 2 \times 2 \times 2 \times 2 \times 2$$

$$= 2^6$$

$$\therefore \quad (2^3)^2 = 2^6$$

Thus squaring a power involves doubling the index.

Also

$$(2^2)^4 = 4 \times 4 \times 4 \times 4$$
$$= 2 \times 2 \times 2 \times 2 \times 2 \times 2 \times 2 \times 2$$
$$= 2^8$$

and generally

$$(2^p)^q = 2^{pq}$$

Summing up—raising a power to a further power involves multiplying the index by that further power.

Taking the Square Root

As $8^2 = 64$, we can equally well say that 8 is the square root of 64,

or we write $\sqrt{64} = 8$

But $64 = 2^6$ and $8 = 2^3$

$$\therefore \quad \sqrt{2^6} = 2^3$$

So taking the square root of a power involves dividing the index by two. (Similarly, taking the cube root involves division by three.)

1.12. Negative Indices

Can we give a meaning to a number raised to a negative power, *e.g.*, 2^{-3}?

Consider 2^5 multiplied by 2^{-3}. Then if the ordinary laws of indices apply we should add 5 to -3. The sum total effect of "5 up" and "3 down" is clearly "2 up".

Hence, if the laws of indices apply

$$2^5 \times 2^{-3} = 2^2$$

$$i.e., \; 32 \times 2^{-3} = 4$$

Thus 2^{-3} is the number which multiplies 32 to give the result 4.

Comparing with $32 \times \frac{1}{8} = 4$ we see that 2^{-3} has the same effect in multiplying as $\frac{1}{8}$ or $\frac{1}{2^3}$.

It is reasonable then to use the notation 2^{-3} for the fraction $\frac{1}{2^3}$ (the " reciprocal " of 2^3).

Similarly 10^{-4} can be interpreted as $\frac{1}{10^4}$, and so on.

1.13. Applications of Indices in Technical Computations

In scientific and technological work we often handle very large or very small quantities.

For example, the distance of the earth from the sun is about 92,900,000 miles. This can be more neatly written as 92·9 million miles or, using indices, as $92{\cdot}9 \times 10^6$ miles ($10^6 = 1$ million). For some purposes we may prefer to put this large number in " standard form " as $9{\cdot}29 \times 10^7$ miles (see Chapter 3).

In contrast, when we investigate the smallest current which will produce an audible sound in a telephone receiver, we discover it to be from 0·000000003 to 0·00000001 ampere. These facts are much easier to handle when this range is written as 3×10^{-9} $\left(\text{or } \frac{3}{10^9}\right)$ to 1×10^{-8} $\left(\text{or } \frac{1}{10^8}\right)$ ampere.

This index notation is particularly valuable when we have to substitute very large or very small values in a formula. For example, in the formula $V = IR$ the units concerned are volts, amperes, and ohms. We frequently deal with tiny fractions of an ampere and millions of ohms in telecommunications practice; the use of indices helps to reduce the risk of errors in computation, for it renders unnecessary the continual writing and counting of rows of noughts.

1.14. Use of Prefixes in the Decimal System

For everyday discussion the following prefixes are used with all scientific units to denote large and small values of quantities:

Large quantities (Greek prefixes).		Small quantities (Latin prefixes).	
Deka	$\times 10$	deci	$\times 10^{-1}$
Hecto	$\times 10^2$	centi	$\times 10^{-2}$
Kilo	$\times 10^3$	milli	$\times 10^{-3}$
Mega	$\times 10^6$	micro	$\times 10^{-6}$
		micro-micro (or pico)	$\times 10^{-12}$

Illustrations of these Prefixes in Use

Description.	Value in scientific units.	In basic units.	Abbreviation.
Length of 1 inch	2·54 centimetres	$2{\cdot}54 \times 10^{-2}$ metres	2·54 cm.
Distance of 1 mile	1·61 kilometres	$1{\cdot}61 \times 10^{3}$ metres	1·61 km.
Rating of electric fire	2 kilowatts	2×10^{3} watts	2 kW.
Operating current of telegraph relay	8 milliamperes	8×10^{-3} amperes	8 mA.
Typical radio frequency	2·1 megacycles per sec.	$2{\cdot}1 \times 10^{6}$ cycles per sec.	2·1 Mc/s.
Insulation resistance of a cable pair	40 megohms	40×10^{6} ohms	40 MΩ
Capacitor in C.B. telephone (capacitance)	2 microfarads	2×10^{-6} farads	2 μF.
Air-spaced capacitor (capacitance)	60 micromicrofarads (or 60 picofarads)	60×10^{-12} farads	60 μμF. (60 pF.)

The student should keep his eyes open for the many other uses of these prefixes.

Note that μ ("mu"), the Greek letter "m", is used as the abbreviation for micro, for the Roman "m" is already used for milli.

1.15. Examples of the Use of Large and Small Units

Example 1(*b*). *Find the weight of 60 metres run of steel rod of rectangular cross-section 8 by 12 cm., if steel weighs 7·8 gm. per cubic centimetre.*

We must clearly obtain the volume of steel first, expressed in cubic centimetres (c.c.). This is the volume of a long rectangular block, whose dimensions must be expressed in centimetres as 6000 by 8 by 12,

i.e., volume of steel $= 6000 \times 8 \times 12$ c.c. $(V = lbh)$

1 c.c. weighs 7·8 gm.

∴ the whole "block" weighs $7{\cdot}8 \times 6000 \times 8 \times 12$ gm.

$= 7{\cdot}8 \times 576{,}000$ gm.

$= 4{,}492{,}800$ gm.

The gram is too small a unit for everyday use, so it is usual to express this result in kilograms (1 kg. is about $2\frac{1}{5}$ lb.) as 4492·8 kg., or we could use the index notation and give the total weight as $4{\cdot}5 \times 10^6$ gm., correct to the first two " significant figures ". (The " megagram " is not used in practice.)

It is worth noting that the density of the steel is probably not exactly 7·8 gm. per c.c., and to include more than 2 significant figures in the answer is a pretence, since digits beyond the first two figures are almost certainly incorrect. The importance of significant figures is discussed at length in para. 1.16.

Example 1(*c*). *A current of* 15 *microamps flows through a resistance of* 3 *megohms. Calculate the voltage across the resistance necessary to maintain this current.*

The Ohm's Law formula $V = IR$ presupposes the quantities V volts, I amperes, R ohms.

In this case $I = 15 \times 10^{-6}$ (15 millionths)

$$= \frac{15}{10^6}$$

and $R = 3 \times 10^6$ (3 million)

applying the formula $V = IR$

$$= \frac{15}{10^6} \times 3 \times 10^6$$

$= 45$ (for clearly 10^6 " cancels ")

∴ 45 volts are required to maintain this current.

Example 1(*d*). *The resonant frequency* f *cycles per second* (*c/s*) *of a tuned circuit containing inductance* L *henrys and capacitance* C *farads is given by the formula* $f = \dfrac{1}{2\pi\sqrt{LC}}$.

If the inductance is 4 *millihenrys* (4 *mH.*) *and the capacitance* 90 *micromicrofarads* (90 μμ*F.*), *calculate the resonant frequency of the circuit.*

This appears formidable, involving as it does the square root of the quantity LC. But if we work step by step, the calculation is quite simple. We will evaluate LC first.

Now $L = 4 \times 10^{-3} = \dfrac{4}{10^3}$ (in henrys)

and $C = 90 \times 10^{-12} = \dfrac{90}{10^{12}}$ (in farads)

so that
$$LC = \frac{4}{10^3} \times \frac{90}{10^{12}}$$
$$= \frac{4 \times 90}{10^3 \times 10^{12}}$$
$$= \frac{360}{10^{15}}$$
$$= \frac{36}{10^{14}} \text{ (dividing numerator and denominator by 10)}$$

But we need $\sqrt{LC}$. Fortunately 36 is a "perfect square", *i.e.*, $\sqrt{36} = 6$. By the index laws

$$\sqrt{10^{14}} = 10^7 \quad \text{(halving the index)}$$

$$\therefore \quad \sqrt{LC} = \frac{6}{10^7} \quad \left(\text{to check this } \frac{6}{10^7} \times \frac{6}{10^7} = \frac{36}{10^{14}}\right)$$

Substituting in the full formula $\left(\pi = \frac{22}{7}\right)$

$$f = \frac{1}{2\pi\sqrt{LC}}$$
$$= \frac{7 \times 10^7}{44 \times 6}$$
$$= \frac{70}{44 \times 6} \times 10^6$$
$$\simeq 0{\cdot}27 \times 10^6 \quad (\simeq \text{ means " approximately equal to "})$$

The resonant frequency is thus 0·27 megacycles per second, or 270,000 c/s. (Accurate to 2 significant figures.)

Note the technique of calculating the awkward part of a formula first.

1.16. Note on Significant Figures

In many practical calculations we have to decide how many figures " count " or are " significant ". This may depend on the accuracy of measurement of the quantities involved or on the tools of calculation themselves (slide-rule, log tables, etc.).

For instance, if a capacitor has a rated capacitance of 0·12 μF., this capacitance may well vary between 0·122 and 0·118 μF. (*i.e.*, $\pm$ 2% deviation approximately). Manufacturers generally specify the percentage tolerance allowed, and on a mass-produced component $\pm$ 5% is not uncommon. With such a component, any calculation based on exactly 0·12 μF. would clearly give a result true to not more than 2-figure accuracy. Thus a calculated frequency of 2300 c/s achieved by using such a capacitor in a tuned circuit might in reality be 2315 c/s or 2279 c/s using two samples of the same rated component. Such variations may be covered by admitting that the frequency we get is 2300 c/s correct to 2 significant figures only.

In radio practice such variations are always anticipated, and in most cases " trimming " capacitors are used to " trim " the frequency to the required value. This is much cheaper than manufacturing very highly accurate components.

All physical measurements are subject to errors, both those of human judgement and those due to the limitations of a machine. With a sensitive chemical balance, for example, we may weigh with 5-figure accuracy (*e.g.*, 15·372 gm.): in other words with care we can expect an accuracy of 1 part in 10,000 ($\pm$ 0·01%). With a Post Office Box (or any Wheatstone Bridge instrument) we may measure a resistance of 293·7 ohms, *i.e.*, 4-figure accuracy. But an ordinary ammeter cannot be reliable beyond 2 significant figures (*e.g.*, 4·2 amperes).

This need not unduly worry us. In fact, for some purposes one significant figure is quite sufficient; *e.g.*, with a " 5-amp. fuse " we are quite content if a specimen fuses at 4·6 or 5·4 amperes.

In determining the last significant figure, if the next figure to its right is 5 or more than 5, increase by one the last figure to be retained.

Thus the number 13,475 should be written

13,480 to 4 significant figures
or 13,500 to 3 significant figures
or 13,000 to 2 significant figures

(since the next figure in the original number is FOUR, not five).

Noughts. Noughts are not normally significant unless they are sandwiched between other digits. Thus 0·00195, 32700, and 30·9 are all correctly given "to 3 significant figures". Only in the last number is the zero significant.

Exceptional cases arise, like the number 29·97, which must be written " 30·0 correct to 3 significant figures ". Both noughts are significant here, and we must state how many figures are significant in such cases.

Practice examples with significant figures will be found in Exercise 1(*b*).

EXERCISE 1(*b*)

A. Complete the following WORD statements, and then write as concisely as possible a FORMULA expressing the same statement in each case in terms of algebra :

1. " The petrol consumption (P gall.) of a car which does 15 m.p.g. for any journey (m miles) is found by . . . "
2. " The volume (V cu. ft.) of a cube is found by . . . "
3. " The weight per unit length (w lb.) of a beam of given length (l ft.) and total weight (W lb.) is found by . . . "
4. " The power rating of an electric fire (P kilowatts) is found by . . . " (V volts, I amperes).
5. " The length (l ft. run) of a cornice to be floated around a rectangular ceiling (p ft. by q ft.) is calculated by . . . "
6. " The speed of a train (V m.p.h.) is found by . . . " (distance s miles, time t hours).

B. Translate each of the following formulæ into words :

7. $V = \frac{1}{3}Ah$. (Volume V cu. yd. of a pyramid h yd. high with base area A sq. yd.)
8. $Q = VC$. (Quantity Q coulombs of electricity held by a capacitor of C farads capacitance when charged to a potential V volts.)
9. $S = 16t^2$. (A stone in t sec. falls S ft.)
10. $I = \frac{E}{R}$. (Current I amperes flowing through a resistance R ohms when an e.m.f. of E volts is applied.)
11. What familiar rule is this last formula equivalent to?
12. $A = 4\pi r^2$. (Surface area A sq. miles of a sphere, radius r miles.)

C. Express the following expressions in algebra in their neatest form:

13. $3a + 2a + 5b - 7b$. 14. $a^3b^2 \times ab^4$. 15. $2pq \times 6p$.
16. $16p^4q \div 6pq^2$. 17. $7a \times 3ab \times a^2b$. 18. $\pi r^2 \times 6l$.

D. Express the following in terms of " standard form " numbers :

19. 519. 20. 240,000. 21. 0·00018. 22. 540 million.
23. 0·000000804.
24. The velocity of sound in dry air is 33130 cm. per sec.
25. The insulation resistance of a cable pair under test is 34 megohms. (Express in ohms.)
26. A mutual inductance of 42 microhenrys. (Express in henrys.)
27. A polarised relay operates to a minimum current of 3 mA. (Express in amperes.)
28. A clearance of 12 thousandths of an inch (in inches).

29. 1 mile expressed in inches.

30. 1 sec. as a decimal fraction of an hour (to 3 significant figures).

E. Write down the following in ordinary numbers:

31. The velocity of Hertzian waves is $2{\cdot}991 \times 10^{10}$ cm. per sec.

32. The average distance from the earth to the moon is $2{\cdot}39 \times 10^{5}$ miles.

33. The conductance of an overhead wire to earth is $1{\cdot}4 \times 10^{-6}$ mhos (or 1·4 micro-mhos).

34. The mass of an electron is approximately $9{\cdot}11 \times 10^{-28}$ gm.

F. Write down the following numbers correct to 3 significant figures:

35. 27·912. 36. 47884. 37. 0·091653. 38. 103130.
39. 0·09078. 40. 0·69967.

G. In the remaining examples results should be given correct to TWO significant figures unless otherwise stated:

41. The earth moves in its yearly journey round the sun at about 3×10^{6} cm. per sec. How many miles does it travel in a day? In a year?

42. Radio waves travel at 186,000 miles per second. The operation of radar equipment depends upon measuring the time taken for a radio "pulse" to travel to the target and for the echo to return. How long will this time be:

(*a*) For a target only 400 yards away?
(*b*) For the moon, assumed 239,000 miles from the earth?

43. When a voltmeter registers 8·4 volts, the current through the instrument is in fact 2·7 mA. Calculate the resistance in ohms of the instrument (formula: $R = \frac{V}{I}$, involving V volts, I amperes, and R ohms).

44. The resistance of a mile of copper wire is approximately 8·8 ohms. Express the resistance of a foot of this wire in the most convenient unit.

45. If the same wire weighs 100 lb. per mile, find the length of an ounce of wire, and the resistance per ounce.

Note that *Loop* resistance (*i.e.*, "Go and Return") per mile in ohms × conductor weight per mile in lb. = 1760, approximately—a useful fact to remember for copper wire.

46. In January 1948 there were approximately 60 million telephones installed in the world. Assuming that 95% of these were equipped with the normal soft-iron diaphragm in the form of a circular disc 2 in. in diameter and 0·009 in. thick, what total weight of iron in tons was then in use for this purpose? (1 cu. in. of iron weighs 0·26 lb.)

47. The velocity of light in a vacuum is $2{\cdot}9985 \times 10^{10}$ cm. per sec. Its velocity in water is 75·2% of this figure. Calculate this velocity to 3 significant figures. (*Hint*: find 75% first and then add 0·2% of the original.)

48. The following news item appeared in the *News Chronicle* of 3rd January 1949.

Quite a distance

A **NEW star** 1,000 times the size of the sun, discovered by Dr. J. Pearce, Victoria B.C. University, has been named after the university. He placed the star roughly at 6,800 light years from the earth. Speed of light is 186,000 miles a second.

A light-year is the distance travelled by a ray of light (at 186,000 miles per sec.) in one year. How many miles are equivalent to one light-year?

Show that this "new" star is approximately $4{\cdot}0 \times 10^{16}$ miles from the earth.

49. How many cubic feet of timber are contained in a pole 30 ft. long, if its mean diameter is 6 in.? (Give your answer to 2 significant figures.)

50. The resistance of a piece of wire has been determined by measuring the voltage across it and the current passing through it, with instruments stated to be accurate to $\pm$ 1% of their readings. The meter readings were 5·17 volts and 2·38 amperes, and the resistance was calculated (dividing voltage by current reading) to be 2·17227 ohms. To what extent do you regard this figure as reliable? How would you state the answer?

51. A piece of resistance wire 112 cm. long and having a resistance of 5·11 ohms per metre is joined in series with another piece 170 cm. long, having a resistance of 10·1 ohms per metre by means of a length of copper wire, 3·52 metres long, having 0·00103 ohms resistance per metre. Find the resistance of the combination correct to 3 significant figures.

52. A 50-volt battery supplies 20 lamps each of 670 ohms resistance, connected in parallel. A voltmeter, of resistance 50,580 ohms, is permanently connected across the battery. Find the total current taken from the battery, correct to sufficient significant figures for practical purposes.

CHAPTER 2

THE LANGUAGE OF GRAPHS

2.1. The Language of Graphs

Very often we are concerned with recording or presenting a group of related facts and figures. We can, of course, make up a table of the data concerned; this is clearer than a rambling essay on the subject. But the eye finds it difficult to connect together a mass of figures, and prefers a complete " picture " or " graphical description " of the facts presented. Compare the tables of values given in the examples below with their graphs, and you will appreciate the clearness of graphical presentation.

In this chapter we shall consider how to choose and construct the type of graph best suited to a particular purpose. Later on, in Chapter 6, we shall see how the shape of a graph often indicates a mathematical relation between two quantities.

2.2. Statistical Graphs

(a) *The Bar Diagram*

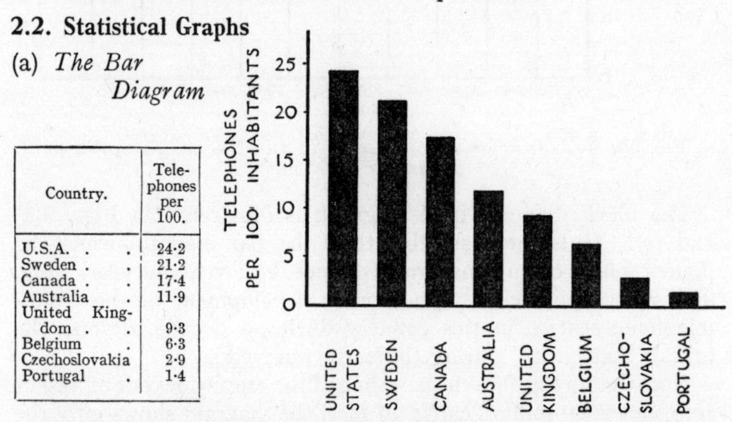

Country.	Telephones per 100.
U.S.A. . .	24·2
Sweden . .	21·2
Canada . .	17·4
Australia .	11·9
United Kingdom . .	9·3
Belgium .	6·3
Czechoslovakia	2·9
Portugal .	1·4

Fig. 8.—Telephones per 100 Inhabitants of Various Countries on 1st January 1949.

This is the simplest kind of graph; thus in Fig. 8 the number of telephones per 100 inhabitants in certain countries is set out both in a table and as a " Bar " graph. The height of each bar measures the required information for the country concerned.

(b) *The Block Diagram*

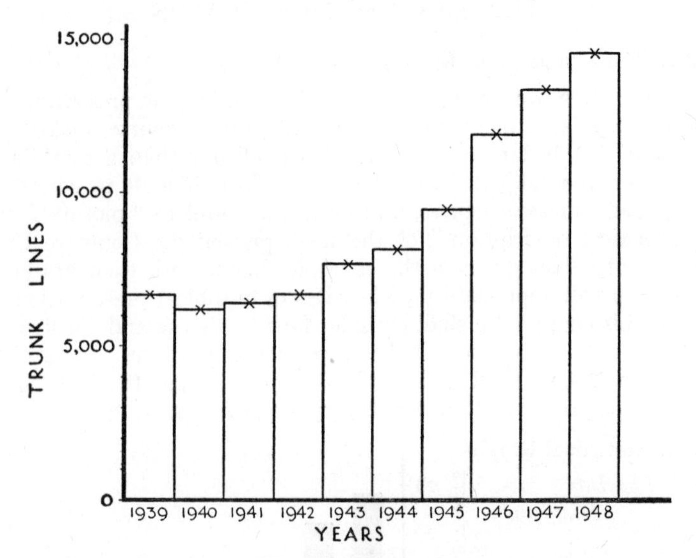

FIG. 9(*a*).—TRUNK TELEPHONE LINES IN THE UNITED KINGDOM AT 31ST MARCH EACH YEAR.

The block diagram type of graph is illustrated in Figs. 9(*a*) and (*b*). It is more suitable than the bar diagram when the figures follow on in a natural sequence, *e.g.*, year by year. Fig. 9(*a*) shows quite clearly the steady development of the trunk-telephone system in this country during a decade, despite the initial " set-back " during the early war years.

Note that in Fig. 9(*b*) the base-line of the graph does not represent zero, but 2000 million calls; in fact, the diagram shows only the tops of the blocks. This device allows us to use a larger scale, and is very often employed in making graphs. It helps to emphasise the *variation* in value rather than the *total* values concerned. For example, our diagram shows much more vividly

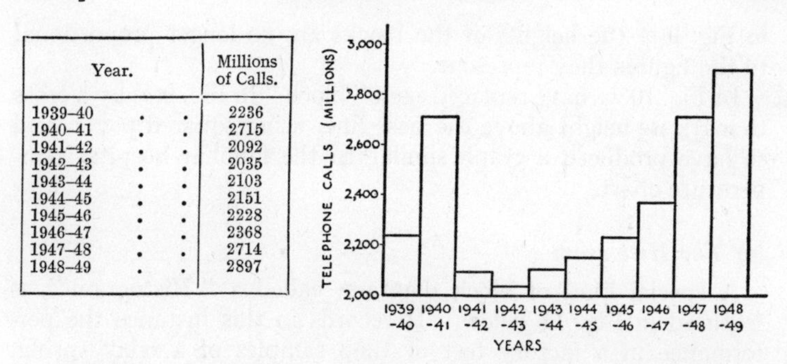

Year.	Millions of Calls.
1939–40	2236
1940–41	2715
1941–42	2092
1942–43	2035
1943–44	2103
1944–45	2151
1945–46	2228
1946–47	2368
1947–48	2714
1948–49	2897

FIG. 9(*b*).—TELEPHONE CALLS PER YEAR (1ST APRIL TO 31ST MARCH) IN THE UNITED KINGDOM.

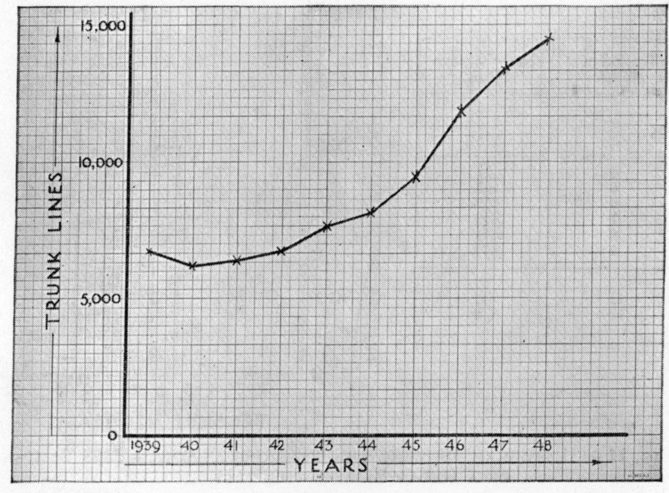

FIG. 10.

the fluctuations in telephone traffic during the war years than would a diagram of the same size but to the reduced scale necessary to include zero calls.

At the same time we may be tempted to overlook the fact that in Fig. 9(*b*) the heights of the blocks are no longer proportional to the figures they represent.

In Fig. 10 we have replaced each " block " in Fig. 9(*a*) by a cross to mark its height above the base-line, using squared paper, and so have produced a graph similar to the familiar hospital temperature chart.

(c) *The Histogram*

A special kind of block diagram, called a " Histogram ", is featured in Fig. 11 below. It records in this instance the performance in a factory test of 1258 samples of a relay spring, which is required to " lift " at a tension of not more than 9 gm.

Tension in grams.	No. of springs lifting.
1	2
2	3
3	5
4	11
5	65
6	299
7	478
8	275
9	78
10	22
11	18
12	2

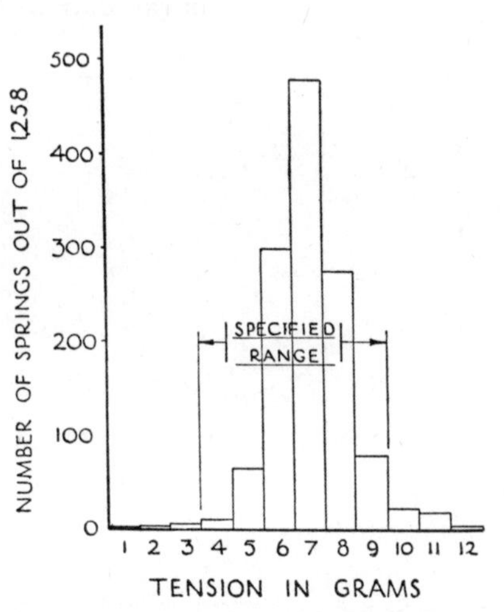

FIG. 11.—HISTOGRAM SHOWING THE TENSIONS AT WHICH 1258 SAMPLES OF A RELAY SPRING WILL JUST LIFT.

and not less than 4 gm. Only 52 of the samples fail to operate within this specified range of " tolerance " (*i.e.*, less than 5% are below standard), so the process is reasonably successful.

Our histogram thus shows a " frequency distribution ", and tells us by how much individual samples of a manufactured product (a relay spring, a capacitor, a resistor, and so on) differ from their rated value or performance.

The histogram is a valuable check on the efficiency of an industrial process, and is frequently used in the quality control of mass-produced components.

2.3. Graphs Showing Continuous Variation

In Fig. 12(*a*) we have reproduced, by kind permission of the Royal Society for the Prevention of Accidents, a poster diagram

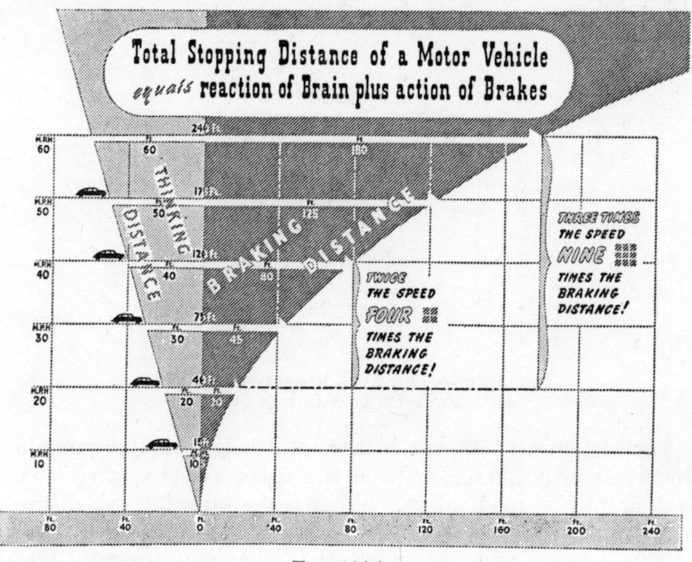

FIG. 12(*a*).

showing how the total stopping distance of a car at different speeds is made up of a " Thinking Distance "—the distance travelled while the brain reacts to a situation—together with the " Braking Distance " in which the brakes can bring the vehicle to a standstill.

It is instructive to note that at 60 m.p.h. the car travels 60 ft. before the brain reacts (and at other speeds in like proportion).

Thus it takes $\frac{1}{5280}$ of an hour, *i.e.*, $\frac{3600}{5280} \simeq \frac{7}{10}$ sec. from the moment the eye notices danger to the instant when the brakes are applied.

The same information has been plotted in more conventional style on squared paper in Fig. 12(*b*). In this case the various distances have been plotted " vertically " against the vehicle speed shown " horizontally ". We have lost some of the vividness of the original diagram, but there is a new gain. For in this second diagram we can find quickly the safe stopping distance at any intermediate speed (*i.e.*, speeds not shown in the poster).

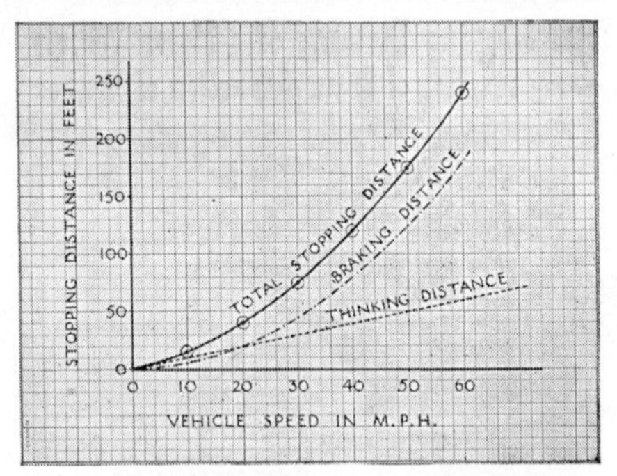

FIG. 12(*b*).—THE SAFE STOPPING DISTANCE OF A VEHICLE AT VARIOUS SPEEDS.

For instance, at 45 m.p.h. the total height is approximately 150 ft. on the graph scale. So at this speed a clear vision of 50 yd. is desirable for road safety. What is the safe stopping distance at 53 m.p.h.? *

This process will also work the other way round. For a motorist driving in fog with a visibility of 10 yd. (*i.e.*, 30 ft.) the greatest safe road speed would be about 17 m.p.h. What would be his safe speed with visibility limited to 15 yd.? †

This facility of " reading off " fresh values from a graph drawn on " graph paper " is very important indeed. We shall discuss this process more fully in the next section.

* 65 yd. (195 ft.).
† 22 m.p.h.

Continuous and Discontinuous Graphs

What is the essential difference between the graphs in Fig. 12(*b*) and those in previous sections? Briefly this: in Fig. 12(*b*) we can give a *definite meaning to every point* on each graph. We call such graphs "*continuous*". In contrast the diagrams in Figs. 8–11 deal with sets of isolated values, and no definite meaning can be assigned to intermediate values. Such graphs are called "*discrete*" or "*discontinuous*".

Even in Figs. 9(*a*) and 10 (trunk-telephone development), which do give an idea of steady development, as shown by the jagged line joining the points in Fig. 10, it would be impossible to state the number of lines at an intermediate date. So these graphs cannot be considered as continuous.

A Practical Point in Plotting

When we are obliged to draw more than one graph on the same graph framework it is wise to vary the manner of plotting the points—either a cross or a circle with a central dot is recommended. Or use different colours to distinguish between your graphs.

2.4. Graphs to Record Observations

Observations can be very simply and conveniently recorded by drawing a line on a suitable chart, that is by plotting a graph. The doctor watches the progress of his patient in this way, by using a temperature chart. And not only is such a graphical record more vivid than a table of figures, but it also lends itself readily to "automatic" recording. How much more complicated the recording barometer or the sunshine recorder would be if they had to produce printed figures instead of a single line on a chart!

The recording voltmeter or ammeter and the photographic record of an oscillograph "trace" are examples of a series of observations automatically recorded as graphs; many other examples will spring to the mind of the engineer, for such records are of vital importance in production and maintenance problems.

Sometimes graphs must be made "by hand" after taking a number of individual observations; this often occurs in the course of fresh research. Provided the quantities measured vary continuously with one another, we may join the points plotted

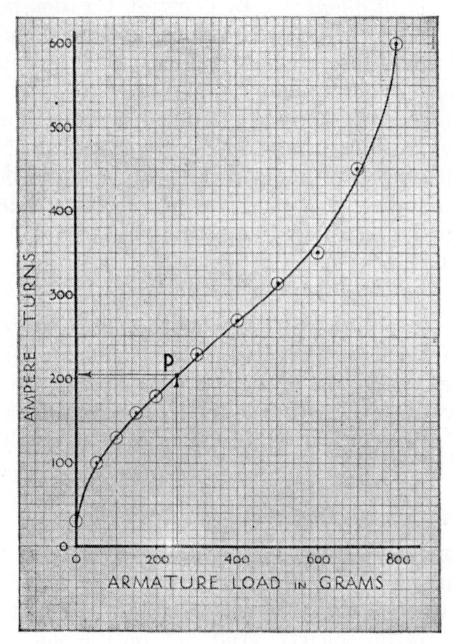

FIG. 13.—THE RELATION BETWEEN ARMATURE LOAD AND AMPERE-TURNS REQUIRED TO OPERATE A RELAY.

from the observations with a continuous line, and then we have a means of finding values of these quantities at intermediate points on the graph.

For example, in Fig. 13 we have the results of an experiment which set out to determine the current required just to operate a certain relay when the armature was loaded with 50, 100, 150, 200, 300, 400, 500, 600, 700, and 800 gm. In order to make the results general for any winding on a relay of the same type we have converted the currents measured to " ampere-turns " by multiplying each current reading by the number of turns in the winding. So we have obtained our graph by plotting " ampere-turns " against the " armature load in grams ".

A graph is for use, not ornament! How are we to make effective use of this particular graph? In this instance we can now estimate the number of ampere-turns required for a given armature load, and *vice versa*.

For example, if we have a relay of this pattern with an armature load of 250 gm., we can forecast the least number of ampere-turns for which the relay would just operate. If we " enter " the load-axis at the " 250 " mark and follow the arrow to meet the graph curve at P, we can then turn left (following the second arrow) and note the value of approximately 205 ampere-turns where this second arrow meets the upright axis. We then predict that this relay would just operate at about 205 ampere-turns. In other words, if we wish the relay to operate when the current exceeds 3 amperes, the winding must contain 68 turns.

This process of estimating fresh values from a graph is called *Interpolation*. With a little practice the guide-arrows can be omitted. Now try for yourself! From the same graph find:

(i) the ampere-turns required for a load of 750 gm.;
(ii) the armature load which just allows operation with $\frac{1}{4}$ ampere, if the winding has 1000 turns.

EXERCISE 2(*a*)

1. In six months of 1945, U.K. taxation yielded the following:

Direct.	£ million.	Indirect.	£ million.
Income Tax	1265	Drink, Tobacco	745
Surtax	74	Purchase Tax	95
Excess Profits Tax	519	Entertainment Tax	46
Death Duties	107	Motors	28
		Other Indirect Taxes	201

Construct two block diagrams to show each of these sets of data—arranged in descending order.

2. Present the following data in a Block Diagram, arranging the materials in descending order of dielectric strength :

Material . .	Ebonite	Glass	Gutta-percha	Mica	Paper (dry)	Paper (oiled)
Dielectric strength in volts per mm.	40,000	16,000	11,000	58,000	9,000	24,000

3. Construct a Block Diagram to present the Quarterly Consumption of a domestic user of Electricity.

Quarter ending .	1948.				1949.			
	Jan.	April	July	Oct.	Jan.	April	July	Oct.
Units used .	275	185	175	103	315	318	105	95

Estimate from your diagram the average quarterly consumption. Now calculate it ! At $\frac{3}{4}d.$ a unit, what is the average annual bill ?

4. In the 1950 Budget, the estimated National expenditure per £ for 1950–51 was given as : Defence 4*s.* 1*d.*, Service of Debt 2*s.* 6*d.*, Grants to Local Authorities 2*s.* 3*d.*, Health Services 2*s.*, Subsidies 2*s.* 1*d.*, Pensions, Family Allowances 1*s.* 11*d.*, Miscellaneous 2*s.* 11*d.*

What was the estimated Surplus ? Express each item as a percentage, and draw a block diagram to show the percentage expended on each item. What is the total area of this block diagram ?

5. The resistance in microhms per c.c. of various metals is as follows :

Metals . .	Aluminium	Copper (hard drawn)	German silver	Platinum	Silver (hard drawn)	Zinc
Microhms/c.c.	2·89	1·62	20·78	8·97	1·62	5·57

Draw a block diagram, giving a visual comparison of these metals in ascending order of resistivity.

6. The minimum fusing current for wires 0·05 in. in diameter was found to be :

Metal . .	Allotin	Aluminium	Copper	Lead	Platinoid	Tin
Current in amperes . .	15	85	118	14	53	19

Draw a block diagram to show the minimum fusing current for each type of wire, arranged in order of easy fusibility.

7. A mass-produced steel spindle is specified to be 0·24 in. in diameter. 193 samples of this product are measured with a micrometer, and graded into bins marked 0·231 to 0·233, 0·233 to 0·235, 0·235 to 0·237, etc., with the following results :

Bin grade .	31–33	33–35	35–37	37–39	39–41	41–43	43–45	45–47	47–49
Number of spindles .	1	4	16	32	69	41	20	8	2

Construct a histogram to show how these products vary. If a tolerance of 5 mils (5 thousandths of an inch) is allowed, what percentage of these samples must be rejected? Do you consider the process is working satisfactorily?

8. A batch of 114 radio valves were given a "life test", with the following results :

Life in thousands of hours .	0–5	5–10	10–15	15–20	20–25	25–30	Over 30
No. of valves .	48	26	19	11	6	3	1

Draw a histogram to show these facts pictorially. What percentage of the valves survive 10,000 hours operation?

9. A certain telegraph relay should not operate to a current of less than 3 mA., and must operate before the current exceeds 8 mA.

In testing a batch of 134 samples, the current, I mA., at which each relay operated was noted as follows :

Operate to I mA. .	1	2	3	4	5	6	7	8	9	10
Number of relays .	0	1	4	9	42	58	12	5	2	1

Construct a histogram to show these data. What information does it provide about the adjustment of these samples? How many samples should be returned for readjustment? What percentage is this?

10. A 2-μF. capacitor is manufactured, allowing $\pm$ 5% tolerance in its capacitance. A batch of 180 samples tested gave the following results :

Capacitance in μF.	1·90–1·94	1·94–1·98	1·98–2·02	2·02–2·06	2·06–2·10	2·10–2·14	2·14–2·18
No. of samples .	3	33	67	48	21	7	1

Comment on the histogram produced.

11. Weekly Faults Record at an Exchange:

	April.				May.			
Week ending . .	8	15	22	29	6	13	20	27
Junction lines . .	2	5	3	0	6	4	4	3
Subs. lines . .	51	42	38	46	59	44	33	36

Draw a graph on squared paper to show the progress of faulting on both junction and subs. lines. Draw a straight line to indicate the average number of subs.-line faults dealt with during this period.

Would you join the points plotted by a broken or a continuous line?

12. The average number of daily calls on a small rural automatic exchange was as follows:

Year . .	1943	1944	1945	1946	1947	1948	1949
Calls .	421	662	857	1008	1132	1191	2020

Have intermediate points on your graph any meaning? Should we join the points plotted? What would you estimate to be the average daily calling rate in 1950?

13. s is the distance in yards in which a train can stop from v m.p.h.

v . .	30	40	50	55	60	65
s . .	100	175	277	336	400	492

From your graph find s when $v = 35$, and find v when $s = 300$.

14. The table gives the times of sunset at the dates shown.

	April.					May.	
Date . . .	1	8	15	22	29	6	13
Time, p.m. . .	6.30	6.42	6.53	7.5	7.16	7.28	7.39

Find the time of sunset on 12th April. At what day does the sun set at 7 p.m.?

15. The temperature ($T°$ C.) of a cooling liquid in a vessel was recorded at 10-sec. intervals, as follows :

t sec. . . .	0	10	20	30	40	50	60
$T°$ C. . . .	86	71	59	49·5	41·5	34·5	28

When would the temperature be exactly 55° C.? When is the liquid cooling the fastest? Why?

16. A similar experiment was conducted with melted beeswax, with these results :

t sec. . . .	0	20	40	60	80	100	120
$T°$ C. . . .	96	81·5	72	65	61	61	61

Can you explain this graph? Find the melting point of this specimen of beeswax, expressed in °F. When did it commence to solidify?

17. The voltage per cell of a lead–acid accumulator discharging at the "12-hr. rate" varies as follows :

Hours on load .	0	1	2	4	6	8	9	10	11	12
Volts per cell .	2·06	2·01	2·0	2·01	2·005	1·995	1·98	1·96	1·91	1·81

Graph the data with a voltage range 1·8 to 2·1. For how long is the battery's voltage within 1% of 2 volts?

18. The electrolyte of a lead–acid accumulator is found to have a specific gravity of 1·215. Every two hours while it is supplying current to a load the specific gravity is checked, as follows :

Time. . .	8 a.m.	10 a.m.	12 noon.	2 p.m.	4 p.m.
Sp. gr. . .	1.215	1.208	1.199	1.190	1.181

Plot a graph giving a range of specific gravity from 1·17 to 1·22. When was the specific gravity exactly 1·195?

19. Weight and resistance per mile of copper conductors :

Wt. in lb./mile .	6½	10	20	40	70	100	150	200
Resistance in ohms/mile	135	87·7	43·9	21·9	12·5	8·78	5·85	4·39

The resistance per mile of No. 20 S.W.G. wire is 40·9 ohms. Find graphically its approximate weight per mile.

20.

Fusing current in amps.	1	5	10	15	20
Diam. of copper wire in in. . . .	0·0021	0·0062	0·0098	0·0129	0·0156
Diam. of aluminium in in. . . .	0·0026	0·0076	0·012	0·0158	0·0191

Draw 2 graphs with the same axes to display this information.

The diameter of No. 32 S.W.G. is 0·0108 in. What would be the fusing current for this gauge of: (*a*) copper, (*b*) aluminium wire?

*2.5. Graphical Aids to Design

In the previous paragraph we had a glimpse of the value of graphs in designing relays and similar components. By plotting either experimental observations or the results of calculations made for sample values we make further similar experiments or calculations unnecessary: for any intermediate values we require can be found from the graph by interpolation.

The data illustrated in Fig. 14 demonstrate another way in which graphs are a valuable aid to the designer. When two (or more) quantities are both varying as a measured quantity changes, we obtain a concise picture of what is happening if we draw two (or more) graphs on the same sheet of squared paper.

In Fig. 14 Graph A tells us how the *magnetic pull on the armature* of a relay varies with the " *travel* ", that is the distance moved by the armature towards the core away from its normal position, *i.e.*, the distance by which the air-gap is shortened. With the same axes Graph B shows how the restoring force exerted by the contact springs also varies with the " travel ". There is a constant current passing through the relay coil during the experiment.

Let us try to interpret in words the behaviour of the relay armature and contact springs as shown in these two graphs.

(*a*) As the armature moves away from its normal position and towards the core, the air-gap is shortened, and the magnetic pull increases, at first slowly and then more rapidly. These changes are shown in Graph A.

(*b*) As the movement of the armature deflects the springs, they exert more and more pressure on the armature, tending to "restore" it, *i.e.*, to push it back to its normal position. The growth of this restoring force in an actual relay is somewhat

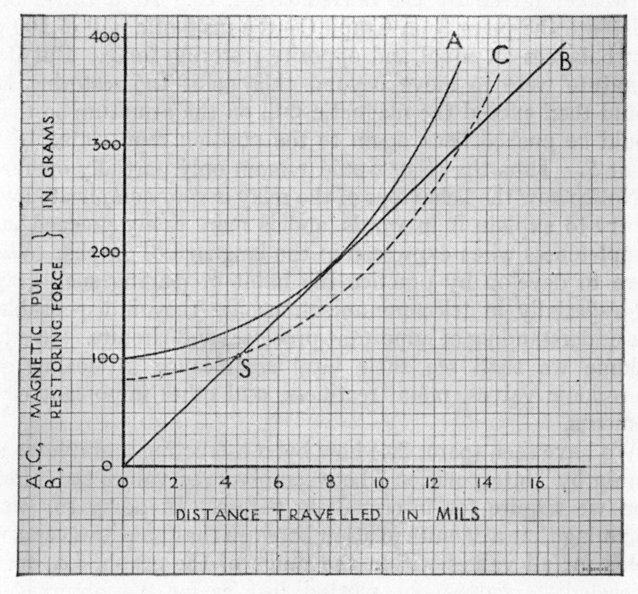

FIG. 14.

irregular, due to the different springs coming into play successively. Our Graph B shows the changes in the restoring force as though these irregularities had been smoothed out.

When the armature commences to move the magnetic pull (Graph A) is about 100 gm. and the opposing force exerted by the springs (Graph B) is much smaller. By the time the armature has moved about $7\frac{1}{2}$ mils the springs are exerting a restoring force of 175 gm. A magnetic pull of at least this amount is required to cause the motion to continue. The magnetic pull is just about equal to it, and once the armature has passed this point (*e.g.*, at 9 mils) the excess of the magnetic pull over the restoring force continually increases, so that the operation of the relay continues until it is completely operated.

We can see from the graph what would happen if the current in the winding were to be reduced—the magnetic pull would then be weaker (given by the dotted Graph C). At a travel of $4\frac{1}{2}$ mils—see the point S on Graph C—the magnetic pull of 100 gm. just balances an equal and opposite restoring force. But if the armature were to move beyond this point the restoring force would then exceed the magnetic pull on the armature so that it would be pushed back again to the position represented by the point S. Thus the relay would remain only partially operated.

Notice how the behaviour of the relay is very complicated when we try to explain it in words, and is more easily grasped in its completeness when expressed in the language of graphs. We all have to develop by practice the habit of reading such graphs, for especially in telecommunications are they indispensable to the understanding of important processes and circuits. In a later volume we shall see their vital significance in dealing with thermionic valves, metal rectifiers, and other important circuit elements.

We have seen (para. 2.4) how by constructing a graph we may obtain by " interpolation " between observed values the information we require for intermediate values. In doing so, we are assuming that a definite relation between the quantities concerned holds for all values within the range of the graph, *i.e.*, for the intermediate values not plotted just as for the observed and plotted ones. In the following examples a very simple relationship is apparent.

Fig. 15 shows a graph plotted from a few observed results to enable us to " read off " quickly the weight in pounds of any given volume of water in cubic feet. In addition to the plotted points, the point (0, 0) which we call the " origin "—it corresponds to " no water " weighing " no pounds "—must also be on the same graph. We see that the graph in this case turns out to be not a curve but a straight line through the origin. In the language of mathematics, we say that the weight is "directly proportional" to the volume of water involved. And whenever one

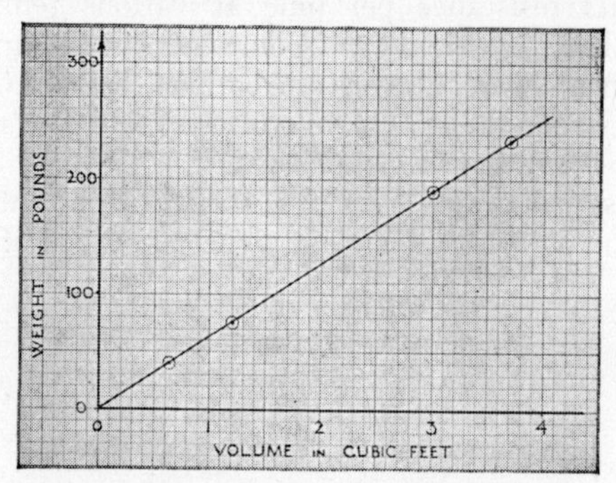

FIG. 15.—A SIMPLE CONVERSION GRAPH.

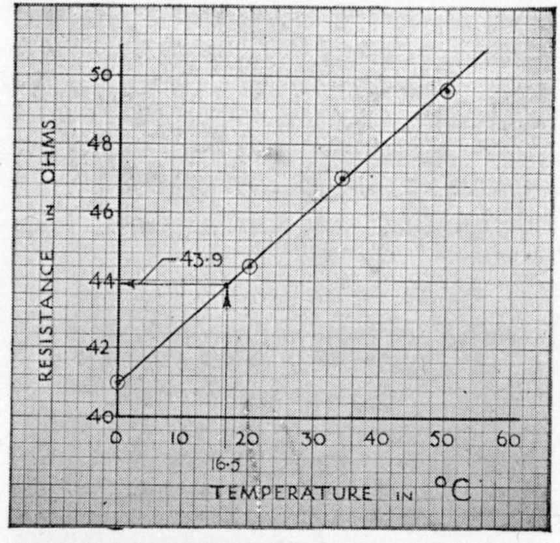

FIG. 16.—THE RESISTANCE OF 1 MILE OF 20-LB. COPPER WIRE AT VARIOUS TEMPERATURES.

quantity is directly proportional to another we always get a straight-line graph like this one.

We sometimes need to know the resistance of a wire at a

particular temperature. In specifying cable we often quote its resistance per mile measured at 16·5° C. Thus, for 20-lb. copper conductor, its resistance per mile at various temperatures is shown graphed in Fig. 16. Some of the temperatures in this experiment were above, and others below, the specified 16·5° C., but from the graph we can read that the resistance at this reference temperature is 43·9 ohms.

In this case the graph is again a straight line, but it does not pass through the origin, for clearly the resistance of the wire at

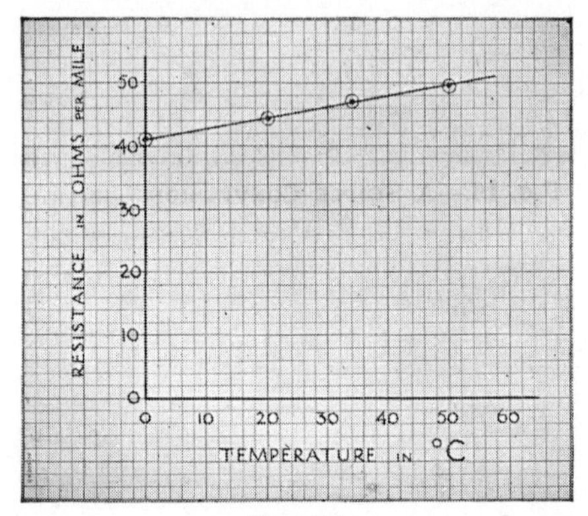

FIG. 17.

0° C. cannot be zero. In fact, in drawing this graph we have had to make use of the same device as in Fig. 9(*b*) so as to emphasise the *change* in resistance with temperature: also to allow an accurate interpolation to find the resistance at 16·5° C. To bring out this last point we have drawn, in Fig. 17, the same graph, but with a true zero on both axes, *i.e.*, to include the " origin ". Notice how the graph is confined to a narrow strip of the graph paper; over 80% of the paper is thus wasted, and the eye finds it difficult to read off values of the resistance at intermediate temperatures.

Although the graph in Fig. 16 does not pass through the origin, so that the resistance of our sample is not directly proportional to the temperature, yet the resistance does increase uniformly as the temperature rises. In fact, by simple division (since the resistance rises by 8·6 ohms between 0° C. and 50° C.) we note that the resistance rises 0·172 ohms for each degree rise in temperature. In general, if we want a formula for finding the resistance at any temperature t° C., we must add to the resistance at 0° C. (41 ohms) an amount 0·172 ohms for every degree rise above 0° C., *i.e.*, a total increase of $0{\cdot}172t$ ohms. Thus, if the resistance at t° C. is called R ohms,

$$R = 0{\cdot}172t + 41$$

As we shall see in Chapter 6, a straight-line relation between two quantities can always be represented algebraically by a similar formula. Such straight-line or "linear" variation occurs in many technical relations; for example, in the behaviour of an elastic spring (plotting the length of the spring against the stretching force); in the variation of the volume of a quantity of gas as the temperature changes (the pressure remaining constant); and supremely in the relation between the voltage across a resistor and the current passing through it (Ohm's Law !).

2.6. Note on the Choice of Scales

To make the most of a graph we must choose the best possible scale for each axis. There are two guiding principles in making our choice:

(*a*) The scales must be as large as possible in order to make the graph as meaningful as we can and at the same time to provide for accurate interpolation. Thus allot where possible the full length of each axis to *the range of values which you have to plot*. Do not attempt to include a zero on either axis unless this is really necessary.

(*b*) The scales chosen must make for easy reading of intermediate values on a decimal scale. That is, the small squares on the graph paper should correspond easily to the subdivisions of the quantities plotted. Generally, we achieve this by taking one small square as representing 1, 2, or 5 units or decimal multiples of these (*e.g.*, 10, 20, or 50; 0·1, 0·2, or 0·5; and so on). Avoid, for example, making one square represent 3 or 4 units, or 3 or 4 small squares represent 1 unit. If necessary, we must be prepared on occasion to use a smaller scale (*i.e.*, to override the consideration of a large scale advocated in (*a*) above) in order to satisfy this requirement of " easy divisibility ".

In conclusion, be careful to use accurately divided graph paper, especially if values are to be read off by interpolation. On many types of " squared paper ", produced by ruling rather than by printing from a block, the " squaring " is often far from being accurate.

EXERCISE 2(*b*)

A. *Straight-line Graphs*

Construct a conversion graph to express :

1. Miles per hour as feet per second (up to 600 m.p.h.).
2. Diameter in inches (up to 1 in.) as diameter in cm. (British Association Screw Thread No. 19 is 0·021 in. in diameter: convert this to mm.)
3. Nautical miles (up to 50 nauts) to km., if 1 naut ≡ 1·853 km. (Convert a boat's speed of 42 km. per hr. to knots.)
4. British Thermal Units (B.Th.U.s) as Calories if 1 Calorie ≡ 3·97 B.Th.U.s (to read up to 10,000 B.Th.U.s).
5. Tons of displacement of a cargo-vessel as cubic feet of hold space : 1 ton of displacement ≡ 35 cu. ft. What is the maximum carrying capacity in cu. ft. of a vessel of 12,500 tons displacement ?

6. The petrol gauge for a French car is graduated 10, 20, 30, 40 litres. Using a graph, devise a stick-over strip to re-graduate the same gauge in gallons. (1 litre = 0·22 gall.)

7. A salary scale commences at £360 per annum, with annual increments of £15, and has a " ceiling " of £690. Assuming 40 years' service, draw a graph from which the salary in any year of service can be read at a glance. In which year of service does the salary first exceed £500? Before the " ceiling " is reached what would be a simple formula for giving the annual salary £S in terms of n years completed service?

8. The electrical resistance R ohms of a wire at a temperature $T°$ C. is given in the following table :

$T =$	10	40	80	120	150	180
$R =$	11·3	12·5	14·2	15·9	17·1	18·4

Draw a graph showing the relation between R and T. Hence determine its equation in the form $R = aT + b$ (see para. 2.5).

9. A domestic consumer of electricity pays an all-in rate of 15*s*. a quarter plus $\frac{3}{4}$*d*. a unit. Draw a graph from which he can read off his quarterly bill for consumption ranging from 1000 to 6000 units.

Make up a formula showing the quarterly bill Q shillings if x units are consumed.

10. An hotel-keeper's overhead expenses (wages, rates, etc.) average £7 a week, and each guest costs him an additional 30*s*. a week in food, laundry, etc. He charges 4 guineas a week.

Draw two graphs with the same axes, one showing expenses, the other receipts for different numbers of guests up to 20.

How many guests does he need to make £10 a week profit?

B. *Miscellaneous Examples*

11. The following are examples of British Association Screw Threads:

No.	0	2	5	10	15	20	25
Diam. (in in.)	0·236	0·185	0·126	0·067	0·035	0·019	0·010
Pitch (in mils)	39·4	31·9	23·2	13·8	8·3	4·7	2·8

A No. 8 thread is 0·087 in. in diameter. What is likely to be its pitch? If a screw of pitch 10 mils is required, what do you estimate its diameter to be?

12. The Standard Wire Gauge (S.W.G.) of steel wire has the following specifications:

S.W.G. No.	1	3	5	7	10	15	20
Diam. (in mils)	300	252	212	176	128	72	36
Lb. per mile	1225	864	612	422	223	70	18

No. 8 gauge is 0·16 in. in diameter. Find its weight per mile. What would be the diameter of steel wire weighing 500 lb. to the mile?

13. A pipe of diameter D in. will discharge a maximum of N gall. of water per minute from a reservoir:

$D =$	2	4	6	8	9	10	11	12
$N =$	5	37	117	262	364	484	624	805

Plot N against D. Estimate: (*a*) maximum discharge for a 5-in.-diameter pipe; (*b*) diameter of a pipe to discharge up to 400 gall. per minute.

14. An apprentice can joint and plumb a 50-pair underground cable in 6 hr., but he takes $7\frac{1}{2}$ hr. to do the same job with a 120-pair cable. Assuming a straight-line graph connects the " number of pairs " (on the base-line) with the " time in hours " to do the job, draw this graph, and estimate the time he would take to joint and plumb 80 pairs.

Estimate the average time he takes to joint one pair in the cable.

15. The time to operate and time to release of a dial impulsing relay is measured for different line resistances, as set out below :

Loop resistance in ohms. . . .	100	200	300	400	500	600	700	800
Operate time in millisecs. . . .	10·3	10·9	11·9	12·7	15·4	18·0	20·6	24·3
Release time in millisecs. . . .	13·1	12·7	12·1	11·2	10·1	8·6	6·7	4·4

With the same graph axes plot operate and release times against loop resistance. For what value of the loop resistance of a subscriber's line is the distortion least? What is the distortion in millisecs. when the resistance is 150 ohms, 750 ohms?

CHAPTER 3

LOGARITHMS AND THE SLIDE-RULE

3.1. Powers of 10—" Standard Form "

In Chapter 1, para. 13, we saw the value of making use of powers of 10 when dealing with very large or very small numbers. When doing so it is a good rule to express all the numbers in *standard form*, that is to move the decimal point so that there is one and only one significant digit to the left of it. For example 2↓7940000 is more manageable as $2{\cdot}794 \times 10^7$ (the decimal point shifted 7 places to the arrow position). And likewise 0·000000001↓29 in standard form appears as $1{\cdot}29 \times 10^{-9}$ (the decimal point having been shifted 9 places in the reverse direction).

We also saw the operation of the following four processes:

A. PROCESS	B. EXAMPLE	C. EFFECT
Multiplication . .	$10^7 \times 10^4 = 10^{11}$	ADD Indices
Division . . .	$\frac{10^8}{10^3} = 10^5$	SUBTRACT Indices
Raising to a power (*e.g.*, cube) . . .	$(10^2)^3 = 10^6$	MULTIPLY Index (by 3)
Taking the (square) root	$\sqrt{10^6} = 10^3$	DIVIDE Index (by 2)

Notice that each process with *numbers* in Column A becomes a *simpler* process with *indices* in Column C.

In this chapter we shall make use of these ideas in reducing the labour of arithmetical calculations. They form the basis of logarithmic tables and of the slide rule.

3.2. Special and Fractional Powers of 10

(*a*) So far we have only met indices which are whole numbers. There are two special powers which require a little explanation.

(i) 10^1 looks unfamiliar, but if we remember that the index indicates how many times 10 is to be included as a factor, we see that it means simply 10 once only, *i.e.*, $10^1 = 10$ itself.

(ii) But what about 10^0? Thinking again of factors this means " don't include it at all "; in other words leave things as they are. Let us see how this works out in a simple operation, *e.g.*, $10^3 \times 10^0$. Adding indices we have $10^{3+0} = 10^3$.

Hence multiplication by 10^0 leaves the multiplicand 10^3 unchanged, just as if it had been multiplied by 1.

Similarly, $$(10^0)^2 = 10^{2\times 0} = 10^0$$

which compares with $1^2 = 1 \times 1 = 1$.

Again $10^3 \div 10^3$ by subtracting indices gives

$$10^{3-3} = 10^0$$

But $$10^3 \div 10^3 = 1$$

So whenever we come across 10^0 we can write $10^0 = 1$.

(As you will see, whatever number we write instead of 10 in the above argument the result is the same; thus in general $a^0 = 1$ for any value of a.)

(*b*) *Fractional Powers.* Consider as an index the simplest fraction—one-half. What is the meaning of $10^{\frac{1}{2}}$?

If we apply the rule for multiplying, then

$$10^{\frac{1}{2}} \times 10^{\frac{1}{2}} = 10^{\frac{1}{2}+\frac{1}{2}} = 10^1 \text{ or } 10 \text{ itself.}$$

So $10^{\frac{1}{2}}$ is the number which multiplied by itself makes 10, *i.e.*, it is the *square root of* 10.

Check this by the rule for raising to a power

$$(10^{\frac{1}{2}})^2 = 10^{\frac{1}{2}\times 2} = 10$$

So again $10^{\frac{1}{2}}$ squared makes 10.

So $10^{\frac{1}{2}} = \sqrt{10} = 3{\cdot}16\ldots$ (to 3 significant figures).

We can apply similar reasoning to any other fraction, *e.g.*, $10^{\frac{1}{4}}$ is the number which multiplied by itself four times $(10^{\frac{1}{4}} \times 10^{\frac{1}{4}} \times 10^{\frac{1}{4}} \times 10^{\frac{1}{4}})$ makes 10. Thus it is the fourth root of $10 = \sqrt[4]{10}$.

But also $10^{\frac{1}{4}} \times 10^{\frac{1}{4}} = 10^{\frac{1}{4}+\frac{1}{4}} = 10^{\frac{1}{2}} \simeq 3{\cdot}16$

so $10^{\frac{1}{4}} \simeq \sqrt{3{\cdot}16}$, *i.e.*, $1{\cdot}78$

Similarly, $10^{\frac{1}{8}} = \sqrt{10^{\frac{1}{4}}} \simeq \sqrt{1{\cdot}78} \simeq 1{\cdot}33$

From these three fractional powers of 10, other numbers like $10^{\frac{3}{8}}$ can be worked out:

e.g.,
$$10^{\frac{3}{8}} = 10^{\frac{1}{4}+\frac{1}{8}} = 10^{\frac{1}{4}} \times 10^{\frac{1}{8}}$$
$$\simeq 1{\cdot}78 \times 1{\cdot}33$$
$$\simeq 2{\cdot}37$$

It is useful to calculate in this way the value of powers of 10 for a whole series of indices lying between 0 and 1. The student should verify the figures given in the following table:

INDEX (x) .	0	$\frac{1}{8}$	$\frac{1}{4}$	$\frac{3}{8}$	$\frac{1}{2}$	$\frac{5}{8}$	$\frac{3}{4}$	$\frac{7}{8}$	1
Index as a decimal .	0	0·125	0·25	0·375	0·5	0·625	0·75	0·875	1
Number (10^x) .	1	1·33	1·78	2·37	3·16	4·22	5·62	7·50	10
Reference .	A	B	C	D	E	F	G	H	J

3.3. Conversion of Powers of 10 to Numbers and *vice versa*

This table gives us a kind of scale connecting the value of a number 10^x with the value of the index x. Such a scale is shown in the table below:

	A	B	C	D	E	F	G	H	J
Index x .	0	0·125	0·25	0·375	0·5	0·625	0·75	0·875	1·0
Number 10^x	1	1·33	1·78	2·37	3·16	4·22	5·62	7·50	10

But the graduations are not complete, and the lengths on the scale are not proportional to the numbers. We should like to be able to read both the indices and the numbers on scales marked decimally. To achieve this we set out two scales with decimal divisions at right angles to one another on a piece of squared paper as in Fig. 18. On the vertical line we mark, on the decimal scale, points A, B, C, D, E, F, G, H, J corresponding to the value of the indices in our table. Along the base line we mark, again on the decimal scale, points corresponding to the values of the numbers,

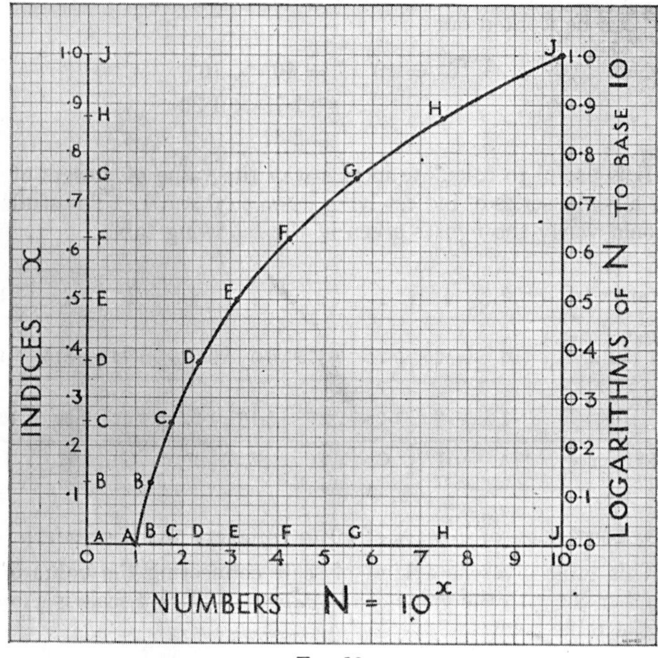

FIG. 18.

identifying the points by the reference letters A to J. Then we find on the grid of squares between the baseline and the vertical line a point which corresponds both to the point on the base line and the corresponding point on the vertical line and mark the point so found with its identifying letter. Now if we join these points by a smooth curve, we can read the value of the number for *any* value of the index and the value of the index which corresponds to *any* number. We have, in fact, plotted a "graph" relating the two quantities.

Make such a graph for yourself to a scale about four times that of Fig. 18 so as to obtain greater accuracy. For example, when the *index* (x) is 0·4 the corresponding *number* (10^x) is approx. 2·52 (the 3rd figure is somewhat of an estimate). That is $10^{0\cdot4} \simeq 2\cdot52$ (a more accurate value is 2·512).

Similarly, we can read off the index which gives the power of 10 equivalent to any number (between 1 and 10). For example, we can express 6 as a power of 10 by noting that 6 on the "Numbers" base line corresponds to 0·78 on the "Indices" vertical line, *i.e.*, $6 = 10^{0\cdot78}$ (a more accurate value is $10^{0\cdot7782}$).

3.4. Use of the Conversion Graph

Now that we can express any number (between 1 and 10) as a power of 10 and convert any power of 10 (between 10^0 and 10^1) into a number, we are in a position to *multiply* numbers by *adding* indices and to *divide* by subtracting indices.

For example, multiply 2·35 by 3·75. Using the conversion graph, we find that, as nearly as we can read it

$$2\cdot35 = 10^{0\cdot372}$$
$$3\cdot75 = 10^{0\cdot575}$$

To multiply we add indices,

$$\therefore \quad 2\cdot35 \times 3\cdot75 = 10^{0\cdot372+0\cdot575} = 10^{0\cdot947}$$

From the graph we find that the *number* corresponding to the index 0·947 is 8·8,

$$\therefore \quad 2\cdot35 \times 3\cdot75 = 8\cdot8 \text{ approx.}$$

Checking by ordinary arithmetic, we get 8·8125.

In a similar way $\dfrac{3\cdot75}{2\cdot35} = \dfrac{10^{0\cdot575}}{10^{0\cdot372}} = 10^{0\cdot575-0\cdot372}$

$$= 10^{0\cdot203} = 1\cdot60$$

Ordinary long division gives the result 1·595.

We can also find squares, say of 2·35. From the graph $2\cdot35 = 10^{0\cdot372}$. To square a power we multiply the index by 2.

So $\qquad 2\cdot35^2 = 10^{0\cdot372\times2} = 10^{0\cdot744}$

Referring again to the graph, we find

$$2{\cdot}35^2 = 10^{0{\cdot}744} = 5{\cdot}55$$

And we can find square roots, say of 3·75. From the graph

$$3{\cdot}75 = 10^{0{\cdot}575}$$

To find the square root of a power we halve the index:

So $$\sqrt{3{\cdot}75} = 10^{\frac{1}{2} \times 0{\cdot}575} = 10^{0{\cdot}29}$$

Referring again to the graph, we find

$$\sqrt{3{\cdot}75} = 10^{0{\cdot}29} = 1{\cdot}95$$

You should try out similar calculations with other numbers and verify your results by ordinary arithmetic.

The use of indices to substitute the simple processes of addition, subtraction, and halving for the more complicated processes of multiplication, division, and extracting the square root is so important that a special name has been given to the indices used in this way. They are called " *Logarithms* " * and accordingly in Fig. 18 we have labelled a second vertical line to the right " Logarithm of N to base 10 ". Thus our graph enables us to find the " logarithm " of any number between 1 and 10, or, knowing the logarithm, to find the corresponding number. But as you will see the vertical line on the right (for logarithms) is an exact reproduction of the vertical line on the left (for indices). *So the logarithm to the base* 10 *of any number is the index of that power of* 10 *which is equal to the number.*

3.5. Tables of Logarithms

Calculations cannot be made very accurately with the graph, because it cannot be read at all accurately; furthermore, since the portions between the plotted points are sketched, it is not very accurate in itself. For greater accuracy we need to use values which are themselves more accurately calculated, and we need a great many more of them. It is possible (by a method which you will learn later) to calculate the power of 10 (or the logarithm to " base 10 ") of any number to any desired degree

* From two Greek words: " Logos "—word, ratio, and " Arithmos " —number, as in " Arithmetic ".

of accuracy. Once we have calculated these logarithms for sufficient numbers between 1 and 10, and arranged them in a systematic table, we have no further need of the graph; we can read off the logarithms and the numbers from the table. Such a table for all numbers from 1·000 to 9·999, advancing 0·001 at a time, is known as a Four-Figure LOGARITHM TABLE (see the Tables at the end of the book).

This table thus covers the same range of numbers as our graph, but with far greater detail and precision.

These tables are very easy and quick to use. For example, to express 2·35 as a power of 10 (to find the " logarithm of 2·35 ", in other words) look for the first two figures 23 in the left-hand column of the table. Along this row in the main body of the table we find under Column 5 the number 3711. This tells us that $2{\cdot}35 = 10^{0{\cdot}3711}$. (Also on the same row, we note that $2{\cdot}30 = 10^{0{\cdot}3617}$ and $2{\cdot}38 = 10^{0{\cdot}3766}$.)

The *fourth* figure of a number, *e.g.*, 6 in 2·356 can be obtained by adding to 3711 the number (11) in the 6th column of the " differences " table, at the right-hand side of the main table, but still on the same row 23,

$$i.e., \quad 2{\cdot}356 = 10^{0{\cdot}3722}$$

Note that all the four-figure numbers in the main body of the table are *decimal* numbers; the decimal points are omitted to make the table more compact.

To summarise: *the tables enable us to express immediately any number between* 1 *and* 10 *as a power of* 10.

3.6. Examples of the Use of Logarithm Tables

(i) *To Multiply* 2·34 *by* 3·19

Using the logarithm tables,

$$2{\cdot}34 = 10^{0{\cdot}3692} \quad \text{(Row 23, Col. 4)}$$
$$3{\cdot}19 = 10^{0{\cdot}5038} \quad \text{(Row 31, Col. 9)}$$
$$\therefore \quad 2{\cdot}34 \times 3{\cdot}19 = 10^{0{\cdot}3692} \times 10^{0{\cdot}5038}$$
$$= 10^{0{\cdot}8730} \text{ (adding `` indices '')}$$

From the body of the log table

$$10^{0{\cdot}8727} = 7{\cdot}46 \text{ (the nearest to } 0{\cdot}8730)$$

A " difference " of 3 appears in the " difference column " 5, on row 74. So the fourth figure is 5.

i.e.,

$$10^{0\cdot8730} = 7\cdot465$$

$$\therefore \quad \underline{2\cdot34 \times 3\cdot19 = 7\cdot465.}$$

Check this on your graph, and by long multiplication.

(ii) *To Divide* 8·729 *by* 2·764.

Now

$$\frac{8\cdot729}{2\cdot764} = \frac{10^{0\cdot9409}}{10^{0\cdot4415}}$$

$$= 10^{0\cdot4994} \text{ (subtracting " indices ")}$$

From the body of the log table

$$10^{0\cdot4983} = 3\cdot15$$

A " difference " of 11 appears in the " Difference Col. 8 " on row 31, so

$$10^{0\cdot4994} = 3\cdot158$$

Thus

$$\underline{8\cdot729 \div 2\cdot764 = 3\cdot158.}$$

(iii) *To Evaluate* $\dfrac{4\cdot67 \times 1\cdot082}{3\cdot57}$

$$= \frac{10^{0\cdot6693} \times 10^{0\cdot0342}}{10^{0\cdot5527}}$$

$$= \frac{10^{0\cdot7035}}{10^{0\cdot5527}} \text{ (adding indices in the numerator)}$$

$$= 10^{0\cdot1508} \text{ (subtracting indices)}$$

$$= \underline{1\cdot415}$$

In this instance

$$10^{0\cdot1492} = 1\cdot41$$

Note that 1492 is 16 short of 1508.

For the fourth figure a difference of 15 appears in Column 5, and a difference of 18 in Column 6.

So 1·415 is nearer to the truth than 1·416.

With complicated fractions we always reduce numerator and denominator to a single power or logarithm, and then subtract the log of the denominator from that of the numerator.

(iv) *Evaluate* $2{\cdot}37 \times (1{\cdot}52)^2$

$$= 10^{0\cdot3747} \times (10^{0\cdot1818})^2$$
$$= 10^{0\cdot3747} \times 10^{0\cdot3636}$$
$$= 10^{0\cdot7383}$$
$$= \underline{5{\cdot}474}$$

N.B. Squaring a power of 10 means *doubling* the index, *i.e.*, doubling the logarithm (see para. 3.1).

(v) *Find the Square Root of* 8·219.

Now $8{\cdot}219 = 10^{0\cdot9148}$

Taking the square root of a power means *halving* the index,

$$\therefore \quad \sqrt{8{\cdot}219} = 10^{0\cdot4574}$$
$$= \underline{2{\cdot}866(5)}$$

(This result is mid-way between 2·866 and 2·867.)

EXERCISE 3(*a*)

A. 1. Using the method of para. 3.2(*b*), find approximately the value of the following powers of 10 :

$$10^{\frac{1}{16}},\ 10^{\frac{9}{16}},\ 10^{1\frac{5}{8}},\ 10^{\frac{3}{2}}$$

2. From your own graph (Fig. 18) read off the values of $10^{0\cdot27}$, $10^{0\cdot35}$, $10^{0\cdot62}$, $10^{0\cdot97}$.

Check your results by the rule for multiplying powers.

3. Express the following as powers of 10, using the log table : 7·3, 4·89, 5·04, 6·346, 1·904, 7·008.

4. Express the following powers of 10 as ordinary decimal numbers, using the log table : $10^{0\cdot8762}$, $10^{0\cdot6069}$, $10^{0\cdot0094}$, $10^{0\cdot9929}$, $10^{0\cdot3070}$.

B. Use the log table to evaluate :

5. $2{\cdot}7 \times 3{\cdot}1$.

6. $8{\cdot}92 \div 5{\cdot}6$.

7. $3{\cdot}402 \times 1{\cdot}964$.

8. $9{\cdot}006 \div 2{\cdot}488$.

9. $\dfrac{2{\cdot}97 \times 1{\cdot}718}{4{\cdot}007}$.

10. $\dfrac{8{\cdot}916}{4{\cdot}29 \times 1{\cdot}374}$.

11. $2{\cdot}93 \times (1{\cdot}76)^2$.

12. $8{\cdot}92 \div (2{\cdot}73)^2$.

C. 13. 1 in. ≡ 2·54 cm. Convert (*a*) 2·19 in. to cm.; (*b*) 7·28 cm. to in.

14. A visiting-card measures 1·24 in. long by 3·32 in. wide. Find the total area of cardboard used in making 1000 such cards.

15. 400-lb. copper wire has a resistance of 2·2 ohms per mile. Show that the resistance of 4200 yd. of such wire is given by $\dfrac{4{\cdot}2}{1{\cdot}76} \times 2{\cdot}2$ ohms, and work this out to the nearest hundredth of an ohm.

16. Find the volume of the contents of a cylindrical zinc container for an element of a dry battery if it is 2·24 in. long and 1·07 in. in radius. (Assume $\pi = 3{\cdot}142 = 10^{0{\cdot}4972}$.)

17. If 1 kg. ≡ 2·205 lb., what is the weight of 7 lb. of solder in kg.?

18. An equipment bay in a telephone exchange is 8 ft. high and 3 ft. 6 in. wide. If 1 metre is equivalent to 1·094 yd., express these dimensions in metres (to nearest cm.)

19. Assuming 1 mile ≡ 1·609 km., how many miles of wire are there on a 5-km. drum?

20. A kilowatt of electrical power is equivalent to 1·34 h.p. in mechanical power.

How many kilowatts do you expect from a small generator, whose rated horse-power is 7·29?

3.7. Extension of the Idea of Logarithms to any Number

Any number can be expressed in " standard form ":

e.g., $$\overset{\downarrow}{2}37{\cdot}6 = 2{\cdot}376 \times 10^2$$

But 2·376 (a " standard number ") can be expressed as $10^{0{\cdot}3758}$ from the log table.

Hence
$$\underline{\overset{\downarrow}{2}37{\cdot}6} = 2{\cdot}376 \times 10^2$$
$$= 10^{0{\cdot}3758} \times 10^2 = 10^{2+0{\cdot}3758}$$
$$= \underline{10^{2{\cdot}3758}} \text{ by rules for multiplying indices,}$$

or the log of 237·6 is 2·3758.

Similarly,
$$0{\cdot}002376 = 2{\cdot}376 \times 10^{-3}$$
$$= 10^{0{\cdot}3758} \times 10^{-3}$$
$$= 10^{-3+0{\cdot}3758}$$

At once we are in a quandary, for

$$-3 + 0{\cdot}3758 \text{ (3 " down " and 0·3758 " up ")}$$

is clearly equivalent to 2·6242 *down* on balance, *i.e.*, $-2{\cdot}6242$. We have had to make a four-figure subtraction in order to find the log of a number less than one.

This is a time-consuming business.

In order to avoid it, we adopt the convention of retaining the separate parts -3 and $+0{\cdot}3758$ and adopt a " bar " symbol $\bar{3}{\cdot}3758$ * to indicate that the whole number part of the log is *negative*, while the decimal portion is positive.

* In words " Bar Three Point Three Seven Five Eight ".

Vocabulary: The decimal part of a logarithm—the number recorded in the log table—is called the " mantissa " *; while the whole-number part, which tells us the position of the decimal point, is known as the " characteristic ".

3.8. Further Examples in the Use of Logarithms

(i)
$$\frac{\overset{\downarrow}{2}\cdot 93 \times \overset{\downarrow}{1}48\cdot 1}{3\overset{\downarrow}{7}\cdot 9} = \frac{10^{0\cdot 4669} \times 10^{2\cdot 1706}}{10^{1\cdot 5786}}$$
$$= \frac{10^{2\cdot 6375}}{10^{1\cdot 5786}}$$
$$= 10^{1\cdot 0589}$$
$$= \underline{1\overset{\downarrow}{1}\cdot 45}$$

The arrow indicates the " standard " position of the decimal point in each number. This helps the eye to see the characteristic, *e.g.*,

$$\overset{\downarrow}{1}48\cdot 1 = 1\cdot 481 \times 10^2$$
$$= 10^{0\cdot 1706} \times 10^2$$
$$= 10^{2\cdot 1706}$$

Thus, for 148·1 the decimal point is 2 *places to the* RIGHT of the standard position; the characteristic of the log is + 2.

(ii)
$$0\cdot 0\overset{\downarrow}{2}97 \times 0\cdot \overset{\downarrow}{1}902 = 10^{\bar{2}\cdot 4728} \times 10^{\bar{1}\cdot 2792}$$
$$= 10^{\bar{3}\cdot 7520}$$
$$= \underline{0\cdot 00\overset{\downarrow}{5}65}$$

In this case the decimal point in each number is to the LEFT of the standard position indicated by the arrow,

e.g.,
$$0\cdot 0297 = \frac{2\cdot 97}{100}$$
$$= 2\cdot 97 \times 10^{-2}$$
$$= 10^{0\cdot 4728} \times 10^{-2}$$
$$= \underline{10^{\bar{2}\cdot 4728}}$$

* A Latin word meaning " makeweight ".

Rough Check

Very roughly

$$0{\cdot}0297 \times 0{\cdot}1902 \simeq 0{\cdot}03 \times 0{\cdot}2$$
$$= \frac{3}{100} \times \frac{2}{10}$$
$$= \frac{6}{1000} \text{ or } \underline{0{\cdot}006}$$

which is roughly (to " 1 significant figure ") the same as 0·00565. A rough check is useful in confirming our placing of the decimal point.

(iii) *To calculate the cube of* 0·0829

$$0{\cdot}0829 = 10^{\bar{2}{\cdot}9186}$$
$$\therefore \quad (0{\cdot}0829)^3 = (10^{\bar{2}{\cdot}9186})^3$$
$$= 10^{\bar{6}+2{\cdot}7558}$$
$$= 10^{\bar{4}{\cdot}7558}$$
$$= \underline{0{\cdot}0005699}$$

Note: Where negative characteristics are involved, it is important to remember that the *mantissa* is always *positive*.

In this case, cubing the number meant multiplying the log $\bar{2}{\cdot}9186$ by 3.

i.e.,
$$3(-2 + 0{\cdot}9186)$$
$$= -6 + 2{\cdot}7558$$
$$= \underline{-4 + 0{\cdot}7558}$$

which we write $\bar{4}{\cdot}7558$

(iv) *The square root of* 4274

Now $4274 = 10^{3{\cdot}6308}$

Taking the square root involves halving the index,

$$\therefore \quad \sqrt{4274} = 10^{\frac{1}{2}(3{\cdot}6308)}$$
$$= 10^{1{\cdot}8154}$$
$$= \underline{65{\cdot}38}$$

So the square root is 65·4 (3 significant figures).

Formal methods of calculating with logarithms are treated in more detail in the next chapter. Use common sense in dealing with negative characteristics; whenever in doubt, make a rough check to confirm your placing of the decimal point.

EXERCISE 3(*b*)

Express the following numbers in standard form:

(Example $3\overset{\downarrow}{9}7{\cdot}2 = 3{\cdot}972 \times 10^2$.)

1. 29·2.
2. 5003.
3. 0·047.
4. 2,370,000.
5. 0·000000818.

Evaluate the following by logs:

6. $\dfrac{172{\cdot}5}{14{\cdot}8}$.
7. $5{\cdot}28 \times (16{\cdot}91)^2$.
8. $\sqrt{15{\cdot}81}$.
9. $\dfrac{0{\cdot}0372 \times 5{\cdot}19}{7{\cdot}34}$.
10. $\dfrac{0{\cdot}1982 \times 17{\cdot}53}{262{\cdot}7 \times 0{\cdot}086}$.
11. $0{\cdot}0092\sqrt{229{\cdot}4}$.
12. $\dfrac{17{\cdot}9 \times (0{\cdot}0188)^3}{73{\cdot}9}$.

13. 1 litre = 0·22 gall. A heavy lorry holds 32 gall. of petrol. Express this in litres to the nearest litre. (Check by simple arithmetic.)

14. 1 B.Th.U. (British Thermal Unit) of heat is equivalent to 0·293 watt-hours of electrical energy. If a kilowatt electric fire burns for 4 hr. 10 min. calculate: (*a*) the number of watt-hours consumed, and (*b*) the equivalent number of B.Th.U.s.

15. If 1 hectare = 2·47 acres, how many hectares does 1 square mile (640 acres) contain?

16. Calculate the area of a circle whose diameter is 4·36 in. (Allow $\frac{\pi}{4} = 0{\cdot}7854$.)

17. Convert the previous answer to sq. cm. (1 in. = 2·54 cm.).

18. Find the radius in feet and inches of a circular flower-bed whose area is 29 sq. yd.

19. A cylinder of diameter 6·2 cm. has a volume 39·7 c.c. Find its length to the nearest mm.

20. Find the weight of a solid lead shot (assumed spherical and 0·42 cm. in diameter) if lead weighs 11·37 gm. per c.c. The volume of a sphere of radius r is $\frac{4}{3}\pi r^3$. (Answer to nearest milligram.)

3.9. Logarithms in Battle Dress—The Slide-rule

For the engineer or the statistician a book of logarithm tables is not handy enough for day-to-day calculation. Many attempts have been made to design a robust pocket-size calculator based on the logarithm tables. The most practicable and convenient device is the " slide-rule ".

You should construct a simple slide-rule for yourself. Even if you own a real rule it is worth while to try this out.

On graph paper draw a convenient base line AA' (see Fig. 19) covering just over 100 small squares. Below AA' mark off in pencil 10 squares at a time over the complete range the distances 0, 0·1, 0·2, . . . etc., up to 1·0.

The zero mark 0 represents the *Logarithm* of 1, as we have seen (*i.e.*, $10^0 = 1$). In INK number this mark as " 1 " both above and below the base line. Similarly, mark as " 2 " the scale distance 0·301 representing the *Logarithm* of 2. Repeat this

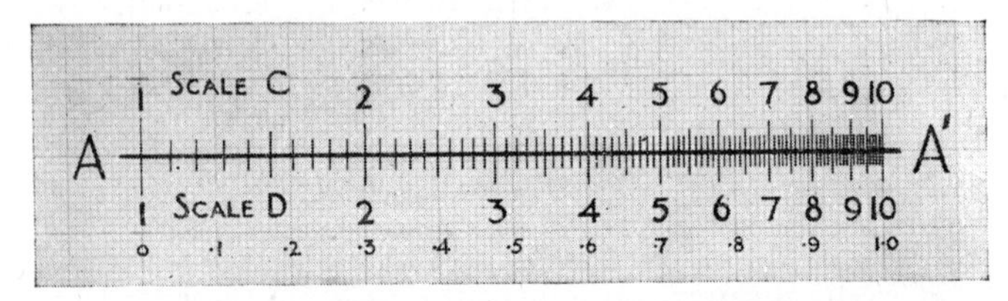

FIG. 19.—THE HOME-MADE SLIDE-RULE.

process with " 3 " (scale distance 0·477), and so on, up to " 10 " (scale distance 1·0, since $10^1 = 10$). The distances are thus *logarithmic*, the logarithm table being used to obtain them.

A little patience is required to insert intermediate values along the scale. For example, the mark " 2·3 " is inserted at a scale distance " 0·362 " ($10^{0·362} = 2·3$).

The distances on this scale (known originally as a " Gunther Scale " after the mathematician who first popularised it) are not proportional to the *numbers* marked, but to their *logarithms*. Distances on this scale thus represent INDICES, and so the addition of distances on the scale is equivalent to the *addition* of *indices*, and therefore to *multiplication* of the numbers shown on the scale.

The first users of these Gunther Scales required a pair of dividers to add these " index-distances ". Adding distances (moving to the *right*) gave the results of *multiplying*. Subtracting distances (moving to the *left*) gave the results of *dividing*.

The inventions of the sliding stock and cursor speeded up these processes, and produced in essentials the modern slide-rule (see Fig. 21).

3.10. Using the Slide-rule

Let us return to our home-made Gunther Scale. Mount this on thin card, and then carefully cut along the line AA'; you will then possess two identical log scales. (If pasting the paper on one side of the card causes it to warp, paste a piece of the same kind of paper on the other side as well.)

(a) *Multiplication*

To multiply two numbers we must add their logarithms and then find the number of which the sum is the logarithm.

On our slide-rule the scale marks represent the numbers, and

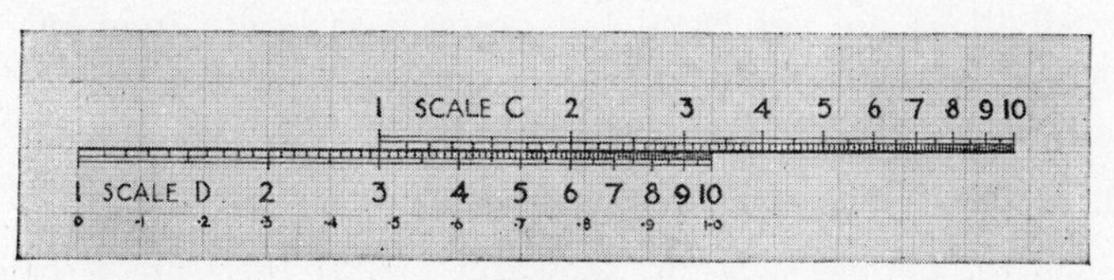

FIG. 20.

the corresponding lengths represent the logarithms. So we have to add the lengths corresponding to the numbers we wish to multiply.

To add lengths we set them end to end. So to multiply 3 by 2 we look for the scale mark " 3 " on one scale (D) and set the beginning of the second scale (C), *i.e.*, the scale mark " 1 ", against it—as in Fig. 20. Then we look on the Scale C for the scale mark " 2 ", and note the scale mark against which it stands on Scale D. The length between this mark and the beginning of Scale D is the sum of the two lengths corresponding to the scale marks " 3 " and " 2 ". So it represents the sum of the logarithms of 3 and 2; that is, it is the logarithm of the product of 3 and 2, and the scale mark represents the product $3 \times 2 = 6$.

(b) *Division*

To divide 6×2 we must subtract from the length representing the logarithm of 6 that representing the logarithm of 2. To subtract lengths we set them side by side and see what balance is left. So we set the scale mark 2 on Scale C against the scale mark 6 on Scale D, and then note the mark on Scale D against which the beginning (*i.e.*, the mark " 1 ") of Scale C stands. The length of this mark from the beginning of Scale D represents the difference between the logarithms of 6 and 2; that is it represents the logarithm of (6 divided by 2), *i.e.*, log 3.

(c) *Two- and Three-figure Numbers*

For numbers having 2 significant figures, such as 2·5, 37, 920, we have definite scale marks. For three-figure numbers, such as 2·53 and 927, we have to imagine that the divisions between scale marks are graduated and make our settings and take our readings from the imagined graduations. Bear in mind that these subdivisions are not equal but decrease in length from left to right.

3.11. One-decade and Two-decade Scale

On the normal slide-rule shown in Fig. 21 there are 4 scales, A, B, C, and D. Scales D and C are one-decade scales extending the full length of the rule, like Scales D and C of our home-made rule. Scales A and B are each made up of two scales similarly marked and divided, but of half the length of Scales C and D.

These two-decade scales help us to deal with numbers having ratios up to 100 to 1, without resetting the rule; with one-decade scales we can deal only with numbers having ratios up to 10 to 1.

For example, we can multiply 9 by 8 by setting the rule to these numbers in the first decade of Scales A and B, and reading off the answer 72 in the second decade of Scale A. But if we try this on the one-decade Scales C and D or on our one-decade home-made rule, the reading against 8 on Scale C is " off the map " on Scale D. In such a case we can set the scale mark 10 at the *end* of Scale C against the mark 9 on Scale D and

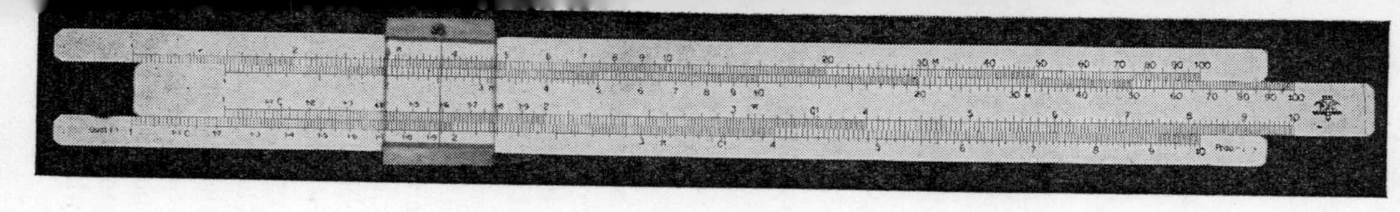

The Normal 10-in. Slide-rule.

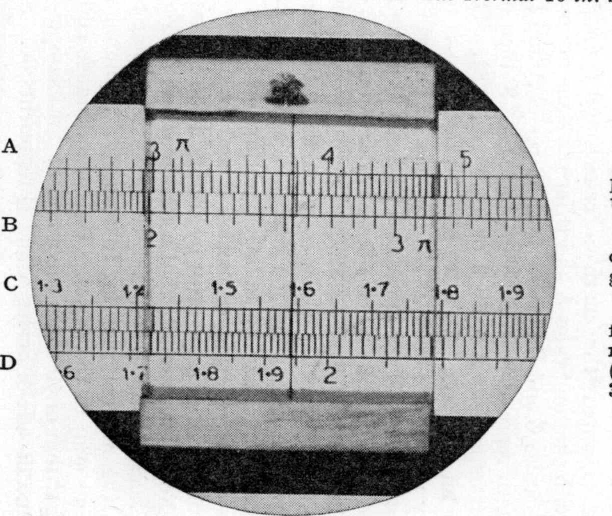

Note : In this figure the rule is set for multiplying 1·49 (on scale A) by 2·53 (on scale B), giving the result 3·77 (on scale A).

The same setting could, of course, be used for dividing 3·77 (on scale A) by 2·53 (on scale B) giving the result 1·49 (on scale A).

The same position of the cursor also illustrates finding the square root of 3·77 (on scale A), by reading down to scale D, giving the result 1·94. (And conversely, the square of 1·94 is seen to be 3·77.)

Enlarged View of Cursor.

FIG. 21.—THE MODERN SLIDE-RULE.

(*By permission of Messrs. A. G. Thornton, Ltd.*)

read off 72 against the mark 8 on Scale C (see Fig. 22). This is equivalent to taking log 9 − log 10 + log 8 = log 7·2.

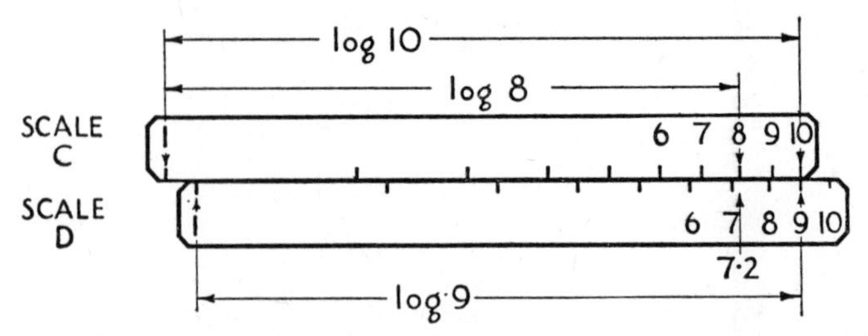

FIG. 22.—TO ILLUSTRATE THE MULTIPLICATION OF 9 BY 8 ON THE SINGLE-DECADE SCALE.

3.12. Use of Two-decade Scales

The two-decade scales (A and B) have another very important use. Compare the length on Scales C and D for, say, the number 9 with that on Scales A and B for the same number. The length on Scales C and D is twice that on Scales A and B. Thus if this length represents on Scales C and D log 9, then on Scales A and B it represents $2 \times \log 9$ or $\log 9^2 = \log 81$. The same is true for all numbers. So if we read straight across to Scale A from any number on Scale D we can read off on Scale A *the square of that number*. Thus opposite 9 on Scale D we find 81 on Scale A and opposite 1·2 on Scale D we find 1·44 on Scale A.

Similarly, if we read straight across from Scale A to Scale D we find a marking which corresponds to the *square root*. Thus if we are calculating on Scales A and B we can introduce a squared quantity by taking a reading on Scales C or D, and if we are calculating on Scales C and D we can introduce the square root of a quantity by reading it on Scales A or B.

Note very carefully that a scale mark on the right-hand decade of Scales A and B is at a distance corresponding to the logarithm of the number plus log 10 from the beginning (1 at the left-hand end) of the scales. Thus the distance corresponds to the logarithm of 10 times the number; for example, 6 on the right-hand decade of Scales A and B represents 60. So when reading off square roots we must regard the numbers as spread over Scales A or B as shown on Fig. 23.

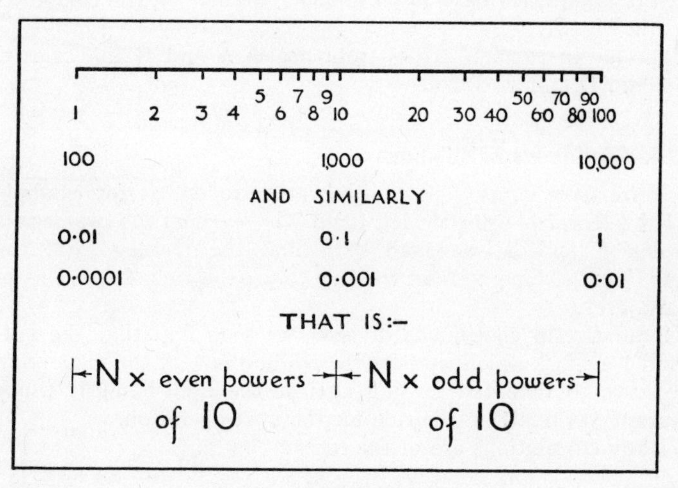

FIG. 23.

Practice

The following examples in Exercise 3(*a*) make excellent first practice in handling a slide-rule; the one-decade scales will be found sufficient for most of these calculations.

Easy: Nos. 5–8, 13, 14, 17, 19, 20
Slightly Harder: Nos. 9–11, 15, 16, 18.

3.13. Use of the Cursor

The cursor is of help:

(*a*) In reading off a number which comes between two divisions on the scale; it provides, so to speak, a movable scale mark when we are using the imagined graduations referred to in para. 3.10.

(*b*) In marking a reading on one scale while we are looking for a reading in another. For example, in dividing 3·77 by 2·53 we can set the cursor to 3·77 on Scale A and then looking at Scale B, bring 2·53 on it under the line on the cursor, as in Fig. 21.

(*c*) In reading across from Scales A and B to Scales C and D and *vice versa*.

3.14. Continuous Calculations

If we have continuous multiplication to do, as for example $5{\cdot}4 \times 6{\cdot}2 \times 7{\cdot}1$, using the left-hand " 1 " even on the two-decade Scales A and B, we shall soon find the reading " off the map " and so have to reset the rule to use the right-hand 1 as in para. 3.11.

Similarly, in continuous division we soon find that the left-hand 1 against which we take the reading is " off the map " and we have to take the reading against the right-hand 1, which necessitates resetting the rule for the next operation.

Many calculations are of the form:

$$\frac{a \times b \times c \times d \times \ldots}{z \times y \times x \times w \times \ldots}$$

and you will find it easier to divide and multiply alternately, rather than to find $a \times b \times c \times d$ and the $z \times y \times x \times w$ and finally divide one by the other. So we first divide a by z and then multiply by b, then divide by y, then multiply by c, and so on.

3.15. Locating the Decimal Point

There are rules by means of which, if one is very methodical in setting the slide-rule, one can determine the position of the decimal point in the result of the calculation.

It is much simpler and more reliable to

(a) *Make a rough check calculation.*—If necessary convert the numbers to standard form (Section 3.1) using only one or at the most two significant figures for each, and fix the decimal point according to the power of 10 arrived at in the rough answer.

For example, to calculate the weight in grams per metre of a wire weighing $6\frac{1}{2}$ lb. per mile:

$$1 \text{ mile} \equiv 1609 \text{ metres}$$

$$\text{Weight of 1 metre} = \frac{6\cdot5}{1609} \text{ lb.}$$

$$1 \text{ lb.} \equiv 453\cdot6 \text{ gm.}$$

$$\text{Weight of 1 metre} = \frac{6\cdot5 \times 453\cdot6}{1609} \text{ gm.}$$

Using the slide-rule, we find that the figures in the answer are 183.

Converting to standard form for a rough check:

$$\text{Weight of 1 metre} = \frac{6 \times 5 \times 10^2}{2 \times 10^3}$$

$$= \frac{3 \times 10^3}{2 \times 10^3} = 1\cdot5 \times 10^0$$

So the wire weighs 1·83 *gm. per metre.*

(b) *Use the correct decades on Scales A and B. Remember always that the left-hand ends of Scales A and B relate to numbers on the even decades*:

e.g.,	0·0001 to 0·0009,	*i.e.,*	1×10^{-4} to 9×10^{-4}
	0·01 to 0·09,	*i.e.,*	1×10^{-2} to 9×10^{-2}
	1 to 9,	*i.e.,*	1×10^{0} to 9×10^{0}
	100 to 900,	*i.e.,*	1×10^{2} to 9×10^{2}

and so on;

and the right-hand ends to numbers on the odd decades:

e.g.,	0·001 to 0·009,	*i.e.,*	1×10^{-3} to 9×10^{-3}
	0·1 to 0·9,	*i.e.,*	1×10^{-1} to 9×10^{-1}
	10 to 90,	*i.e.,*	1×10^{1} to 9×10^{1}
	1000 to 9000,	*i.e.,*	1×10^{3} to 9×10^{3}

and so on.

If this distinction is made when setting the slide-rule for any calculation, an indication is obtained of whether the result is to be assigned to an even or an odd decade. Referring to the above example, we set the rule as follows:

6·5 (Scale A l.h.) divided by 1809 (Scale B r.h.) and multiplied by 453·6 (Scale B l.h.) gives the reading 162 on Scale A l.h.—indicating that an even decade must be assigned. We have seen that the decade is the 10^0 or units decade.

If the calculation involves square roots this distinction is *essential*, since $\sqrt{64}$, for example, is 8 and $\sqrt{6{\cdot}4}$ is 2·53. For $\sqrt{64}$ we must use the right-hand end, and for $\sqrt{6{\cdot}4}$ the left-hand end.

Example 3(*a*). *Calculate the resonant frequency* (f) *of a circuit containing an inductance* (L) *of* 360 *microhenrys and a capacitance* (C) *of* 0·0005 *microfarads from the formula*:

$$f = \frac{1}{2\pi} \cdot \frac{1}{\sqrt{LC}}$$

$$f = \frac{1}{2 \times \frac{22}{7}\sqrt{360 \times 10^{-6} \times 0{\cdot}0005 \times 10^{-6}}}$$

$$= \frac{7}{2 \times 22 \times \sqrt{3{\cdot}6 \times 10^{-4} \times 5 \times 10^{-10}}}$$

$$= \frac{7}{44 \times 10^{-7} + \sqrt{3{\cdot}6 \times 5}} = \frac{7}{44 \times 10^{-7} \times \sqrt{18}}$$

Since we have to introduce square roots, we must start on Scales C and D of our slide-rule; so we divide 7 (Scale D) by 4·4 (Scale C), giving us the figure 1·59 on Scale D. To divide this by $\sqrt{18}$ we set (with the help of the cursor) 18 on *Scale B* (RIGHT-HAND end since 18 is in an odd decade) against 1·59 on Scale D and read the result 3·74 on Scale D against the end (Scale mark 1) of Scale C.

Making our rough check to locate the decimal point:

$$\frac{7}{44 \times 10^{-7} \times \sqrt{18}}$$

$$\simeq \frac{7}{4{\cdot}4 \times 10^{-6} \times 4}$$

$$\simeq \frac{7 \times 10^{6}}{2 \times 10}$$

$$\simeq 3{\cdot}5 \times 10^{5}$$

So the resonant frequency is $3{\cdot}74 \times 10^5 = 374$ kilocycles per second.

Had we used the figure 18 on the left-hand end of Scale B the result would have been quite different and quite wrong. (Try it !)

3.16. Significant Figures

You will have noticed that no matter how many significant figures are given in the data you are using, you cannot take into account when using the slide-rule more than 3 figures at the "9" end of the scales (*e.g.*, 9·64) or more than 4 figures at the "1" end (*e.g.*, 1·078). It is not possible to rely upon the last figure entirely—it might in the examples given be "5" instead of "4" or "7" instead of "8". So slide-rule calculations cannot at best be more accurate than about 2 in 1000 or 1 in 500. If you need greater accuracy than this, you must use Tables of Logarithms, choosing a 4-figure table or a 5-figure table according to the number of significant figures given in your data and required in your answer. In most cases 4-figure tables are adequate, giving 3-figure accuracy in results.

Note too that the logarithmic scales on a slide-rule have a uniform degree of accuracy throughout. On an ordinary proportionate scale, say a foot rule, an error of $\frac{1}{100}$ in. represents 1 in 50 when measuring half an inch and 1 in 1000 when measuring 10 in. But on a slide-rule a given error in reading (*i.e.*, a given difference between the true and the observed distance) represents the same difference between the two logarithms wherever it occurs and, therefore, the same *ratio* between the true and the observed numbers whatever their size, *i.e.*, the same degree of accuracy throughout.

It is these two facts, namely:

(i) slide-rule calculations are accurate to about 1 in 500, which is adequate for most purposes;

(ii) the degree of accuracy does not vary with the magnitude of the quantities;

that make the engineer so fond of his slide-rule!

EXERCISE 3(*c*)

Mainly in the use of the slide-rule. (For formal practice, the examples in Exercise 3(*b*) can now be attempted with the slide-rule.)

Examples marked "H" can be done using the home-made slide-rule.

1. If 1 oz. = 28·352 gm. express 1 kg. in lb. (to 3 significant figures).

2. A cricket ball weighs 5·6 oz.; express in kg. the weight of a gross of cricket balls.

3. If 1 h.p. = 33,000 ft.-lb. per minute,
= 42·41 B.Th.U. per minute,

how many ft.-lb. of energy are equivalent to 1 B.Th.U.?

4. Open-wire telephone lines often use 200-lb. (per mile) copper conductors. Express this gauge of conductor in kg. per km. (1 kg. = 2·205 lb.; 1 km. = 0·62 mile.)

How would you construct a conversion chart to "translate" lb. per mile to kg. per km.?

5. In English-speaking countries we describe petrol consumption in "miles per gallon", while on the Contintent "litres per kilometre" are quoted. Translate: (*a*) 30 m.p.g. to its continental equivalent; (*b*) 0·15 litres per km. to m.p.g. (1 gallon = 4·54 litres; 1 ml. = 1·609 km.)

6. What is the resistance of a coil of 5320 turns, the average diameter of a turn being 0·62 in., if the wire used has a resistance of 2230 ohms per km.? (1 metre = 39·37 in.)

7. Table A gives the weight per 1000 yd. of three sizes of copper wire in Standard Wire Gauge. Compile a similar Table B in metric units using: (i) a slide-rule; (ii) 4-figure logarithms.

TABLE A

S.W.G.	Diameter in in.	Weight of copper in lb. per 1000 yd.
20	0·036	11·77
26	0·018	2·943
36	0·0076	0·5246

TABLE B

S.W.G.	Diameter in mm.	Weight of copper in kg. per km.

(1 in. = 25·4 mm. 1 kg. = 2·205 lb. 1 km. = 1094 yd.)

8. Convert a map scale of 4 miles to the inch to "km. to the cm." (2 significant figures). (H.)

9. A continental map is quoted as 1 : 25,000. (1 cm. on map ≡ 25,000 cm. on the ground.)

For an English tourist, convert the scale to "inches to the mile" (2 significant figures). (H.)

10. A model is in a scale of $\frac{1}{2}$ in. $\equiv$ 1 ft. Express this in " cm. to the metre ". (H.)

11. One bay of a continental single-channel-carrier telephone equipment has dimensions 55 $\times$ 200 $\times$ 35 in cm. Give the same information in feet and inches. (H.)

12. 1 ampere flowing for 1 sec. in an electroplating vat will deposit 0·329 mg. of copper. Calculate how many kg. will be deposited by a current of 250 amperes in a working-day of 8 hr., allowing 15 min. each end of the day for starting up and shutting down.

13. If the coating in Question 12 is to be 5 mils thick on components each having a surface area of 4·5 sq. in., how many can be treated in one day? (1 cu. in. of copper weighs 0·32 lb.; 1 lb. = 453 gm.)

14. A certain selector switch in an automatic exchange carries an average of 12 calls per busy hour, and an associated set of contact springs operates twice for each call. If the springs wear out in 250,000 operations, how many years of service should they provide? (Assume daily traffic 8 times that in the busy hour, and 300 working-days per annum.) (H.)

15. In the laboratory tests a similar set of springs (Question 14) are operated automatically every 2 sec. How long will it take to test for 250,000 operations, running day and night with 3 half-hour breaks for observation each day? (H.)

16. The following measurements were taken on various samples of copper wire. Find for each specimen the resistance in ohms per mile, and the weight in pounds per mile.

No. of sample.	Lengths in in.	Weight in lb.	Resistance in ohms.
1	40	0·014	0·026
2	54	0·033	0·019
3	24	0·0039	0·032

17. In large automatic exchanges there are on the average about 6·5 relays per line. It can be assumed that the pull which the relay exercises when energised is about 850 gm. Supposing all the relays in an exchange with 7500 lines were energised at once, what would be the total pull in tons? (1 lb. = 453 gm.)

CHAPTER 4

THE DISCOVERY AND MANIPULATION OF FORMULÆ

4.1. Sources of the Formula

For the practical man a formula may arise in three ways:

(a) *As the Generalisation of a Working Rule*

The mensuration formulæ in Chapter 1 are examples of this. In para. 4.2 we shall consider some further instances of such formulæ, which are usually based on intuition—thus we feel " in our bones " that the area of a rectangle must be " length times breadth ".

(b) *As a Concise Expression for an Experimental Law*

Ohm's Law is the illustration most familiar to the student of telecommunications. Such a law could not be argued from intuition—it is based on observation and experiment. Furthermore, it is *not* true for every situation which we meet; it is well to recognise the limitations of Ohm's Law, as well as appreciate its great usefulness. The student should be encouraged to " try for himself " and not accept blindly this or any other mathematical law. To this end some typical experiments are suggested, in para. 4.3. Such experiments should form an integral part of the student's work in mathematics.

(c) *By Mathematical Deduction from some Fundamental Law*

The important formulæ for the periods of oscillation of a swinging pendulum $\left(t = 2\pi\sqrt{\frac{l}{g}}\right)$ and of an oscillating electrical circuit $(t = 2\pi\sqrt{LC})$ are formulæ of this type. They can be verified by experiment, but are not sufficiently simple to be discovered by experiment alone. The mathematical processes involved are beyond the scope of this present volume.

4.2. Discovering Further Mensuration Formulæ

Example 4(*a*). *Find a formula for the total surface area* (A *sq. in.*) *of a closed cylindrical can of radius* r *in. and length* l *in.*

If we imagine the can to be made of thin tinplate, we see that the area of tinplate required to make the whole can consists of three portions—the two flat ends, which are just circles of tinplate, and the curved part of the can, which would normally be made from a rectangle of tinplate. This

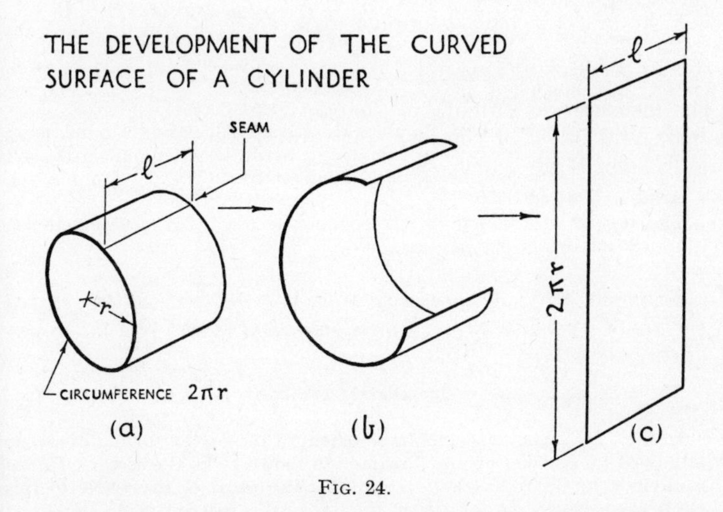

FIG. 24.

" development " of the curved surface of the can as a rectangle l in. long and $2\pi r$ in. wide (*i.e.*, of width equal to the circumference of the can) is shown diagrammatically in Figs. 24(*a*), 24(*b*), 24(*c*).

Thus the area of the curved part $= l \times 2\pi r$
$= 2\pi rl$ sq. in.

And the area of the two ends $= 2 \times \pi r^2$
$= 2\pi r^2$ sq. in.

Combining these two observations into one statement as a formula, if A sq. in. is the total area of tinplate, then

$$A = 2\pi rl + 2\pi r^2$$

This formula as it stands would involve two separate multiplications. But notice that the two terms on the right-hand side of the formula have a common factor $2\pi r$. In fact, we have l times $2\pi r$ added to r times $2\pi r$, giving us $(l + r)$ times $2\pi r$ altogether. So we could write

$$A = 2\pi r(l + r)$$

which has " streamlined " the formula to provide us with one single multiplication. It is indeed the neatest form of the required formula.

Let us verify this by calculation, using simple numbers :

Suppose $l = 8$ (in.), $r = 7$ (in.), using $\pi = \frac{22}{7}$

(i) $A = 2\pi rl + 2\pi r^2$
$= 2 \times \frac{22}{7} \times 7 \times 8 + 2 \times \frac{22}{7} \times 7^2$
$= 44 \times 8 + 44 \times 7$
$= 352 + 308$
$= \underline{660}$ (sq. in.)

or

(ii) $A = 2\pi r(l + r)$
$= 2 \times \frac{22}{7} \times 7 \times 15$
$= 44 \times 15$
$= \underline{660}$ (sq. in.)

Even with such very simple numbers the second calculation is much more compact. The saving in time and energy would be even more striking if the measurements had been more complicated. (Try it with $l = 4{\cdot}3$, $r = 2{\cdot}9$, $\pi = 3{\cdot}14$.)

This process of " taking out the common factor " ($2\pi r$ in this instance) is a most important one (see para. 4.4).

Example 4(*b*). *Obtain a formula for the area of the curved surface of a cone in terms of the slant height* (S *in.*) *and the radius of the base* (r *in.*).

The area we require is like that of the material of a " dunce's cap " (Fig. 25(*a*)).

In Figs. 25(*b*) and 25(*c*) the development of the curved surface of a cone (obtained by opening up the " seam ") is shown to be the sector OPP' of a circle S in. in radius and centre O. The area of the whole of this circle would be πS^2 sq. in. What fraction of this will our sector OPP' be?

We note that the arc PP' in Fig. 25(*c*) is the same length as the circumference of the base of the cone (*i.e.*, $2\pi r$ in.) in Fig. 25(*a*).

Thus $$\frac{\text{the length of arc } PP'}{\text{circumference of whole circle}} = \frac{2\pi r}{2\pi S} = \frac{r}{S}$$

Common-sense intuition tells us that the *area* of the sector OPP' will be this same fraction of the area of the whole circle of radius S in.

i.e., Area of sector OPP' in sq. in. $= \frac{r}{S} \times \pi S^2$

$$= \frac{r \times \pi \times S \times S}{S}$$

$= \pi r S$ (for S " cancels ")

But OPP' is the development of the curved surface (area A sq. in.) of the original cone, so we have the surprisingly simple formula $A = \pi r S$ for the surface area of a cone.

Note: S is the slant height, not the vertical height of the cone. Such a slant height as the " seam " OP is sometimes called a *generator* of the cone

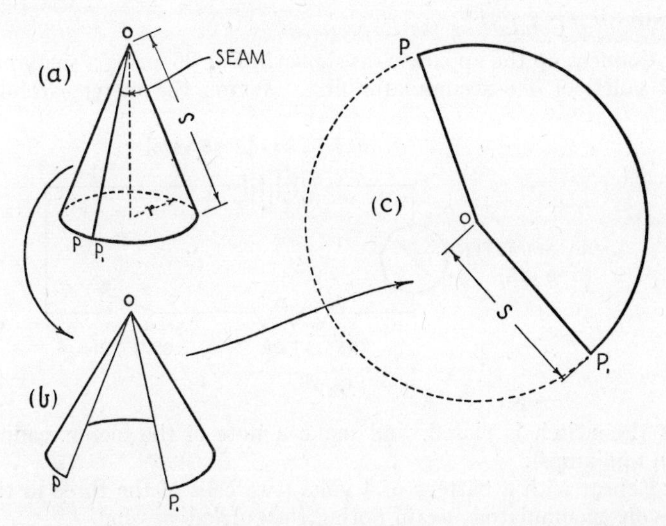

FIG. 25.—THE DEVELOPMENT OF THE CURVED SURFACE OF A CONE.

for if it were to revolve about the vertical height as axis it would generate the curved surface of the cone.

Note that we have " discovered " the above formulæ by the exercise of reason and intuition, and not as the result of experiment.

4.3. Discovering Formulæ through Experiment

EXPERIMENT A. *To Verify Ohm's Law and Achieve a Formula to Describe It*

Apparatus Required

Three fully charged 6-volt accumulators and " wander " lead; a resistor—between 2000 and 3000 ohms for preference (a rheostat would be suitable); an ammeter to read 0–10 in amperes or similar range; a simple switch; insulated wire for connections.

Aim

To investigate how the current passing through a resistor changes (or varies) with different voltages applied.

Method of Conducting the Experiment

Connect up the apparatus as shown in Fig. 26, using a single cell (2 volts) of one accumulator first. Watch the meter carefully

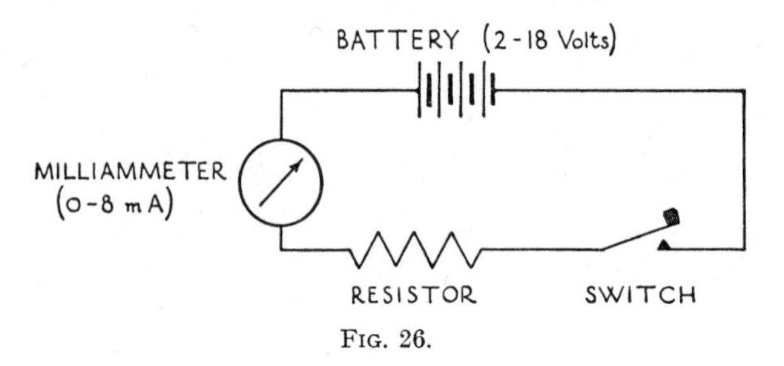

FIG. 26.

as the switch is closed, and make a note of the meter reading (in milliamps).

Repeat with a battery of 4 volts (two cells of the three in the 6-volt accumulator), again noting the current passing.

Continue to obtain readings, stepping up the voltage 2 volts at a time until almost a full reading of the ammeter is achieved or the full 18 volts has been tested. In every observation the switch should be closed only when taking readings.

The readings are best noted in the form of a table; a typical one is shown below:

E.m.f. of battery V volts	2	4	6	8	10	12	14	16	18
Ammeter reading I mA. .	0·85	1·6	2·5	3·3	4·1	5·0	5·8	6·6	7·5
Ratio V/I .	2·35	2·50	2·40	2·42	2·44	2·40	2·41	2·42	2·40

The third line of the table gives the ratio of the voltage to the current for each pair of readings (obtained by slide-rule.)

Trying to Interpret the Results

(i) *Inspecting the figures themselves* it appears that the ratio $\frac{V}{I}$ is about 2·4, whatever the voltage applied may be. If this is so—if, that is, $\frac{V}{I}$ is constant—then we can say that the current passing is directly proportional to the voltage applied. If we double the voltage we double the current passing, and so on.

(ii) *Plotting our results on graph paper* (see Fig. 27), we see that

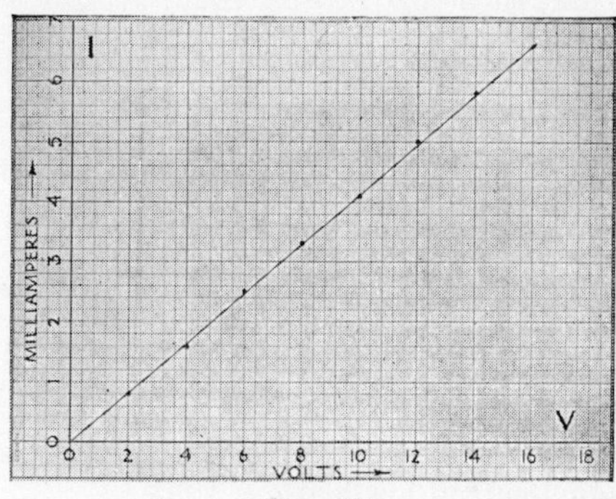

FIG. 27.

the points on the graph corresponding to our pairs of observations lie very nearly on a straight line through the origin. (The origin is the point on the graph paper where $V = 0$ and $I = 0$.)

This confirms our idea that a definite law connects V and I. As we shall see in a later chapter, a " straight-line " relationship between two quantities indicates a simple " linear " law. A straight line through the origin indicates a law of direct proportion, *i.e.*, $\frac{V}{I}$ is constant or $V \propto I$.

Drawing the straight line of " best fit " through our points, we see it passes through the point where $V = 16$ and $I = 6·6$. In other words, as the voltage increases from nothing up to 16 volts the current increases from nothing up to 6·6 mA.

Numerically then, $\frac{V}{I} = \frac{16}{6 \cdot 6} = 2 \cdot 42$ (slide-rule) and this ratio is the same for any other point we care to take on our straight line: for instance, when $V = 9$, $I = 3 \cdot 7$, then

$$\frac{V}{I} = \frac{9}{3 \cdot 7} = 2 \cdot 43 \text{ (slide-rule)}$$

(iii) *Giving a name to this constant* $\frac{\mathrm{V}}{\mathrm{I}}$.

Clearly, if a large voltage is required to force a small current through a resistor, then that resistor could be said to have a high resistance to the passage of current. And, of course, $\frac{V}{I}$ would be large.

So it is sensible to measure the resistance of the resistor by using this ratio $\frac{V}{I}$, for it is independent of the size of the voltage, and depends only on the character of the resistor itself.

If we express V in volts and I in amperes, the ratio $\frac{V}{I}$ is said to be in ohms, after George Simon Ohm, a German physicist and pioneer in electrical work, who was a professor at Munich University from 1849 till his death in 1854.

This relation between the voltage and current, expressed symbolically as $R = \frac{V}{I}$ or $V = IR$, is called OHM'S LAW.

With our sample figures

$$V = 16 \text{ (volts)} \quad I = 6 \cdot 6 \times 10^{-3} \text{ (amperes)}$$

Then the resistance R ohms in our circuit is given by

$$\begin{aligned} R &= \frac{V}{I} \\ &= \frac{16}{6 \cdot 6 \times 10^{-3}} \\ &= \frac{16}{6 \cdot 6} \times 10^{3} \\ &= \underline{2420} \text{ (ohms)} \end{aligned}$$

To 2 significant figures, the resistance in the circuit (including that of the ammeter) is thus 2400 ohms.

A Note to the Sceptical Student

It should be noted that the plotted points in the graph above do not exactly fit the straight line. It would be suspicious if they did! For consider the following sources of possible error:

(*a*) The meter itself is read by eye—the smaller the current the greater is the percentage inaccuracy in reading it. Thus an error of 0·05 in measuring 1 mA. is a 5% error, while 0·05 in 5 mA. is only 1% error. Two-figure accuracy is the best we can hope for.

(*b*) We have ignored the internal resistance of the accumulators. This is negligible if the resistor has a comparatively high resistance.

(*c*) We have assumed that the individual cells of the three accumulators have the same e.m.f. of 2 volts, or at least an equal voltage. There is likely to be considerable variation here.

EXPERIMENT B. *To Investigate the Movement of a Swinging Pendulum*

We are all fascinated by the rhythm of an oscillating pendulum. Galileo is said to have been inspired by the swinging of a massive bronze lamp in the Cathedral at Pisa. The present writers never fail to enjoy the measured pulse of the Foucault Pendulum at the Science Museum.

How does the time of the pendulum's swing vary with its length? Does the amount of swing (its " amplitude ") alter the tempo?

Galileo used his own pulse to time the oscillations of that bronze lamp, and years later used the fruits of his observations in constructing one of the first astronomical clocks.

To perform a similar experiment, the simplest apparatus will be sufficient.

Aim of Pendulum Experiment

To discover how the time of swing (or the " period " of oscillation) of a simple pendulum varies with its length.

Method of Conducting the Experiment

Improvise a simple pendulum, using a small heavy "bob" (a lead plummet is ideal) suspended from a long, light silken cord. The longer this cord the more accurate will be the results. Beginning with a length of say, 9 ft., set the pendulum swinging gently and record the time taken for 50 complete swings (to and fro). Divide by 50 to obtain the period in seconds. Then shorten the length of the cord, and again find the period of oscillation. Draw up a table of values—a specimen set are shown below:

Length of pendulum, l ft. .	9	8	7	6	5	4	3
Time for 50 swings in sec. .	165	156	146	135	124	111	97
Period (one complete swing), t sec. . .	3·30	3·12	2·92	2·70	2·48	2·22	1·94

Trying to Interpret the Results

(i) *Examination of figures in our table.* It does not appear from the table that the time t sec. is directly proportional to the length l ft.; thus considering the cases $l = 4$, $l = 8$, doubling the length certainly does not double the period of swing. However, as the length increases, the period does indeed steadily increase.

On comparison of the times for $l = 4$ ($t = 2{\cdot}22$) and $l = 9$ ($t = 3{\cdot}30$) we might spot that these lengths are in the ratio 4/9, while the times are very nearly in the proportion 2/3. Note that $2/3 = \sqrt{4/9}$. Is this pure accident, or is it possible that the period is proportional to the square root of the length? Let us test another pair of readings, say $l = 7$, $t = 2{\cdot}92$ and $l = 4$, $t = 2{\cdot}22$.

Using a slide-rule, $$\frac{2{\cdot}92}{2{\cdot}22} \simeq 1{\cdot}316$$

and $$\sqrt{\frac{7}{4}} \simeq 1{\cdot}319$$

So the ratio of the times is very nearly equal once more to the square root of the ratio of the lengths.

If we assume this to be true in every case, we then postulate a law in words—" the period of a pendulum is proportional to the square root of its length ". In algebra $t \propto \sqrt{l}$ or better still $t = K\sqrt{l}$, where K is a numerical constant.

Taking a typical pair of results: $l = 9$, $t = 3{\cdot}3$

$$3{\cdot}3 = K \times \sqrt{9}$$
$$= K \times 3$$

or

$$\underline{K \simeq 1{\cdot}1}$$

So we could write as the result of our experiment a formula for the time of swing of a simple pendulum as

$$t = 1{\cdot}1\sqrt{l}$$

This should give us reasonable 2-figure accuracy. (Test this for other readings in the table, using your slide-rule.)

(ii) *Plotting results on graph paper.* Fig. 28(a) shows the result

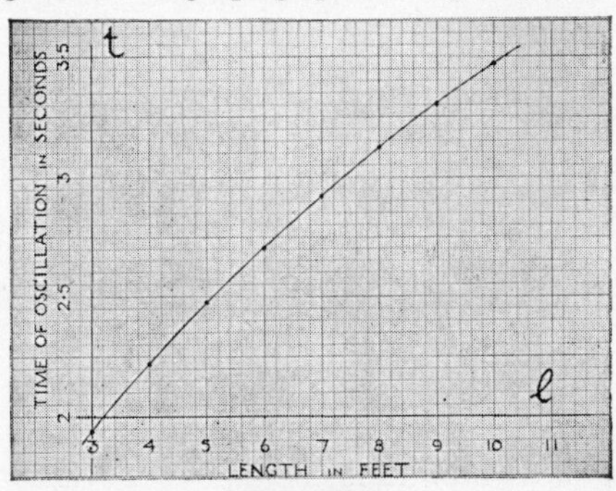

FIG. 28(a).

of plotting t against l. The graph shows a gentle curve—if t were simply proportional to l we should have had a straight line.

Fig. 28(b) shows t plotted against $\sqrt{l}$ for the values shown in the table below:

l . . .	9	8	7	6	5	4	3
$\sqrt{l}$ . . .	3	2·83	2·65	2·45	2·24	2	1·73
t . . .	3·30	3·12	2·92	2·70	2·48	2·22	1·94

It will be seen that the points plotted on this graph lie in a straight line through the origin—which indicates that t is directly proportional to $\sqrt{l}$,

$$i.e., \quad t = K\sqrt{l}$$

where K is a numerical constant.

From the graph we can find K by choosing any one convenient pair of values of t and $\sqrt{l}$; thus when $\sqrt{l} = 2$, $t = 2{\cdot}22$, and $2{\cdot}22 = K \times 2$ or $K = 1{\cdot}1$, to 2 significant figures.

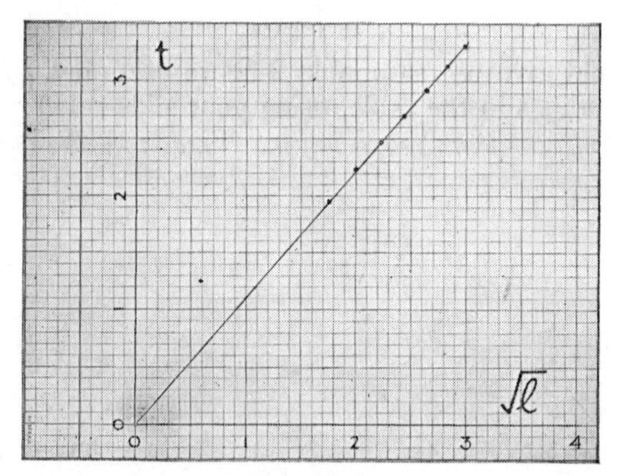

FIG. 28(*b*).

The drawing of a graph thus checks our previous conclusion.

Work out $\dfrac{t}{\sqrt{l}}$ for each pair of observations and check in every case that this ratio is 1·1 to 2 significant figures.

(iii) *Summarising in a clear-cut formula.* We have seen that the period t sec. of a simple pendulum of length l ft. obeys an experimental law of the form $t = K\sqrt{l}$, where $K \simeq 1{\cdot}1$.

In para. 4.1(*c*) above we suggested that the true formula was $t = 2\pi\sqrt{\dfrac{l}{g}}$, where g ft./sec./sec. is the acceleration of a falling body ($g \simeq 32{\cdot}2$).

But note that we can rewrite this formula as

$$t = \frac{2\pi}{\sqrt{g}} \times \sqrt{l},$$

and if $g \simeq 32{\cdot}2$ and $\pi \simeq 3{\cdot}14$, then

the " constant " $$\frac{2\pi}{\sqrt{g}} = \frac{6{\cdot}28}{\sqrt{32{\cdot}2}}$$
$$= 1{\cdot}11 \text{ (by slide-rule).}$$

So that a more exact meaning can be given to the " constant " we have called K.

The experiment we have performed can be said to *verify* the formula $t = 2\pi\sqrt{\frac{l}{g}}$, which was first mathematically *deduced* from Newton's Laws of Motion.

Experiment C. *To Investigate the Stretching of a Helical Spring under Different Loads*

Apparatus Required

A length of unstrained helical (" spiral ") spring (30 cm. or more is advisable to obtain sufficient accuracy); a suitable weight-carrier or scale-pan to be attached to one end of the spring; a metre rule; a set of weights, *e.g.*, 50–500 gm.

(Some laboratories may have special apparatus for such a test, including vernier scales. But improvised apparatus is very much cheaper, and often as effective !)

Method of Conducting the Experiment

Fix the metre rule vertically against a firm support and fix the free end of the specimen of helical spring at the side of the rule so that the spring and weight-carrier hang freely, but sufficiently near the graduated edge of the rule so that the length of the spring may be readily noted during the experiment.

Record the length of the spring itself (*i.e.*, from the fixed end to the point of its attachment to the weight-carrier) for different loads placed on the weight-carrier. The results of such an experiment are set out in tabular and graphical form in Fig. 29 below:

TOTALS		INCREASES	
LOAD GRAMS	LENGTH CMS	LOAD GRAMS	LENGTH CMS
0	34·0	-	-
50	35·7	50	1·7
100	37·5	50	1·8
150	39·3	50	1·8
200	41·2	50	1·9
250	43·0	50	1·8
300	44·8	50	1·8
350	46·5	50	1·7
400	48·3	50	1·8

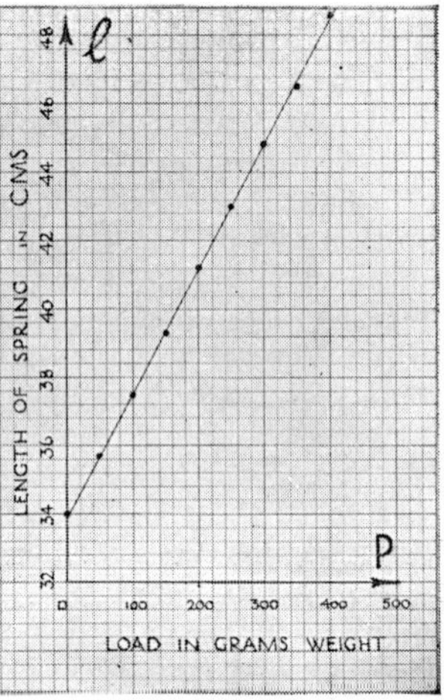

FIG. 29.

Trying to Interpret the Results

(i) By examination of the numerical data we observe that the length of the spring appears to increase by equal amounts for equal increases in the load which stretches it. In other words, the extension of the spring is directly proportional to the tension (the stretching force).

(ii) On examining the graph, the points plotted lie on a straight line (within the limits of any errors in measurement). This straight line cuts the l-axis at the 34-cm. mark—this represents the unloaded length of the spring. We see from the graph too that any increase in length is proportional to the load applied.

(iii) Endeavouring to reduce the above word-statements to a formula, we can write

$$\text{Extension} \propto \text{Tension}$$

As the unloaded length is 34 cm., for any total length l cm. the extension may be written $(l - 34)$ cm., and we could write

$$l - 34 = KP$$

where P gm. is the load and K a " constant " for this particular spring.

We can deduce K by common-sense argument, *i.e.*, for every increase of 50 gm. in the load the length increases by approximately 1·8 cm.

So $$1{\cdot}8 = K \times 50$$

or $$K = \frac{1{\cdot}8}{50}$$

$$= \underline{0{\cdot}036}$$

So, as a formula, $l - 34 = 0{\cdot}036P$

or $$\underline{l = 0{\cdot}036P + 34}$$

This is called a linear law, as it is represented graphically by a straight line.

In general, for any specimen of helical spring, we expect a linear law $l = aP + b$ to hold for the length of the spring and the stretching force P.

The constant a (0·036 in our specimen) is called the " slope " of the straight line, for it describes the steepness of the line on the graph. The " tougher " the spring, the smaller will the slope a be. The constant b represents the value of l when $P = 0$ ($b = 34$ in our specimen), and on the graph is the value of l where the straight line cuts the l-axis. We call b the " intercept " on the l-axis.

This linear law for the stretching of a spring or elastic string was first propounded by Robert Hooke, a pupil of Robert Boyle, and is generally known as Hooke's Law. Hooke became Professor of Geometry at Gresham's College, London, in 1665, and after the Great Fire of 1666 he acted as surveyor to the city authorities. His work in the study of gravitation paved the way for Newton's great discoveries a few years later.

4.4. The Value of Algebraic Factors in Simplifying Formulæ

(a) " *The Common Factor* "

In Example 4(*a*) above we observed how a formula

$$A = 2\pi rl + 2\pi r^2$$

was more compact and more suitable for calculation when the common factor $2\pi r$ was brought to the fore, and we wrote

$$A = 2\pi r(l + r)$$

This process of " taking out the common factor " is of very wide application. As a general pattern, we could say that

$$ab + ac \equiv a(b + c)$$

where a is the common factor of the two terms ab and ac.

Note the geometrical analogy (Fig. 30) of adding together the areas of two rectangles having a common width to form one long rectangle.

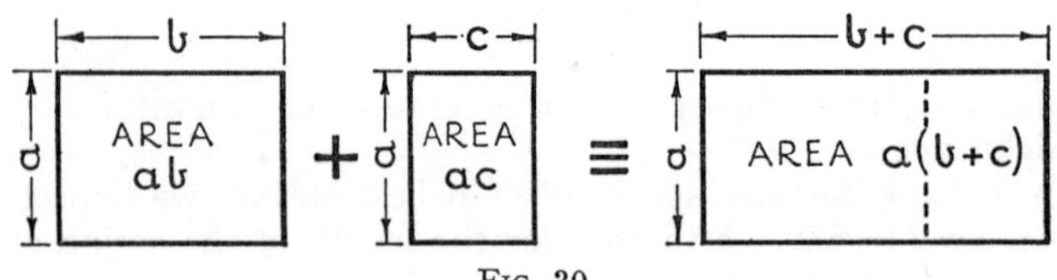

FIG. 30.

Electrical Analogy—Resistances in Series

In Fig. 31 we have 3 resistances r_1, r_2, and r_3 ohms; we wish to assess their total effect in the circuit when connected in series across the terminals of a battery of e.m.f. E volts.

Clearly the same current i amperes will flow through each resistance—it is " common " to all three resistances. Using Ohm's Law:

The voltage drop across r_1 is ir_1 volts.
" " r_2 " ir_2 "
" " r_3 " ir_3 "

So the voltage drop between the battery terminals must be $ir_1 + ir_2 + ir_3$ volts. But if we neglect the internal resistance of the battery as compared with r_1, r_2, r_3 this total drop in voltage must be equal to the total e.m.f. of the battery.

$$\therefore \quad E = ir_1 + ir_2 + ir_3$$
$$= i(r_1 + r_2 + r_3)$$

when we take out the common factor i from the whole expression.

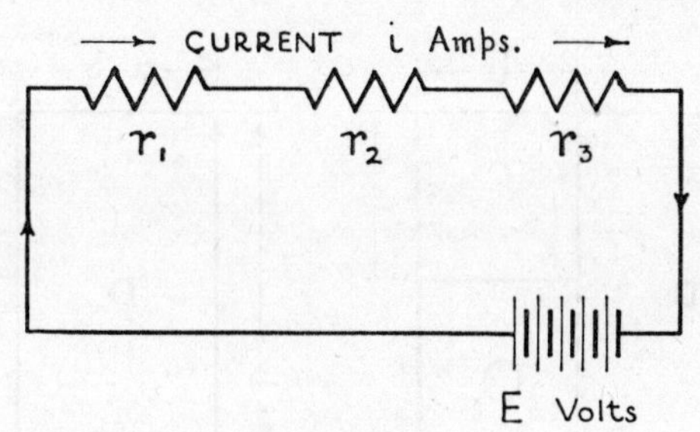

FIG. 31.

If R ohms is the total effective resistance between the battery terminals

$$E = iR$$

so that
$$R = r_1 + r_2 + r_3$$

which is the well-known formula for the sum of three resistances in series.

(See Exercise 4(*a*), No. 34, for the case of resistances in parallel.)

Note how a common element in the circuit—the current passing—gives rise to a common factor when we express the circuit's behaviour in the language of Algebra.

(b) "*The Difference of Two Squares*"

Very frequently we come across the difference of two squares in technical work. For instance, the area A sq. in. of a flat ring with external and internal radii R and r in. is clearly

$$A = \pi R^2 - \pi r^2$$
$$= \pi(R^2 - r^2) \qquad \text{(taking out the common factor } \pi)$$

So this formula involves the difference of the squares of the

D

two radii. We now want to discover simple factors of such an expression.

To show that $a^2 - b^2 = (a + b)(a - b)$

The L-shaped figure in Fig. 32(*a*) represents the area $(a^2 - b^2)$ sq. ft.—*i.e.*, a square of side *a* ft. from which a smaller square of

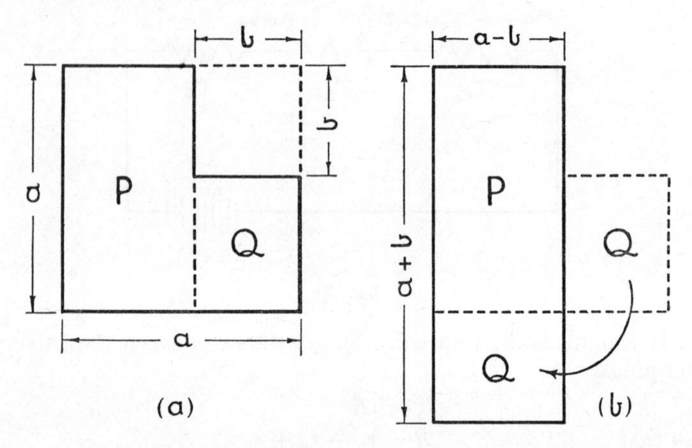

FIG. 32.—THE DIFFERENCE OF TWO SQUARES.

side *b* ft. has been removed. A dotted line has been drawn to divide the area into a larger rectangle *P* [*a* ft. by $(a - b)$ ft.] and a smaller rectangle *Q* [$(a - b)$ ft. by *b* ft.].

In Fig. 32(*b*) rectangle *Q* has been moved into a new position below *P* so that one complete rectangle replaces the original L-shape. This latter rectangle has the combined length $(a + b)$ ft. and the common width $(a - b)$ ft.

So	$a^2 - b^2$	$\equiv$	$(a + b)(a - b)$
	Area enclosed in solid lines in Fig. 32(*a*)		Area enclosed in solid lines in Fig. 32(*b*)

$(a + b)$ and $(a - b)$ are said to be the " factors " of $(a^2 - b^2)$. In words, " the difference of the squares of two numbers is equal to the product of their sum and difference ".

Thus for the flat ring mentioned above, the area of the flat surface

$$A = \pi(R^2 - r^2)$$
$$= \underline{\pi(R + r)(R - r)}$$

e.g., if the external and internal radii of a metal washer are 0·68 and 0·44 cm. respectively, its surface area

$$= \pi \times (1{\cdot}12) \times (0{\cdot}24) \text{ sq. cm.}$$
$$= 0{\cdot}84 \text{ sq. cm. (by slide-rule multiplication)}$$

Compare the results of using $A = \pi(R^2 - r^2)$ instead of the last calculation:

Then
$$A = \pi[(0{\cdot}68)^2 - (0{\cdot}44)^2] \text{ sq. cm.}$$
$$= \pi[0{\cdot}462 - 0{\cdot}194]$$
$$= \pi \times 0{\cdot}268$$
$$= 0{\cdot}84 \text{ sq. cm. (slide-rule)}$$

The calculation by logs or slide-rule is much more tedious and clumsy if we neglect to use the factors of the difference of two squares.

Note that in many practical instances a common factor must be extracted first before the pattern $a^2 - b^2$ is revealed, *e.g.*, $12x^3 - 3xy^2$ does not look like the difference of two squares until we extract the common factor $3x$.

Then
$$12x^3 - 3xy^2$$
$$= 3x(4x^2 - y^2)$$
$$= 3x(2x + y)(2x - y)$$

Here we spot that $4x^2$ is simply the square of $2x$, so that the term $4x^2 - y^2$ is of the pattern $a^2 - b^2$, where $2x$ corresponds to a and y corresponds to b.

(c) " *The Perfect Square* "

A very important pattern in algebra is the " perfect square ", which is illustrated in Fig. 33.

If we multiply $(a + b)$ by itself

$$(a + b)^2 = (a + b)(a + b)$$
$$= a(a + b) + b(a + b)$$
$$= a^2 + ab + ab + b^2$$
$$= \underline{a^2 + 2ab + b^2}$$

Fig. 33 shows how this identity can be displayed geometrically. The term $(a + b)^2$ represents the area of a square $PQRS$ of side $(a + b)$ units. It will be seen that this square divides easily into

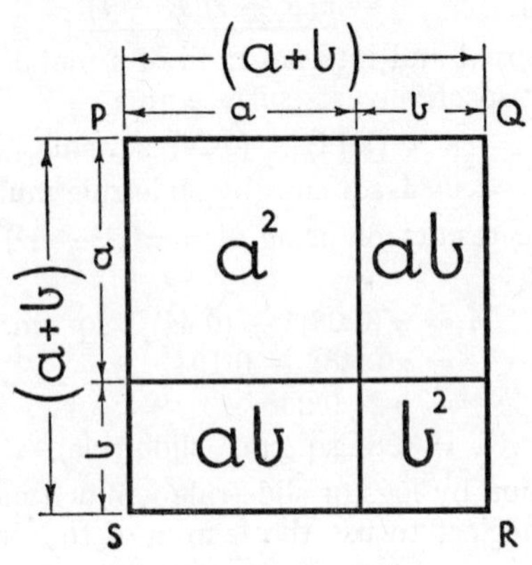

FIG. 33.—THE PERFECT SQUARE.

two squares of area a^2 and b^2 sq. units respectively, and two equal rectangles, each of area ab sq. units.

In words:

The Square of the Sum of Two Numbers
= Sum of their two Squares + Twice their Product

By multiplication, we can similarly expand $(a - b)^2$

i.e.,
$$\begin{aligned}(a - b)^2 &= (a - b)(a - b)\\ &= a(a - b) - b(a - b)\\ &= a^2 - ab - ab + b^2\\ &= \underline{a^2 - 2ab + b^2}\end{aligned}$$

In words:

The Square of the Difference of Two Numbers
= Sum of their Squares − Twice their Product

Example 4(c). *Calculate the volume of concrete required to line the sides of a manhole to a thickness of three inches, if the manhole is to have internal dimensions* 3 *ft.* 6 *in. by* 3 *ft.* 6 *in. by* 6 *ft. deep.*

Let us concoct a simple formula for finding the volume V cu. ft. of the concrete, where the internal and external dimensions are as shown in Fig. 34.

Then $$\begin{aligned} V &= a^2d - b^2d \text{ (expressed in cu. ft.)} \\ &= d(a^2 - b^2) \\ &= \underline{d(a + b)(a - b)} \end{aligned}$$

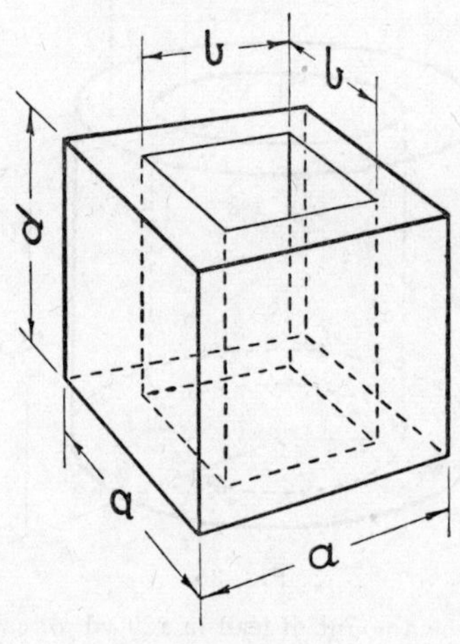

FIG. 34.

In this particular example $d = 6$, $a = 4$, $b = 3\frac{1}{2}$ (all expressed in feet).

So our volume $$\begin{aligned} V &= 6 \times 7\tfrac{1}{2} \times \tfrac{1}{2} \\ &= 6 \times \tfrac{15}{2} \times \tfrac{1}{2} \\ &= 22\tfrac{1}{2} \text{ (cu. ft.)} \end{aligned}$$

Thus nearly a cubic yard of concrete would be required (1 cu. yd. $= 3 \times 3 \times 3$ cu. ft. $= 27$ cu. ft.).

Check this last result by finding the amount of concrete required by a different method, *e.g.*, dividing into 4 " slabs ".

Example 4(d). The Volume of a Hollow Cylinder. *Fig.* 35 *represents a hollow cylinder (such as a length of conduit or a portion of the lead sheath of an underground cable)* 1 *in. long with external and internal radii* r *in. and* R *in. respectively.*

We can find the volume (V cu. in.) of material in the hollow cylinder by subtracting the volume of a cylinder of radius r in. from the same length cylinder of radius R in.

i.e., $V = \pi R^2 l - \pi r^2 l$

$= \pi l(R^2 - r^2)$ (taking out the common factor)

$= \pi l(R + r)(R - r)$ (difference of two squares)

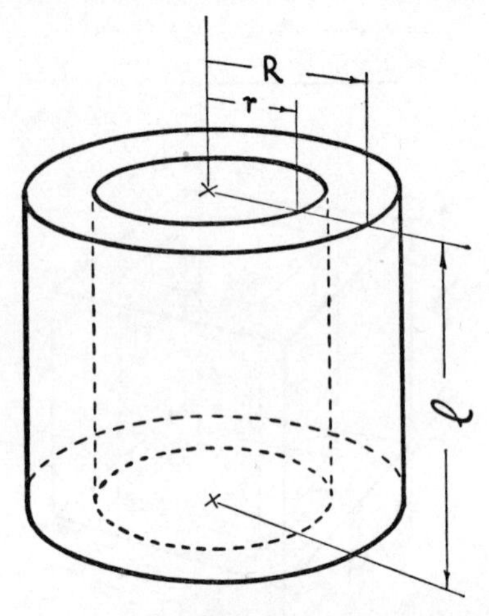

FIG. 35.

So, for example, the amount of lead in 220 yd. of cable, if the external and internal radii of the lead sheath be 0·64 in. and 0·48 in., would be :

$$\frac{\pi \times 660(1{\cdot}12)(0{\cdot}16)}{144} \text{ cu. ft.}$$

$$= \frac{3{\cdot}14 \times 6{\cdot}6 \times 1{\cdot}12 \times 16}{144} \text{ cu. ft.}$$

$$\simeq 2{\cdot}63 \text{ cu. ft. (slide-rule)}$$

As lead is 11·36 times as heavy as water and 1 cu. ft. of water weighs 62·5 lb., it follows that 220 yd. of this cable contains 11·36 × 62·5 × 2·63 lb., *i.e.,* 16·7 cwt. of lead.

Example 4(*e*). Use of the "Perfect Square". *An immersion heater is designed for* 200 *volts. If the voltage rises by* 10%, *what percentage increase is there in its heating effect* (i.e., *its running costs*)?

The energy (heating effect) of electric current is proportional to the *square* of the voltage.

When operating at the rated 200 volts the heating effect is proportional to

200^2. When the voltage rises by 20 volts (*i.e.*, 10% of 200 volts) the new heating effect is proportional to $(200 + 20)^2$.

Using $$(a + b)^2 = a^2 + 2ab + b^2$$

where $a = 200$, $b = 20$.

Then $$(200 + 20)^2 = 200^2 + 2 \times 200 \times 20 + 20^2$$
$$= 4 \times 10^4 + 8 \times 10^3 + 4 \times 10^2$$

Note that 20^2 is small compared with the " twice the product " term

$$2 \times 200 \times 20$$

So, approximately, the percentage increase in heating

$$= \frac{\overset{2}{\not{8}} \times 10^3}{\not{4} \times 10^4} \times 100\% \text{ (ignoring } 20^2\text{)}$$
$$= \underline{20\%}$$

Generalise the above argument, to show that if designed for V volts, an increase of $p\%$ in voltage will mean approximately $2p\%$ increase in the running costs of the heater.

(d) " *Linear Factors of a Quadratic* "

An expression such as $x^2 - 5x + 6$ is neither a perfect square nor a difference of squares. It is an example of a more general quadratic expression of the form $ax^2 + bx + c$, where a, b, c are numbers.

Such expressions may *sometimes, but not always*, be factorised by trial into the product of two linear factors, *i.e.*, in this case

$$x^2 - 5x + 6 = (x \qquad)(x \qquad)$$

where numbers may be found to fill the blank spaces in the brackets, giving when multiplied out a number term $+6$ and two x-terms which together make $-5x$.

By trial, such a pair of numbers is -3, -2, for

$$(x - 3)(x - 2) = x^2 - 3x - 2x + 6 = x^2 - 5x + 6$$

It is interesting to compare this with $x^2 - 5x - 6$, which factorises as $(x - 6)(x + 1)$. Whereas an equally simple-looking quadratic such as $x^2 + 3x + 4$ has no factors. The existence of such linear factors must indeed be regarded as the exception rather than the rule!

The use of linear factors, where they exist, to solve quadratic equations is discussed in Chapter 6 (p. 161).

EXERCISE 4(*a*)

A. *Formal Practice*

Express the following in factors (in the neatest form for computation):

1. $3x^2 + xy$.
2. $6p^2 - 4pq$.
3. $\pi a^2 - \pi b^2$.
4. $r^2 - 4s^2$.
5. $3p^2q - 27q^3$.
6. $63x^4 - 7x^2$.
7. $\pi at - \pi bt$.
8. $2\pi a^2 + \pi ab$.
9. $16p^2q^2 - r^4$.
10. $2l^3m^3 - 8lm$.

Illustrate the following products by diagrams involving areas of rectangles and squares:

11. $a(b + c) = ab + ac$.
12. $(l + m)(p + q) = lp + lq + mp + mq$.
13. $(x + 3)(x + 4) = x^2 + 7x + 12$.
14. $(x + a)(x + b) = ?$.
15. $(x + 7)(y + 2) = ?$.

Express the following without brackets and check the correctness of your expression by substituting the numerical value(s) suggested:

16. $(x - 1)(x - 2)$; $x = 9$.
17. $(2p - 1)(p + 4)$; $p = 3$.
18. $(y + 1)^2$; $y = 5$.
19. $(p - q)^2$; $p = 7, q = 4$.
20. $(3x - 4)^2$; $x = 2$.

" Completing the Square." Insert the missing term or terms in the following statements:

21. $(x + 3)^2 = x^2 + \quad + 9$.
22. $(x - 7)^2 = x^2 - 14x \quad$.
23. $x^2 + 10x + \quad = (x \quad)^2$.
24. $y^2 \quad + 16 = (y - \quad)^2$.
25. $p^2 - 6p \quad = (\quad)^2$.
26. $x^2 + 7x \quad = (\quad)^2$.
27. $x^2 \quad + 49 = (x - \quad)^2$.
28. $r^2 - 2{\cdot}4r \quad + \quad = (r - \quad)^2$.
29. $x^2 - 2px \quad = (\quad)^2$.
30. $x^2 + qx \quad = (\quad)^2$.

(See Appendix to Chapter 6 for the application of this " Completing the Square " technique.)

B. *Building up Formulæ, and Using Them.* (Assume $\pi = 3{\cdot}14$.)

31. Express the cross-sectional area A sq. in. of a steel tube in terms of its internal radius a in, and the thickness t in. of its wall.

Hence find the weight per foot run of such a tube, with internal diameter 3·4 in. and wall thickness 0·42 in. (A cu. ft. of water weighs $62\frac{1}{2}$ lb. and the specific gravity of the specimen of steel is 7·8.)

32. A certain cigarette is $2\frac{3}{4}$ in. long and $\frac{5}{16}$ in. in diameter. Calculate the area of paper required to make 1000 such cigarettes, if $\frac{1}{8}$ in. overlap is allowed for gumming.

How much tobacco (by volume) is saved by incorporating a filter-tip $\frac{5}{8}$ in. long in this cigarette. What is the *percentage* saving?

33. A steel spindle consists of a central cylindrical portion, length l cm. and radius r cm., with conical ends of the same radius. If the height of each cone is h cm., obtain a concise formula for the volume V c.c. of metal in the spindle. Find the weight of such a spindle, given that $l = 4·2$, $r = 0·9$, $h = 1·6$, and steel weighs 7·8 gm. per c.c.

34. Three resistors, with resistances R_1, R_2, R_3 ohms are connected in parallel across a common supply voltage of V volts. Write down the current in amperes which flows through each resistor, and thus the *total* current taken from the supply. If the three resistors were replaced by one with resistance R to give the same total current drain, express R in terms of R_1, R_2, R_3. Find the single resistance which is equivalent to 12, 18, and 24 ohms connected in parallel.

35. Find the total surface area of a solid cone, if the development of its curved surface is a quadrant of a circle, 6 in. in diameter.

36. A spinning-top is made in the shape of an inverted cone (radius r in., height h in., slant height l in.) capped by a hemisphere, also of radius r in.

Express as a formula: (*a*) the volume V cu. in.; (*b*) the surface area A sq. in. of the top, in forms best suited to calculation. Find the volume and surface area when $r = 2·4$, $h = 3·6$. (Note $l^2 = r^2 + h^2$ by Pythagoras' theorem.)

37. A valve envelope has a total length of 3 in. and is 1 in. in diameter. It is made of thin glass with a hemispherical end. Calculate the area of the outer surface of the tube, and its cubical content.

Show that if the total length of the tube is l in., its radius r in., the outer surface A sq. in. is given by $A = 2\pi rl$ (*i.e.*, the same as the curved surface of a cylinder of length l in.).

Also obtain a neat formula for the cubical content V cu. in. in terms of l and r.

38. Find the volume of metal contained in a gross of washers, each of which is $\frac{3}{16}$ in. in thickness, with internal and external diameters 0·42 and 1·24 in.

39. Find the weight in cwt. or 200 lead sleeves used in plumbing underground joints, each sleeve being a hollow cylinder 2·6 in. in external diameter, and 16 in. long, the lead being 0·22 in. in average thickness. (A cu. ft. of water weighs 62·5 lb. and the specific gravity of lead is 11·36.)

40. A tubular metal pole used for permanent line work is 9 m. high, and its average external cross-section is a circle 12·4 cm. in diameter. If the metal is 3 mm. thick, find the volume of metal in the pole, and its weight in kg. if 1 c.c. of the metal weighs 7·6 gm.

C. *Some Experiments which the Student May Wish to Try*

41. Test the relation between the voltage across, and the current passing through, the following components:

(*a*) A small battery-type light bulb or 6-volt switchboard lamp;
(*b*) a tubular (ceramic) resistor (100 ohms);
(*c*) a wire-wound 100-ohm resistor;
(*d*) a 100-ohm relay winding;
[Step up the voltage until each type of resistor is too hot to touch.]
(*e*) a metal rectifier element.

Test each of the above with voltage reversed, and plot a graph for the current–voltage " characteristic " response of each one. Hence form a judgement as to the practical limitations of Ohm's Law. Which type of resistor gives the best linear response? Which response is most definitely *not* linear?

42. Repeat the previous type of experiment for the flow of current through an electrolyte, *e.g.*, applying different voltages across two copper electrodes immersed in an electrolyte of copper sulphate solution (*i.e.*, a simple electro-plating experiment).

43. Test the *compression* of a really stout helical (" spiral ") spring for different compressing forces. Does Hooke's Law appear to hold for compression as well as for stretching?

44. Experiment with the (longitudinal) oscillations of a fixed length of helical spring (not such heavy gauge as in No. 43) when different weights are attached to the lower end. What possible formula connects the period of oscillation with the weight?

45. Repeat No. 44 with the same weight and different lengths of the same specimen of spring. Does the period vary with the length? If so, how?

46. Tension up a bay-length of " 100-lb." copper wire, and note the tension recorded on the spring-balance attachment of the ratchet and tongs. Pluck the wire, and find the time of a complete " wave " to and fro (" Ping-pong "!). Alter the tension, and again note the time of travel of the " wave ".

Plot the results on a graph, and see if a " law " connects the time of travel and the tension in the wire.

47. Devise an experiment (using a spring balance or weights) to find the least force required to hold the telephone dial against the finger stop when dialling numbers 1, 2, 3 . . . These numbers are measures of the " angular displacement " of the dial spring. Test whether a straight-line law connects this angular displacement with the least force required.

D. *Linear Factors*

By trial, factorise where possible:

48. $x^2 + x - 6$.
49. $x^2 - 7x + 12$.
50. $x^2 + 5x + 4$.
51. $3x^2 - 2x - 1$.
52. $2y^2 - 5y + 2$.
53. $4p^2 + p - 3$.
54. $x^2 - 7x - 6$.
55. $2x^2 + 7x - 4$.

4.5. Changing the Subject of a Formula

We can regard a formula as showing in the clearest possible way the relationship between two or more quantities.

Thus $V = IR$ explains how voltage is related to current and resistance—with V in the centre of the stage as the subject of this formula. Thus, knowing I and R, we can easily find V. Such a situation occurs, for instance, in checking the grid bias on a valve in a circuit which includes a " cathode resistor ".

But sometimes we need to shift the emphasis to some other quantity—R for example—and make this the subject of our formula. This did indeed occur in Experiment A above. Starting with the fact that $IR = V$, we shift the spotlight to R and write, by common-sense deduction,

$$R = \frac{V}{I}$$

This process of remoulding a formula and shifting the emphasis to another quantity (or " variable ") is called " changing the subject of the formula ".

An analogy from human relationships may help. In a certain family John is Kate's brother, and Kate has a daughter Mary. Then we say " John is Mary's uncle "—John being the subject here. But if we shift our attention to the rising generation, we can express the same family relationship in a different form as " Mary is John's niece ". The relationship is the same, but the subject of our statement is changed.

So the formulæ $V = IR$ and $R = \frac{V}{I}$ express the same law of relationship (Ohm's Law) but emphasise first " voltage " and then " resistance " as alternative subjects.

What is this same relationship if we make I the subject?

Example 4(*f*). *Express*: (a) *the height* h *ft.*, (b) *the radius* r *ft. of a cylinder in a formula, involving also its volume* V *cu. ft.*

We already know the formula for the volume of a cylinder; *i.e.*, $V = \pi r^2 h$.

(*a*) We wish first of all to remodel this basic formula to give us h in terms of V and r. We generally begin by re-stating the original formula with the new " subject " (h) on the left side; *i.e.*, $\pi r^2 h = V$. But we can argue that if πr^2 times h gives us V, then h alone must be V divided by πr^2.

$$\therefore \; h = \frac{V}{\pi r^2} \quad \text{(equivalent to dividing both sides of the formula by } \pi r^2\text{)}$$

This is our formula to calculate h if we know V and r.

(*b*) Similarly,

$$r^2 = \frac{V}{\pi h} \quad \text{(dividing both sides of original formula by } \pi h\text{)}$$

so that our radius r must be equal to the square root of $\frac{V}{\pi h}$,

or

$$r = \sqrt{\frac{V}{\pi h}}$$

This expresses r in terms of V and h.

Sometimes a formula is rather harder to recast. Thus in the important formula for the resonant frequency f c/s of a tuned circuit containing capacitance C farads and inductance L henrys, the quantities L and C are " locked " within a square root:

i.e.,

$$f = \frac{1}{2\pi\sqrt{LC}}.$$

Example 4(g). *Find the capacitance required to give a resonant frequency of* 24 *kc/s when in circuit with an inductance of* 120 μH.

We now need to express C in terms of f and L, where

$$f = \frac{1}{2\pi\sqrt{LC}}$$

But C is " locked up " inside the square root. We can release C only if we consider the *square* of the frequency;

i.e.,

$$f^2 = \frac{1}{4\pi^2 LC} \quad \text{(equivalent to squaring both sides of our formula)}$$

If we compare this with (say) $f^2 = \frac{1}{64}$, we note that in this numerical example we may write with equal truth $64f^2 = 1$.

So in our more complicated statement above, we deduce that

$$4\pi^2 LCf^2 = 1.$$

Rearranging, $4\pi^2 f^2 LC = 1$

And therefore C itself must be 1 divided by $4\pi^2 f^2 L$

i.e.,
$$C = \frac{1}{4\pi^2 f^2 L}$$

In this particular case

$$L = 120 \times 10^{-6} \text{ (henrys)}$$
$$f = 24 \times 10^3 \text{ (c/s)}.$$

Then
$$f^2 L = 24^2 \times 10^6 \times 120 \times 10^{-6}$$
$$= \frac{576 \times 10^6 \times 120}{10^6}$$
$$= 576 \times 120$$

and
$$C = \frac{1}{4\pi^2 f^2 L}$$
$$= \frac{1}{4\pi^2 \times 576 \times 120} \text{ (expressed in farads)}$$

Expressing the result in microfarads, the required capacitance

$$= \frac{10^6}{4\pi^2 \times 576 \times 120} \mu\text{F}.$$

As a rough approximation, this is nearly

$$= \frac{10^6}{40 \times 600 \times 100} \qquad (\pi^2 \simeq 10)$$
$$= \frac{10^6}{24 \times 10^5}$$
$$= \frac{5}{12}$$
$$= 0{\cdot}4 \ \mu\text{F}. \quad \text{(to ONE significant figure)}$$

Using a slide-rule, the required value is 0·37 μF. The same calculation using logarithms is set out in detail in the next paragraph (4.6.).

Note that the slide-rule accuracy shown here is usually quite sufficient, as the rating of a mass-produced capacitor would seldom have more than 2-figure accuracy. Indeed a small " trimming " capacitor would normally be fitted to provide a fine adjustment of the frequency.

4.6. Tabular Methods of Using Logarithms to Evaluate Formulæ

There are some calculations in which a slide-rule either cannot be used or is insufficiently accurate for our purpose. To achieve speed and accuracy with logarithms we need to set out our work in tabulated form, as suggested below.

As an illustration, consider the calculation involved in Example 4(*g*) above, where

$$C = \frac{10^6}{4 \times (3{\cdot}1416)^2 \times (2{\cdot}4 \times 10^4)^2 \times (1{\cdot}2 \times 10^{-4})} \quad (\mu\text{F.})$$

$$= \frac{10^6}{10^{0\cdot6021} \times (10^{0\cdot4972})^2 \times (10^{4\cdot3802})^2 \times 10^{\bar{4}\cdot0792}}$$

$$= \frac{10^6}{10^{0\cdot6021} \times 10^{0\cdot9944} \times 10^{8\cdot7604} \times 10^{\bar{4}\cdot0792}}$$

$$= \frac{10^6}{10^{6\cdot4361}}$$

$$= 10^{\bar{1}\cdot5639}$$

$$= 0{\cdot}3664 \quad (\mu\text{F.})$$

Thus the required capacitance is 0·366 μF. (to 3 significant figures).

There are two serious drawbacks to the above mode of working:

(*a*) it is most clumsy and tedious;

(*b*) it involves adding figures across the page instead of in columns, to which the eye is accustomed.

We therefore drop the base 10 from our log calculations and tabulate our work as follows:

	No.	Log.	Log.
Numerator . . .	10^6	6·0000	
	4 π^2 f^2 L	0·6021 0·9944 8·7604 $\bar{4}$·0792	 0·4972 × 2 4·3802 × 2
Denominator . . .		6·4361	
C in μF.	0·3664	$\bar{1}$·5639	

←

Note the use of an auxiliary log column to deal with the logs of numbers to be squared, cubed, square-rooted, etc. Log π is taken as 0·4972. The arrow below the last line of working indicates the use of anti-logs to find C.

Accuracy.—In working with 4-figure logs the 4th significant figure in the final result is bound to be suspect. The result should therefore be given as "a capacitance of 0·366 μF. (to 3 significant figures)".

Example 4(*h*). *Find the diameter of a sphere whose volume is 64 c.c.*

The volume of a sphere (V c.c.) of radius r cm. is given by the formula

$$V = \tfrac{4}{3}\pi r^3$$

$$\therefore \quad \tfrac{4}{3}\pi r^3 = V$$

$$\therefore \quad 4\pi r^3 = 3V$$

i.e.,

$$r^3 = \frac{3V}{4\pi}$$

so to obtain r we must take the *cube root* of $\dfrac{3V}{4\pi}$

$$\therefore \quad r = \sqrt[3]{\frac{3V}{4\pi}}$$

The use of logarithms provides the most ready way of finding a cube root, since the cube root of a number merely involves dividing its log by 3. Tabulating our calculation :

	No.	Log.	Log.	
	3 V (64)	0·4771 1·8062		
Numerator . .		2·2833		
	4 π		0·6021 0·4972	
Denominator . .		1·0993		
	r^3	1·1840		
Taking the cube root	$r = 2·481$	0·3947		Dividing log by 3

←

So the radius of this sphere is 2·48 cm. to 3 significant figures, *i.e.*, it is very nearly 5 cm. in diameter.

In this example we see the second log column used for working out the denominator, thus leaving a clear space for the subtraction in the first log column.

It is interesting to observe that a cube having the same volume as this sphere (64 c.c.) would have sides 4 cm. (*i.e.*, $4^3 = 64$). Thus the ratio of the diameter of a sphere to the side of a cube of equal volume is very nearly 5 : 4.

Example 4(*i*). *Find the period* (t *sec.*) *of oscillation of a compound pendulum, using the formula* $t = 2\pi\sqrt{\dfrac{l^2 + k^2}{lg}}$ *where* $l = 24{\cdot}2$ (*cm.*), $k = 6{\cdot}8$ (*cm.*), *and* $g = 981$ (*cm./sec./sec.*).

Here we are confronted with the *sum* of two squares $l^2 + k^2$, which cannot be factorised. So we must work out this awkward expression first.

Then
$$l^2 + k^2 = (24{\cdot}2)^2 + (6{\cdot}8)^2 = 585{\cdot}6 + 46{\cdot}2 = \underline{631{\cdot}8}$$

	No.	Log.
l .	24·2	1·3838
l^2 .	585·6	2·7676
	←	—
k .	6·8	0·8325
k^2 .	46·24	1·6650
	←	—

Using the complete formula

$$t = 2\pi\sqrt{\frac{631{\cdot}8}{24{\cdot}2 \times 981}}$$

	No.	Log.	Log.
$l^2 + k^2$.	631·8	2·8005	
l . .	24·2		1·3838
g . .	981		2·9917
lg		4·3755	←
		2)$\bar{2}$·4250	
		$\bar{1}$·2125	
	2	0·3010	
	π	0·4972	
	1·025	0·0107	

←

Taking square root divide log by 2

Thus the pendulum makes one complete oscillation (to and fro) in 1·03 sec. (3 significant figures).

Check the above, using your slide-rule.

4.7. Snags with Negative Characteristics

(a) *Square and Cube Roots*

In the previous example (4(*i*)) we had to find the square root of an expression which involved halving $\bar{2}\cdot4250$. Fortunately, as the negative characteristic was even, this was simple, for

$$\tfrac{1}{2}(-2 + 0\cdot4250) = -1 + 0\cdot2125$$

so we could write $\bar{1}\cdot2125$ as the log of our square root. But what would have happened if the negative characteristic had been ODD?

Consider for instance $\sqrt{0\cdot00296}$. In this case log (0·00296) = $\bar{3}\cdot4713$. Remembering the meaning of $\bar{3}\cdot4713$ as $-3 + 0\cdot4713$; halving this as it stands, we obtain $-1\cdot5 + 0\cdot2356$!

To avoid such complications we try to arrange matters so as to *keep the negative part of the logarithm a whole number.* So we deliberately "make up" the -3 to the next exact multiple of 2, *i.e.*, to -4 and compensate the decimal part of the log by adding $+1$ to it. We therefore consider $-3 + 0\cdot4713$ as equivalent to $-4 + 1\cdot4713$, which divides by 2 to give $-2 + 0\cdot7356$.

Symbolically $\tfrac{1}{2}(\bar{3}\cdot4713) = \tfrac{1}{2}(\bar{4} + 1\cdot4713)$
$= \bar{2}\cdot7356$

Similarly, to find the cube root of 0·0000187:

No.	Log.	
0·0000187	$\bar{5}\cdot2718$	
	$3)\overline{\bar{6} + 1\cdot2718}$	
0·02654	$\bar{2}\cdot4239$	Taking the cube root

←

i.e., $\sqrt[3]{0\cdot0000187} = 0\cdot0265$ (to 3 significant figures)

(b) *Awkward Subtractions*

The other difficulty commonly experienced is in subtracting logs involving negative characteristics. As before, if we remember that the whole-number part ONLY of the logarithm is negative, we can reason out such problems as they occur.

Thus, in evaluating $\frac{52\cdot7}{0\cdot069}$ $\left(\text{Roughly } \frac{50}{0\cdot07} \simeq \frac{5000}{7} \simeq 700\right)$

No.	Log.	
52·7 0·069	1·7218 $\bar{2}\cdot8388$	
763·8	2·8830	Subtracting

⟵

we are faced with subtracting $\bar{2}\cdot8388$ from 1·7218. When we reach the " tenths " column we have to take 9 from 7, and we probably say " 9 from 10 leaves 1 and 7 makes 8—carry 1 ". This means " carry + 1 ", *i.e.*, we add + 1 to − 2 in the characteristic on the second line. This brings this characteristic to the value − 1. The final step is to take − 1 away from + 1. In detail, this gives

$$+1-(-1)=+1+1=+2$$

as the characteristic of the log of the answer. The rough check assures us that the decimal point has been correctly sited, and this is, of course, the " job " of the characteristic.

Our result is thus 764 correct to 3 significant figures.

(The discussion above uses the " equal addition " method of subtraction—any other method may be used in a similar commonsense argument.)

4.8. A Word of Warning—and a " Short List "

Formulæ make good servants but bad masters! Resist the temptation to collect a mass of technical formulæ—they will only clog the mind in the attempt to assimilate them. It is important to know thoroughly the essential minimum of " foundation formulæ " and then to be able to work out from these the applied formula required in each practical problem as it arises.

The following might be regarded as a suitable " short list " of such fundamental formulæ. Build compound formulæ on these foundations as necessity dictates.

(a) *Mensuration Formulæ*

AREAS

Plane surfaces			Curved surfaces		
Rectangle	Triangle	Circle	Cylinder	Cone	Sphere
$A = lb$	$\frac{1}{2}bh$	πr^2	$2\pi rl$	πrs	$4\pi r^2$

VOLUMES

Rectangular block	Prism	Cylinder	Pyramid	Cone	Sphere
$V = lbh$	Area of cross-section × Length	$\pi r^2 l$	$\frac{1}{3}$ Area of base × Height	$\frac{1}{3}\pi r^2 h$	$\frac{4}{3}\pi r^3$

(b) *Formulæ in Mechanics*

Period of Simple Pendulum:

$$t = 2\pi\sqrt{\frac{l}{g}}$$

(t sec. for one complete oscillation, length l cm., g cm./sec./sec. the "acceleration due to gravity".)

Hooke's Law for a stretched spring or string:

$$T = \frac{\lambda x}{a}$$

(T tension, x extension, a unstretched length, and λ a numerical constant depending on the particular sample of spring.)

(c) *Formulæ in Electrical Work*

Ohm's Law: $V = IR$ (V volts, I amperes, R ohms).
Power Formula: $P = VI$ (P watts).
Charge on a Capacitor: $Q = VC$ (Q coulombs, C farads).

Resonant Frequency of L–C circuit: $f = \dfrac{1}{2\pi\sqrt{LC}}$ (f c/s, C farads, L henrys).

Relation between wavelength (λ cm.) and frequency (f c/s):

$$\lambda f = \text{constant}$$
$$= 3 \times 10^{10} \text{ (approx.)}$$

More accurately, this " constant " is the speed of propagation of Hertzian waves, *i.e.*, $2{\cdot}991 \times 10^{10}$ cm./sec.

EXERCISE 4(*b*)

Use logarithms to evaluate the following, tabulating your work carefully. Use a rough check in each instance.

1. $\dfrac{(0{\cdot}3872)^2}{0{\cdot}0194}$.
2. $\sqrt{0{\cdot}8182}$.
3. $(0{\cdot}03268)^3$.
4. $\sqrt[3]{0{\cdot}02677}$.
5. $\sqrt{\dfrac{0{\cdot}5678 \times 0{\cdot}07938}{0{\cdot}9596}}$.
6. $\sqrt[3]{\dfrac{0{\cdot}00791}{1{\cdot}93 \times 0{\cdot}026}}$.
7. The radius of a ball whose volume is 0·84 c.c.
8. $pv^{1{\cdot}4}$, where $p = 31{\cdot}7$, $v = 18{\cdot}93$.
9. $t = 2\pi\sqrt{\dfrac{l}{g}}$, where $\pi = 3{\cdot}142$, $l = 27{\cdot}6$ cm., $g = 981$ cm./sec./sec. (Result: period t sec.).
10. The speed v f.p.s. in the formula $v^2 = u^2 + 2gs$, where $u = 12{\cdot}4$ f.p.s., $g = 32{\cdot}2$ ft./sec./sec., and $s = 34{\cdot}6$ ft.
11. Write down the circumference (c cm.) and the area (A sq. cm.) of a circle in terms of π and the radius r cm. Now write a formula expressing its area in terms of c and π only. If the area of a circle is 24 sq. in., find its circumference.
12. Find the length (l in.) of a solid cylinder, 6 in. in radius (r) if the total surface area is 600 sq. in. (A). (Make up a formula giving l in terms of r, A.)
13. What is the internal radius at the top of a conical vessel, of internal depth 6·4 in., if its capacity is to be one-tenth of a cubic foot?
14. If $f = \dfrac{1}{2\pi\sqrt{LC}}$ connects the frequency of an oscillating circuit with its inductance L henrys and its capacitance C farads, make a formula presenting L in terms of f and C. Find the inductance of the inductor required to provide oscillation at 2300 c/s with a capacitor rated as 0·1 μF.
15. Work out a formula to express the internal radius of a lead cable

sheath (r in.) in terms of the volume of lead V cu. in. per foot run of cable and the thickness t in. of the sheath.

If the lead sheath is 0·09 in. thick, find the internal diameter of the sheath if there is 1 cu. ft. of lead in every 40 yd. of cable.

16. The "dip" d ft. in the middle of a bay of an open-wire route is given by the formula $d^2 = \frac{3}{8}l(S - l)$, where the poles are l ft. apart, and there is S ft. of wire in each bay.

Write a formula expressing S in terms of d and l. If there is a "dip" of 15 in. in each 35-yd. span of a permanent-line route, by how much per cent does the length of each wire exceed the length of the route?

17. The resistance R ohms of a specimen of wire is given by $R = \rho \cdot \frac{4l}{\pi d^2}$, where ρ is the resistivity of the metal in ohms per cm. cube, l cm. is the length, and d cm. is the diameter of the wire.

Express: (a) ρ in terms of R, l, and d; (b) d in terms of R, ρ, and l.

If the resistance of a kilometre of copper wire is 25·6 ohms, and the particular wire is No. 20 S.W.G., of diameter 0·914 mm., calculate the resistivity of copper.

With this information, what is the diameter of copper wire whose resistance is 1 ohm per 100 metres?

18. Calculate to the nearest ton the weight of water which falls on an acre of ground during a rainfall of 1 in. (1 gall. contains 227 cu. in. and weighs 10 lb.)

19. A motorist maintains his front tyres at a pressure of 28 lb. per sq. in. How many kg. per sq. cm. should he ask for at a French garage? (1 metre ≡ 39·37 in. 1 kg. ≡ 2·2 lb.)

20. A cylindrical resistor is capable of dissipating 2 watts per sq. cm. of its surface without an excessive rise in temperature. What is the permissible loading (in watts) of a resistor of this type if its diameter is 1·3 cm. and length 5·35 cm.? (Include the ends as dissipating surfaces.)

21. Find the greatest internal diameter of an iron water pipe, 1 in. thick, if it should not exceed 20 lb. in weight per foot run. (1 cu. ft. of water weighs $62\frac{1}{2}$ lb.: iron is 7·7 times as heavy.) Check your log calculation with a slide-rule.

22. Find the weight of lead and copper respectively in 220 yd. of a 400-pr. underground cable, if the lead sheath is 0·09 in. thick and 1·75 in. in external diameter, whereas each copper conductor is 0·025 in. in diameter (10 lb. gauge). The specific gravity of lead is 11·36 and of copper 8·89; 1 cu. ft. of water weighs $62\frac{1}{2}$ lb.

Ignoring the weight of paper in the cable, what percentage by weight consists of lead?

23. The mechanical pull of an electromagnet P gm. is given by the formula $P = \frac{B^2A}{8\pi g}$, where B lines per sq. cm. is the flux density (strength of magnetic field) in the air gap, A sq. cm. is the cross-sectional area of this gap, and $g = 981$ cm./sec./sec.

Find a formula for expressing B in terms of P, A, and g. In an experiment the mechanical pull is measured as 6·4 kg., and the air gap is 13·74 sq. cm. in cross-section. Find the magnetic field strength in lines per sq. cm.

24. The cutting speed of a tool (V ft. per min.) and the life of the tool (t hr.) are related by an approximate formula $Vt^{\frac{1}{8}} = C$, where C is a constant for a particular tool. In one test a tool cutting at 110 ft. per minute was found to last for 40 hr. continuous working. Estimate at what speed a similar tool would last 128 hr.

Find the probable life of this tool when it has to cut at 80 ft. per minute.

25. A relation commonly found between the anode current, I_a mA, and the anode voltage, V_a, of a diode valve is $I_a = K \, . \, V_a^{\frac{3}{2}}$.

For a certain valve the constant $K \simeq 0{\cdot}05$. Calculate the anode current for anode voltages of 50, 100, 150, 200, 250, and graph I_a against V_a. From the graph find the anode voltage which would provide an anode current of 75 mA.

26. The capacitance of a capacitor having air as dielectric is given in electrostatic units (e.s.u.) by $C = \frac{A}{4\pi d}$, where A sq. cm. is the effective area of each plate, and d cm. the distance between the plates. If 1 farad $\equiv$ 9×10^{11} e.s.u., how large must each plate be for a 5-$\mu\mu$F. capacitor with an air-gap of 2 mm.?

27. Two aluminium bus-bars 15 metres long and 20 mm. thick are to be installed for the main power supply of an exchange. The resistance of the two in series must not exceed 0·0005 ohms. The standard widths available are 6–12 cm. in steps of 2 cm. Which width would you choose?

$$\text{(Resistance} = \text{Specific Resistance} \times \frac{\text{Length (in cm.)}}{\text{Cross-sectional Area (in sq. cm.)}}.$$

The Specific Resistance of aluminium = 2·9 microhms per cm. cube.)

28. An electromagnetic wave travels at 186,000 miles per sec. What is this speed of propagation in cm. per sec.? (1 mile $\simeq$ 1610 metres.)

In wireless transmission the wavelength λ cm. and the frequency f c/s are such that the product λf equals the speed of propagation in cm. per sec. Convert the following wavelengths to frequencies:

Programme.	*Wavelengths in metres.*	
Home Service (London) . . .	330	—
Light Programme	1,500	247
Third "	464	194

29. The frequencies in megacycles per second of B.B.C. television programmes are as follows:

	Crystal Palace.	*Sutton Coldfield.*
Vision	45	61·75
Sound	41·5	58·25

Convert these to wavelengths.

30. The frequency f c/s of a tuned circuit containing inductance L henrys and capacitance C farads is given by the formula $f = \frac{1}{2\pi\sqrt{LC}}$.

If an inductance of 5 mH. is used in such a tuned circuit, work out the the values of capacitance to tune into the two frequencies of the Light Programme (see Question 28).

REVISION PAPERS A

A.1

1. Express the following fractions with the same (lowest) denominator; arrange them in descending order of magnitude; verify by converting each to a decimal : $\frac{3}{7}$, $\frac{19}{42}$, $\frac{5}{16}$, $\frac{4}{9}$.

2. Find the H.C.F. and L.C.M. of the following :

   (*a*) 160, 224, 128;
   (*b*) $12a^3b^2c$, $18a^4b^4$, $42ab^2c^4$.

3. A piston-ring for a cylinder 12 cm. in diameter has itself a diameter of 12·5 cm. before being cut. How much must be taken out of its circumference so that it will just fit the cylinder ?

4. What multiplying factor converts a charge of pence per ounce to a charge of £'s per cwt. ? Write a simple formula to demonstrate this. If the tax on tobacco were 7½*d*. an ounce, what is this per cwt. ?

5. The current through a resistor is 0·05 ampere to within ± 5%, and the voltage across it is 100 volts to within ± 2%. By how much either way may the resistance differ from 2000 ohms ? What " tolerances " do these figures represent, expressed as percentages to 2 significant figures ?

6. The following table gives the man-hours per telephone spent each year in the maintenance of subscribers' apparatus :

Year . . .	1927–28	1928–29	1929–30	1930–31	1931–32	1932–33	1933–34
Man-hours per telephone . .	2·5	2·2	1·9	1·7	1·6	1·6	1·5

Make a block diagram to show this information. What would you estimate to be the man-hours required in 1934–35 ?

A.2

1. (*a*) Express 3 ft. 8½ in. in feet to 2 places of decimals.
   (*b*) Convert 2·89 lb. to lb. and oz. (nearest oz.).

2. The following table shows how it is proposed to use the land in the satellite town of Harlow :

	Acres.
Farms, allotments	2112
Residential	1748
Woods, parks, playing-fields	844
Industrial plant	556
Shops and town centre	115
Schools, hospitals, public buildings . . .	422
Roads	175
Miscellaneous	348
Total	6320

Calculate each item as a *percentage* of the total acreage (using a slide-rule) and construct a block diagram to show the " percentage utilisation of land ". What advantage is there in using percentages?

3. If the large and small wheels of a railway locomotive have radii 32 and 14 in. respectively, how many more revolutions will be made by the small wheel than by the large in a journey of 178 miles?

4. A dealer buys some Government Surplus dry batteries at $2\frac{1}{2}d.$ each and twice the number at $4\frac{1}{2}d.$ each. He sells all the sound ones at $6s.$ per dozen. Find his profit per cent if 10% of the batteries are " dud ".

5. If scrap brass is selling for £75 per ton, what is it worth per lb.?

Make a formula for the value of p lb. of scrap brass: (*a*) at £75 per ton, and (*b*) at £B per ton.

Find the value as scrap of 150 yd. of copper wire, weighing 300 lb. per mile, if copper is quoted at £124 per ton.

6. The insulation resistance of a cable pair in a newly laid cable was measured daily, with the following results:

Date: March . .	1	2	3	4	5
Insulation in megohms	40	38	32	21	4

Plot these results on a graph, and comment upon what you see. Why should you use a graph, and not a block diagram?

A.3

1. Convert £4 15*s*. 2*d*. to pounds (4 significant figures). What is its dollar equivalent, if £1 = 2·80 dollars?

2. The formula $d^2 = \frac{3}{8}l(s - l)$ gives the " dip " d ft. at the middle of a wire s ft. long supported at two points on the same level l ft. apart. What is the meaning of $(s - l)$ in words?

If a 30-yd. span of interruption cable has a dip of 2 ft. 6 in., find the length of cable used in this span.

3. Find the volume of metal in a pipe, of external diameter 11 in. and 1 in. thickness, which is 10 ft. 6 in. long. (Take $\pi = \frac{22}{7}$.)

4. Show that 2^{10} is very nearly 10^3. Hence find a good rough approximation to the logarithm of 2 to base 10. Check by the log table.

5. The price of steel rises by 30% after an estimate has been given for a certain project. By what percentage must the quantity of steel used be reduced, if the cost of steel used in the project may only rise by 20%?

6. A householder's gas meter records the quarterly consumption of gas as follows:

Quarter ending .	1948.		1949.				1950.	
	July	Oct.	Jan.	Apr.	July	Oct.	Jan.	Apr.
Therms consumed	32·33	27·56	34·45	38·69	34·45	29·68	39·75	38·69

Plot these values in a graph. Have intermediate points on the line joining these plotted points any meaning? Estimate from the graph the *average* quarterly consumption, and confirm by calculation. Comment on the seasonal variation in consumption, if any.

A.4

1. How would you convert dollars per short ton (2000 lb.) to £'s per ton (2240 lb.)? Draw a graph to enable this conversion to be made for prices up to 500 dollars per short ton. Give the American equivalent of a quotation for scrap copper at £130 per ton. (£1 = 2·80 dollars.)

2. In a town of 32,540 inhabitants it is estimated that 17% are employed in textiles and 24% in engineering. Find the numbers engaged in these two trades. If the average wage in the engineering trade is 53% higher than in textiles, what is the ratio of total wages paid for engineering to the corresponding figure for textile work?

3. The area A sq. cm. of a curved surface of a cone, of height h cm. and base radius r cm., is given by the formula $A = \pi r\sqrt{r^2 + h^2}$. Express h in terms of r and A, and find the height of a conical cap of area 42·8 sq. cm. if the diameter of its base is 6·4 cm.

4. The resistance of enamel-insulated Eureka wire, size 30 S.W.G., is 5575 ohms per 1000 yd. and it weighs 1·5 lb. per 1000 yd.

How many pounds of wire will be required to wind 1000 resistors each of 40 ohms, neglecting waste?

5. Find the resistance of a cylindrical coil using 250 turns of the wire mentioned in Question 4, if the average diameter of the turns is 0·32 in.

6. In a circuit proposed for metering subscribers' calls the service meter

is required to operate on 29 mA. and *not* on 26 mA. A hundred samples of the meter available were tested, with the following results :

Current to " just operate "	24	25	26	27	28	29
No. of meters . .	3	18	20	23	15	11
Current to " just operate "	30	31	32	33	34	35
No. of meters . . .	4	0	2	1	2	1

Construct a histogram to show the response of this batch of meters. What percentage of these meters were suitable ? Say whether the proposed circuit will work satisfactorily with this type of meter.

A.5

1. Evaluate $\dfrac{24{\cdot}72 \times 156{\cdot}8}{(0{\cdot}6194)^2 \times (13{\cdot}85)^3}$ using logarithms.

2. Use logs to find K in the formula $K = 25\pi\sqrt{\dfrac{12gM}{Wl^3}}$, where $\pi = 3{\cdot}142$, $g = 32{\cdot}2$, $M = 572$, $W = 4{\cdot}53$, $l = 56$. Express l in terms of K, M, g, and W.

3. A radio dealer buys a gross of valves and sells three-quarters of them at a profit of 35%. If he disposes of the remainder at a loss of 7%, what was his net profit per cent. on the whole transaction ?

4. The pull P gm. of an electromagnet is given by $P = \dfrac{B^2A}{8\pi g}$, where B lines per sq. cm. is the flux density in the air-gap, A sq. cm. is the cross-sectional area of this air-gap, $g = 981$ cm./sec./sec., $\pi = 3{\cdot}142$. If A is 0·4 sq. cm., calculate the pull for $B = 1000$, 2000, 5000, 8000 lines per sq. cm., and plot these values on a graph (P against B). From this graph find (*a*) the pull of the electromagnet when $B = 3500$, (*b*) the flux-density if the measured pull is 512 gm.

5. Three channels of a V.F. Telegraph equipment are transmitting steady " tones " of frequencies 540, 660, 780 c/s. How many times a second would all three tones be " in step " if they could all be started together ?

6. The price of a 2000-ohm relay varies according to the number of contact springs provided :

No. of contact springs .	3	7	11	13
Price (each) . . .	17*s*. 3*d*.	19*s*. 6*d*.	22*s*. 3*d*.	23*s*. 5*d*.

Plot these pairs of values on a graph, suitably titled, and from this graph estimate the price of such a relay having 9 contact springs. Should a continuous or a dotted line join the points plotted ?

A.6

1. The weights of potassium, chlorine, and oxygen in a sample of potassium chlorate are in the ratio 31·97 : 28·87 : 39·15. Assuming that all the gas is collected, what weight of oxygen can be obtained from 14·5 gm. of potassium chlorate?

2. The walls and ceiling of a room measuring 17 ft. $9\frac{1}{2}$ in. by 12 ft. $8\frac{1}{2}$ in. by 10 ft. 2 in. are to be covered with exactly fitted square panels. What is the largest size of panel which can be used?

Also find the total number of such panels required.

3. During the 4 weeks ended 19th April 1950 the Post Office handled 17,900,000 toll and trunk calls, compared with 16,930,000 in the corresponding period of 1949.

What percentage increase does this represent?

4. For a certain quantity of gas, its pressure p lb. per sq. in., its volume v cu. ft., and its temperature $T°$ C. are connected by the formula

$$pv = 17{\cdot}8(T + 273).$$

Express T in terms of p and v, and find the temperature of the gas at the instant when its pressure is 48 lb. per sq. in. and its volume 123 cu. ft.

5. In a town with 5400 telephones installed, 43% are residential and the remainder business subscribers. If the average revenue from a business telephone is 3·6 times that for a residential telephone, what percentage of the telephone revenue is derived from business subscribers?

If the number of voters in the town is 63,210, what percentage are residential subscribers?

6. A mixture of aniline and water is heated and the temperature, $T°$ C., of the vapour distilling from the hot mixture is measured at 2-min. intervals with the following results:

t min.	0	2	4	6	8	10	12	14	16	18	20	22	24
$T°$ C.	12	63	99	100	100	100	100	100	104	140	183	184	184

Plot a graph of T against t. What conclusions can be drawn from it? What is the likely temperature 17 min. from the start?

CHAPTER 5

EQUATIONS

5.1.

Historically, algebra began with the solving of equations, as an exercise for the philosopher rather than as a useful pursuit. The word algebra itself is derived from the first two words of the Arabic *al jebr wa'l mugābalah* (" integration and equation "), and it was through the Arabs that this study first reached Western Europe.

In our technological age formulæ and equations are of value in the practical understanding and interpretation of everyday problems in science and industry. It is most important for the modern engineer to be skilled in translating a statement in words into mathematical language (and *vice versa*), either as an algebraic expression, *i.e.*, a " formula ", or as a relation between known and unknown numbers—as an " equation ". This mastery of the interpretation of facts in terms of mathematics is more vital than acquiring tremendous skill in handling algebra itself.

Equations may arise in two ways: in response to a problem, or in the course of using a formula or other mathematical tool.

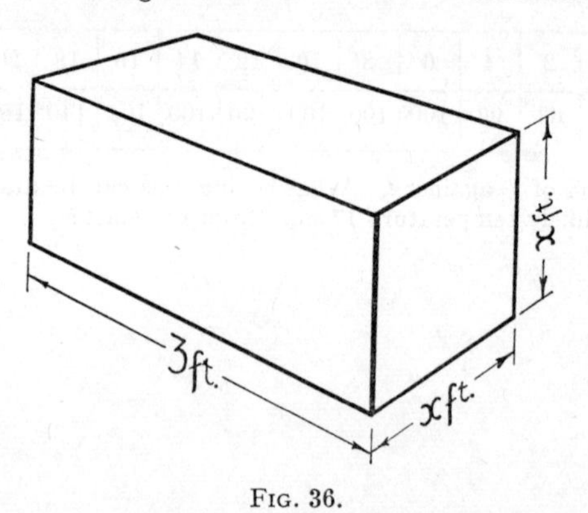

FIG. 36.

5.2. Solving Simple Problems

Example 5(*a*). *G.P.O. regulations state that the combined length and girth of a parcel may not exceed* 6 *ft. Find the capacity of the biggest parcel* 3 *ft. long and square in section which may be sent by parcel post.* (*This might conceivably arise in packing* 3-*ft. rods for parcel post.*)

Let us sketch this parcel (Fig. 36) and suppose the unknown cross-section to be x ft. by x ft.

Then the girth will clearly be $4x$ ft. So the combined length and girth is $(3 + 4x)$ ft.

If this is the largest possible parcel, this last measurement will be exactly equal to 6 ft.

so that $$3 + 4x = 6$$

This is an " equation " connecting the unknown number x with the known numbers 3, 4, and 6.

We can " solve " this equation (*i.e.*, find x) by common-sense thinking. (Later we will develop formal rules.) We argue thus :

If $3 + 4x = 6$ (*i.e.*, 3 more than $4x$ equals 6), then without the 3 added to it, $4x$ would be 3 (3 less than 6), *i.e.*, $4x = 3$.

But if 4 times the unknown number is 3, the unknown number itself must be 3 divided by 4 or $x = \frac{3}{4}$.

Hence the required dimension is $\frac{3}{4}$ ft. (or 9 in.).

We interpret this result in terms of the original problem, as " the dimensions of the largest parcel permissible will be 3 ft. by 9 in. by 9 in." (Its capacity is thus $1\frac{11}{16}$ cu. ft.)

(Reason out the same problem by ordinary arithmetic, and then note that the same mental steps are involved as in solving by algebra.)

Example 5(*b*). *A river flows at* 2 *m.p.h. What is the speed in still water of a boat which goes three times as fast downstream as upstream?*

We usually use the letter v (for " velocity ") to denote a speed. Suppose then that the boat would travel at v m.p.h. in still water.

For clearness we can tabulate our data :

Speed in still water.	Speed upstream.	Speed downstream.
v m.p.h.	$(v - 2)$ m.p.h.	$(v + 2)$ m.p.h.

But $3 \times \text{Speed Upstream} = \text{Speed Downstream}$

or using algebra,
$$3(v - 2) = v + 2$$
i.e.,
$$3v - 6 = v + 2$$
But if 6 less than $3v$ equals $v + 2$, it is clear that $3v$ itself will equal $v + 8$ (*i.e.,* 6 more),

i.e.,
$$3v = v + 8$$
But if 8 with v added equals $3v$, then 8 by itself will equal $3v - v$

or
$$3v - v = 8$$
i.e.,
$$2v = 8$$
And if twice the unknown number is 8, clearly
$$\underline{v = 4}$$
Thus, the speed of the boat in still water is 4 m.p.h.

Common-sense Check.—In this event, the speed downstream would be 6 m.p.h. and upstream 2 m.p.h. The first is clearly 3 times the second speed.

5.3. Equations Arising from the Use of Formulæ

Example 5(c). *To convert a measurement of temperature on the Centigrade scale to the corresponding Fahrenheit reading, we may use the formula*
$$F = 32 + 1{\cdot}8C$$
where F *is the number of degrees on the Fahrenheit scale and* C *the reading on the Centigrade scale.*

What would be the reading on a Centigrade thermometer corresponding to " blood-heat ", i.e., *98° F.?*

Using our formula, $32 + 1{\cdot}8C$ must have the special value 98;

i.e.,
$$32 + 1{\cdot}8C = 98$$
This is the equation which gives us this particular value of C. It is fairly simple to " unwrap " or solve the equation and so find this value.

For clearly $1{\cdot}8C$ by itself will be 32 less than 98;
$$\therefore \quad 1{\cdot}8C = 98 - 32$$
$$= 66$$
But if 1·8 times an unknown number is 66, it follows that the number itself must be 66 divided by 1·8;

i.e.,
$$C = \frac{66}{1{\cdot}8} = \frac{\overset{110}{\cancel{660}}}{\underset{3}{\cancel{18}}}$$
$$= 36\tfrac{2}{3}$$

So the reading for blood-heat on a Centigrade thermometer will be about 36·7° C. (Substitute this for C in the formula, and check that this does indeed correspond to 98° F.)

An equation arises, then, whenever we know all the quantities in a formula save one. We saw in Chapter 4 how this unknown could in many instances be made the "subject" of a fresh formula—but this is not always possible. In the above instance we could in fact have made C the subject of a fresh formula; one expressing the Centigrade reading in terms of the Fahrenheit.

Thus $$32 + 1{\cdot}8C = F$$

gives $$1{\cdot}8C = F - 32$$

hence $$C = \frac{F - 32}{1{\cdot}8}$$

which is a "new" formula for converting F degrees Fahrenheit to C degrees Centigrade.

Thus, for blood-heat, $F = 98$: so substituting in the latest formula,

$$C = \frac{98 - 32}{1{\cdot}8}$$
$$= \frac{66}{1{\cdot}8}$$
$$= 36{\cdot}7 \text{ (3 significant figures)}$$

Note that the arithmetic involved is exactly the same as in solving the equation above. It is in fact the same process, with the number 98 introduced at a later stage of the working.

Example 5(*d*). *Find the length of a closed cylindrical can if its diameter is* 4 *in. and* 88 *sq. in. of tinplate is used in making it.*

Sketch figures are worth drawing here, to avoid careless errors in thinking. Fig. 37(*b*) shows the flat "development" of the closed can of Fig. 37(*a*), *i.e.*, a rectangle with unknown length l in. and width equal to the circumference of the cylinder (4π in.) together with two circles each 2 in. in radius.

We could write a general formula for the surface area A sq. in. of a closed can (radius r in.) as

$$A = \underset{\text{(curved surface)}}{2\pi rl} + \underset{\text{(2 circular ends)}}{2\pi r^2}$$
$$= 2\pi r(l + r)$$

If we take $\pi = \frac{22}{7}$, $r = 2$, $A = 88$ we get

$$88 = \tfrac{88}{7}(l + 2)$$

(or taking an eighty-eighth of each side)

$$1 = \tfrac{1}{7}(l + 2) \qquad \text{(one-seventh of } (l + 2) \text{ is 1)}$$
$$\therefore \quad 7 = l + 2$$

i.e., 2 more than l is 7

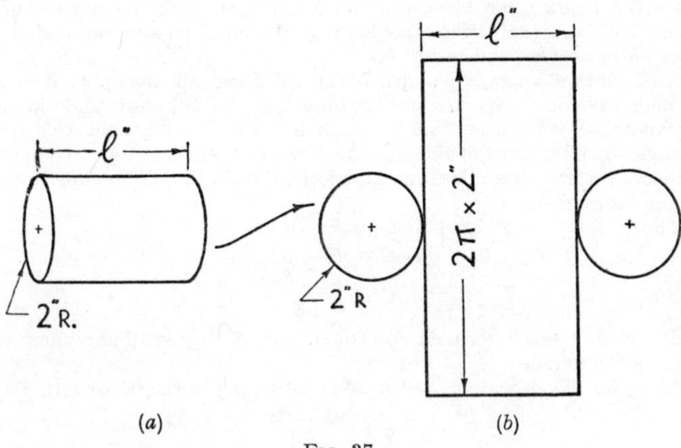

FIG. 37.

Hence $\underline{l = 5}$

So the required cylinder is 5 in. long.

Check : Area of 2 ends is $2 \times \frac{22}{7} \times 4 = \frac{176}{7}$ sq. in.

" curved surface $4 \times \frac{22}{7} \times 5 = \frac{440}{7}$ "

Total surface area is $\frac{616}{7} = 88$ "

Example 5(e). *Suppose that in the previous example we had been given the length of the cylinder (4 in., let us say) and had been asked to determine its diameter, if the total surface area were still to be 88 sq. in.*

Then in our general formula

$$A = 2\pi rl + 2\pi r^2$$

$A = 88$, $\pi = \frac{22}{7}$ approx., $l = 4$.

So we can write

$$88 = 2 \times \frac{22}{7} \times r \times 4 + 2 \times \frac{22}{7} \times r^2$$

i.e.,

$$88 = \frac{176r}{7} + \frac{44r^2}{7}$$

Dividing both sides by 44,

$$2 = \frac{4r}{7} + \frac{r^2}{7}$$

Multiplying both sides by 7,

$$14 = 4r + r^2$$

which we usually write as $r^2 + 4r = 14$.

In this case our equation includes not only the unknown r, but also its square r^2. It is called a " quadratic " equation, and cannot be solved by the common-sense methods we are using in this chapter. In the next chapter we shall see how to tackle such an equation graphically (see para. 6.5).

This is an instance where there is no neat way of changing the subject of a formula; *i.e.*, we should have difficulty in giving r in terms of A, l, and π in the general formula or by completing the square, as on p. 160.

5.4. Simple Rules for Solving Equations

In the above examples we have appealed to common sense rather than rule. To save time and energy it is worth formulating a few simple rules for solving equations.

We shall do well to regard an equation as a " balance " or pair of scales—each side of the equation (*i.e.*, on either side of the " equals " sign) corresponding to the weight (known or unknown) in each " scale-pan ". We distinguish between the two sides by referring to the " left-hand side " (l.h.s.) and the " right-hand side " (r.h.s.).

We may summarise the rules of the equation-solving game as follows:

(1) We may $\begin{cases}\textit{add}\\\textit{subtract}\end{cases}$ any number $\begin{cases}\textit{to}\\\textit{from}\end{cases}$ both sides of any equation without disturbing its balance.

(2) We may $\begin{cases}\textit{multiply}\\\textit{divide}\end{cases}$ each side of any equation by the same number without disturbing its balance.

More rarely needed,

(3) We may $\begin{cases}\textit{square}\\\textit{take the square root of}\end{cases}$ both sides of an equation without disturbing its balance.

In this book, with one exception (" completing the square ") we shall confine ourselves to using the first two rules.

In all three rules the word " number " is intended to include unknown numbers, which are signified by letters.

Until the process becomes automatic, it is advisable to insert " stage directions " above each line of working (" b.s." means " both sides of equation ").

Taking as an illustration the equation from Example 5(*c*) above:

$$32 + 1{\cdot}8C = 98$$

Subtract 32 from b.s., then

$$1{\cdot}8C = 66$$

Divide b.s. by 1·8, then

$$C = \frac{66}{1{\cdot}8}$$

i.e.,

$$\underline{C = 36\tfrac{2}{3}}$$

So that the " blood-heat " temperature 98° F. corresponds to $36\frac{2}{3}$° C.

Note how the answer to a problem is expressed as a simple English sentence. Avoid such senseless rigmarole as " Ans. $= 36\frac{2}{3}$ " !

Check : Put $C = 36\frac{2}{3}$ in the original equation and then

$$\begin{aligned} \text{l.h.s.} &= 32 + (1{\cdot}8 \times 36\tfrac{2}{3}) \text{ and r.h.s.} = 98 \\ &= 32 + 66 \\ &= 98 \end{aligned}$$

∴ the solution is correct.

Example 5(*f*). *Solve the equation* $16 + 4x = 71 - 7x$.

In equations of this type, with the unknown involved on both sides, we have to " collect " the unknown numbers to one side of the equation, and the ordinary numbers to the other. In this instance it is simpler to collect " unknowns " on the l.h.s.

We proceed as follows :,

$$16 + 4x = 71 - 7x$$

Subtract 16 from b.s. (to remove 16 from l.h.s.)

Then $4x = 55 - 7x$

Add $7x$ to b.s. (to bring r.h.s. up to 55)

Then $11x = 55$

Divide b.s. by 11

$$\therefore \quad \underline{x = 5}$$

Check : Substitute $x = 5$ in the original equation.

Then l.h.s. $= 16 + 20$ and r.h.s. $= 71 - 35$

$= 36 \qquad = 36$

So our solution is correct.

5.5. Equations which Involve Brackets

In such equations it is prudent to multiply out the brackets first, and then the equation can be solved by the rules suggested in the previous paragraph.

Example 5(*g*). $16 + 2(5x - 7) = 3(x + 10)$.

Multiply out the brackets on b.s.

$$16 + 10x - 14 = 3x + 30$$

Simplify the l.h.s. before proceeding,

$$2 + 10x = 3x + 30$$

Subtract 2 from b.s.,

$$10x = 3x + 28$$

Subtract $3x$ from b.s.

$$7x = 28$$

Divide b.s. by 7,

$$\therefore \underline{x = 4}$$

Check : Substitute $x = 4$ in original equation

l.h.s. $= 16 + 2 \times (13)$ r.h.s. $= 3 \times (14)$

$= 16 + 26 \qquad = 42$

$= 42$

So our solution is correct.

5.6. Equations which Involve Fractions

A very simple rule to follow in such cases is to multiply both sides of the equation by the smallest number which will "clear" the equation of fractions. This number will be the L.C.M. of the denominators involved on both sides of the equation.

Example 5(*h*). $\frac{x}{4} + \frac{7}{12} = \frac{3x}{8} - \frac{4}{9}$.

Multiply b.s. by 72 (the L.C.M. of 4, 12, 8, 9).

Then $$\frac{72x}{4} + \frac{72 \times 7}{12} = \frac{72 \times 3x}{8} - \frac{72 \times 4}{9}*$$

i.e., $$18x + 42 = 27x - 32$$

It is easier here to " collect " the unknown quantities on the r.h.s. of the equation.

Subtracting $18x$ from b.s.

then $$42 = 9x - 32$$

Add 32 to b.s.

and $$74 = 9x$$

Divide b.s. by 9,

$$\tfrac{74}{9} = x$$

And in this case $$x = 8\tfrac{2}{9}$$

Check in original equation :

$$\text{l.h.s.} = 2\tfrac{1}{18} + \tfrac{7}{12} = 2\tfrac{23}{36}$$

$$\text{r.h.s.} = \tfrac{37}{12} - \tfrac{4}{9} = 3\tfrac{1}{12} - \tfrac{4}{9} = 2\tfrac{23}{36}$$

Sometimes the denominators may include unknown as well as known numbers, as in the example which follows.

Example 5(*i*). *Robinson cycles at an average speed of 24 m.p.h., and starts 20 minutes before Cleaver, who overtakes him 108 miles from the start. Find Cleaver's average speed.*

We will suppose Cleaver's speed is v m.p.h. and work in hours and m.p.h.

Now Cleaver takes $\frac{108}{v}$ hr. for a journey of 108 miles, while Robinson takes $\frac{108}{24}$ or $\frac{9}{2}$ hr.

But Robinson had $\frac{1}{3}$ hr. start (20 min.)

$$\therefore \quad \frac{108}{v} + \frac{1}{3} = \frac{9}{2}$$

The denominators here are v, 3, 2 (L.C.M. = $6v$).
Multiply b.s. by $6v$,

Then $$\frac{6v \times 108}{v} + \frac{6v}{3} = \frac{6v \times 9}{2}$$

or $$648 + 2v = 27v$$

Subtract $2v$ from b.s.

Then $$648 = 25v$$

---

* This line may be omitted with practice.

Divide b.s. by 25

and $\frac{648}{25} = v$

Whence $\underline{v = 25{\cdot}92}$

Note : To divide by 25 : Multiply by 4 and divide by 100, *i.e.*, shift the decimal point 2 places to the left.

Thus Cleaver's average speed was 26 m.p.h. (to nearest m.p.h.)
(Check this with the original data.)

Example 5(*j*). *Solve the equation*

$$\tfrac{1}{3}(x - 3) - \tfrac{1}{4}(5 - x) = \tfrac{1}{12}(9 - 2x).$$

Here we have both fractions and brackets. We clear the fractions first : *i.e.*, multiply b.s. by 12.

Then $\frac{12}{3}(x - 3) - \frac{12}{4}(5 - x) = \frac{12}{12}(9 - 2x)$

$\therefore \quad 4(x - 3) - 3(5 - x) = 9 - 2x$

Now multiply out the brackets

and $4x - 12 - 15 + 3x = 9 - 2x$

or $7x - 27 = 9 - 2x$

Add $2x$ to b.s.

$$9x - 27 = 9$$

Add 27 to b.s.

$$9x = 36$$

$$\therefore \quad \underline{x = 4}$$

Check : l.h.s. $= \frac{1}{3}(4 - 3) - \frac{1}{4}(5 - 4)$ r.h.s. $= \frac{1}{12}(9 - 8)$

$= \frac{1}{3} \times 1 - \frac{1}{4} \times 1$ $\qquad = \frac{1}{12} \times 1$

$= \frac{1}{12}$ $\qquad = \frac{1}{12}$

Note that in the fourth line of working above we were faced with multiplying out $-3(5 - x)$.

How can we check that this is in fact the same as $-15 + 3x$?

The common-sense check is to replace x by a simple number such as 2.

Then clearly $-3(5 - 2)$ is the same as $-3(3)$ or -9. Also $-15 + 3x$, when x has the value 2, becomes $-15 + 6$, which is again -9 (" down 15 " and " up 6 " has the net result of " down 9 ").

When multiplying a bracket by a negative number, it is a safe rule to say that " the sign of each term inside the bracket is changed ".

i.e., When $+5$ is multiplied by -3 the result is -15.

When $-x$ is multiplied by -3 the result is $+3x$.

In every case, when in doubt, check by substituting ordinary numbers for the unknowns.

5.7. Summary : Linear Equations in One Unknown

All the equations solved so far have involved one unknown only, and are called " linear " equations, as they can be solved by drawing straight-line graphs (see Chapter 6).

In solving such equations the following simple " rules " will be helpful. They are matters of style rather than necessity, but they help towards accuracy as well.

(i) Put ONE statement only on each line.

(ii) Put ONE " equals " sign only on each line.

(iii) Keep the " equals " signs under one another. This leaves no doubt at any stage which side of the equation is which !

(iv) If answering a problem in words, express the result in a simple English sentence.

(v) *Check* in the original equation that your solution " works ". In a problem check with the original data.

(vi) Remember always the unwritten rule of mathematics—*common sense.*

EXERCISE 5(*a*)

Check each solution carefully.

A. Solve the following equations :

1. $7x - 5 = 16$.
2. $15x + 22 = 27$.
3. $5x + 14 = 2x + 50$.
4. $6p - 47 = p - 12$.
5. $5y - 11 = 2 + 3y$.
6. $5s + 7 = 19 - 4s$.
7. $5 - 4x = 3x - 23$.
8. $64 - 7r = 52 - 3r$.
9. $8x - 75 - 3x = 0$.
10. $17 - 33q = 23 - 49q$.

In the following equations, multiply brackets first :

11. $4(3x - 7) = 5x + 7$.
12. $3 + 2(x + 5) = 31$.
13. $17 - 2(3x + 2) = 4x + 7$.
14. $13 - 3(y - 4) = 4$.
15. $4p - (3p + 1) = p - 1$.
16. $2x - 11 = 3x - (2x + 3)$.
17. $5y - 6(y - 1) = 2(y - 3)$.
18. $42 - (4q - 7) = 16 - q$.
19. $8 + 3(4 - 5x) = 6x + 13$.
20. $x - 3(4 - 7x) = 7x + 3$.

In the next few equations, clear the fractions first :

21. $\frac{2m}{3} - 5 = \frac{m + 3}{5}$.
22. $\frac{y}{3} = \frac{2y - 1}{6} + \frac{y - 1}{4}$.

23. $\frac{5}{6} = \frac{p-2}{4} - \frac{11}{12}$. 24. $3 - \frac{2x+7}{3} = \frac{x-5}{2}$.

25. $4x - \frac{3x-128}{8} = \frac{5-x}{6}$.

B. Problems which can involve the solution of linear equations.

26. How many dry cells, each having an e.m.f. of 1·5 volts, must be joined in series with a 50-volt H.T. battery to raise its voltage to 80 volts? (Assume N cells required.)

27. A specimen of wire has a resistance of 5 ohms per metre. What length of this wire (x metres, say) must be added to a bobbin, which already measures 87·5 ohms, to bring its resistance up to 100 ohms?

28. Jack cycles from A to B, a distance of 69 miles, at an average speed of 12 m.p.h. Jill leaves B for A at 2 p.m., 15 min. after Jack left A. If she cycles at 10 m.p.h. when will they meet? (Suppose it is t hr. after 2 p.m.)

29. Resistances R_1 and R_2 ohms in parallel have a combined resistance of R ohms given by $\frac{1}{R} = \frac{1}{R_1} + \frac{1}{R_2}$. What resistance must be in parallel with 20 ohms to give a combined resistance of 12 ohms?

30. If the temperature remains constant, the pressure p and volume v of a quantity of gas are related by the formula $pv =$ constant (Boyle's Law). A diving-bell contains 1200 cu. ft. of air at atmospheric pressure (30 in. of mercury). It is lowered into a harbour until the same quantity of air occupies only 960 cu. ft. What is the pressure inside the diving bell at this depth? (Suppose it is x in. of mercury.)

31. A battery of internal resistance r ohms and e.m.f. E volts supplies a current I amperes to a load of R ohms. Then $I = \frac{E}{R+r}$. If a 24-volt battery passes 2·5 amperes through a load of 9·4 ohms, find the internal resistance of the battery.

32. If a railway wagon weighing W_1 tons moving at U m.p.h. is shunted on to a stationary wagon (W_2 tons) their common speed after impact is V m.p.h. given by $V = \frac{W_1 U}{W_1 + W_2}$.

An empty wagon moving at 8 m.p.h. is shunted on to a loaded wagon, and the two move off together at 3 m.p.h. If each wagon weighs 6 tons empty, find the load in the second wagon.

33. A radio dealer buys 20 R.F. chokes and sells 15 of them at a profit of 30%. The remainder he sells at 3*s.* each. If his total profit is 13*s.* 9*d.*, find the original buying price of one choke.

34. The installation costs of suspended cable on existing pole routes and of new underground cable are assumed to be £700 and £1050 per mile respectively. A new cable route 25 miles in length is to be partly suspended and partly buried at an estimated total installation cost of £19,775.

Find the length of aerial cable required for this project. (Suppose x miles aerial, and $(25 - x)$ miles buried cable.)

35. An overhead route over mountainous country is estimated to require 160 poles. After the project has started, some pole-holes have to be blasted out of the rock at an extra 50% cost per pole. If the total erection costs are increased by $7\frac{1}{2}\%$, find the number of holes blasted. (You are advised to assume a nominal basic cost per hole, and that x holes require blasting.)

5.8. Simultaneous Equations

Equations may sometimes arise which contain more than one unknown quantity. We shall illustrate this by discussing a practical problem in which two unknown quantities are involved.

Example 5(*k*). *In the accompanying diagram* (*Fig.* 38) *a* 30-*volt battery* B *with an internal resistance of* 2 *ohms is shown being charged from* 108-*volt D.C. mains* A, *through a resistance of* 27 *ohms. Both sources of e.m.f. supply current to an external load* L. *We want to find how much current is supplied by the mains and how much by the battery when the external load is equivalent to* 6 *ohms. We thus have* two *unknown currents to determine.*

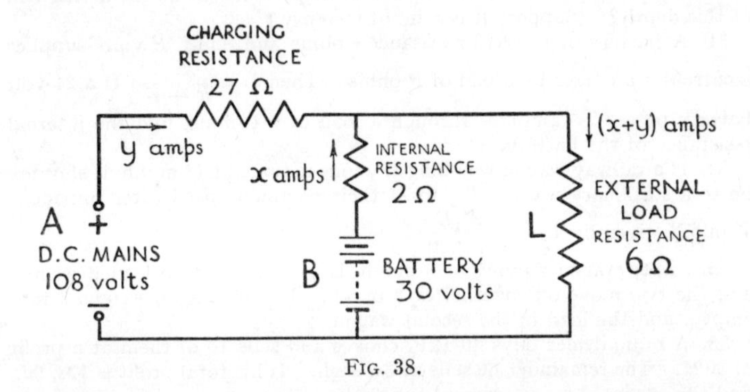

Fig. 38.

Suppose the battery supplies x amperes and the mains y amperes to the load. Then a total of $(x + y)$ amperes will be flowing through the load.

Using " Kirchhoff's 2nd Law ", in the closed circuit B–L the total " *IR* drops " must equal the total e.m.f. in that circuit.

In our shorthand of algebra,

$2x$	$+$	$6(x + y)$	$=$	30
Voltage drop across 2 ohms		Voltage drop across 6 ohms		Total e.m.f. in the circuit

Simplifying this equation,

$$2x + 6x + 6y = 30$$

i.e.,

$$8x + 6y = 30$$

Dividing b.s. by the common factor 2,

$$\underline{4x + 3y = 15} \quad . \quad . \quad . \quad . \quad . \quad . \quad \text{(i)}$$

This is a " linear " equation in the two unknowns x and y. Notice that we have not yet used the voltage of the mains, so that equation (i) would still be true whatever the value of this voltage.

Now consider the closed circuit A–L, which brings in the mains supply. Again employing " Kirchhoff ":

$27y$	$+$	$6(x + y)$	$=$	108
Voltage drop across 27 ohms		Voltage drop across 6 ohms		Total e.m.f. in the circuit

Test for yourself that this simplifies to

$$\underline{2x + 11y = 36} \quad . \quad . \quad . \quad . \quad . \quad . \quad . \quad \text{(ii)}$$

We now have two equations which connect our unknown x and y: there is one unique pair of numbers which simultaneously " fit " both equations. We shall see how to obtain these in the next section.

Before we go on to solve our equations, note how we have brought each of them to its simplest form, by multiplying out any brackets, and dividing both sides of each equation by a " common factor " to reduce the numerical " bulk ". *Always* reduce simultaneous equations to their lowest terms in this way before attempting to solve them !

5.9. Solution of Simultaneous Equations by Method of " Elimination "

Let us examine more closely the equations we obtained in 5.8:

i.e.,

$$4x + 3y = 15 \quad . \quad . \quad . \quad . \quad . \quad . \quad \text{(i)}$$

and

$$2x + 11y = 36 \quad . \quad . \quad . \quad . \quad . \quad . \quad \text{(ii)}$$

We have to find a pair of unknown numbers x and y which fulfil these two conditions. As the equations stand it is difficult to separate the x's from the y's. There are different amounts of each in the two equations. But if we had an equal amount of one unknown (y, for instance) in each equation it would be quite simple to compare the two equations and remove an equal

" dose " of this unknown from the field of operations—to " eliminate " it in fact!

In this case we will set out to eliminate y, and leave the field clear to find x. To get an equal weight of y in each equation we will multiply both sides of the first equation by 11, and both sides of the second by 3, and we shall have an equal amount, $33y$, in both cases.

We set out these operations as follows:

Multiply (i) by 11,

$$44x + 33y = 165 \quad . \quad . \quad . \quad . \quad . \quad \text{(iii)}$$

Multiply (ii) by 3,

$$6x + 33y = 108 \quad . \quad . \quad . \quad . \quad . \quad \text{(iv)}$$

(Of course, equations (iii), (iv) are just (i), (ii) in disguise.)

Now, if we subtract the l.h.s. of (iv) from the l.h.s. of (iii), and the r.h.s. of (iv) from the r.h.s. of (iii) the differences must also balance—so we write:

Subtract (iv) from (iii), $38x = 57$

$$\therefore \quad x = \tfrac{57}{38}$$

i.e., $$\underline{x = 1 \cdot 5}$$

If we substitute this value of x in either of the original equations (we shall use (i) here) we obtain an equation to find y.

Substitute $x = 1 \cdot 5$ in (i),

$$4 \times 1 \cdot 5 + 3y = 15$$

i.e., $$6 + 3y = 15$$

$$\therefore \quad 3y = 9$$

and $$\underline{y = 3}$$

In terms of the original problem,

the battery supplies 1·5 *amperes to the load, while the mains supply* 3 *amperes.*

Check: We substituted in (i) to find y; we therefore check our results in the other equation.

i.e., Test for $x = 1{\cdot}5$, $y = 3$ in (ii),

$$\text{l.h.s.} = 2 \times 1{\cdot}5 + 11 \times 3 \qquad \text{r.h.s.} = 36.$$
$$= 3 + 33$$
$$= 36$$

Now test yourself by solving the same equations, *beginning by eliminating* x.

5.10. Further Examples of Simultaneous Equations

Whenever the equations have whole numbers as the coefficients of x and y, it is always best to use the method of elimination outlined above. For more unwieldy equations perhaps the neatest method of solution is the graphical one shown in Chapter 6. Sometimes, the elimination demands the adding of two equations, as shown below:

Example 5(*l*). *To solve the equations*

$$2p + q = 10 \quad . \quad . \quad . \quad . \quad . \quad (1)$$
$$3p - 2q = 1 \quad . \quad . \quad . \quad . \quad . \quad (2)$$

We will eliminate q in this instance. All we need here is $2q$ in both cases, so (1) must be multiplied by 2, but (2) is left as it is.

Multiply (1) by 2, $\quad 4p + 2q = 20 \quad . \quad . \quad . \quad . \quad . \quad . \quad . \quad (3)$

and $\quad 3p - 2q = 1 \quad . \quad . \quad . \quad . \quad . \quad . \quad . \quad (2)$

Equation (3) has a surplus of $2q$, while (2) has a deficit of the same amount.

So if we add (3) to (2), $\quad 7p = 21$

$$\therefore \quad \underline{p = 3}$$

Substitute in (1), $\quad 2 \times 3 + q = 10$

$$6 + q = 10$$
$$\therefore \quad \underline{q = 4}$$

So the solution is $\quad \underline{p = 3,\ q = 4}$

Check: In (2), $\quad \text{l.h.s.} = 3 \times 3 - 2 \times 4 \qquad \text{r.h.s.} = 1$

$$= 9 - 8$$
$$= 1$$

Example 5(*m*). *The area shown in Fig.* 39 *is* 72 *sq. ft. and the perimeter is* 48 *ft. Find the unknown dimensions.*

We have two unknown measurements to determine, and two separate pieces of information. We therefore should be able to construct a pair of equations involving both x and y, *i.e.*, two simultaneous equations.

(*a*) The area of 72 sq. ft. can be split up into 2 rectangles each 3 ft. by x ft., and a single rectangle 4 ft. by y ft.

Hence $\qquad 2 \times 3x + 4y = 72$ (the whole area)

or $\qquad 6x + 4y = 72$

Divide b.s. by the common factor 2,

$$\underline{3x + 2y = 36} \quad \ldots\ldots \quad (1)$$

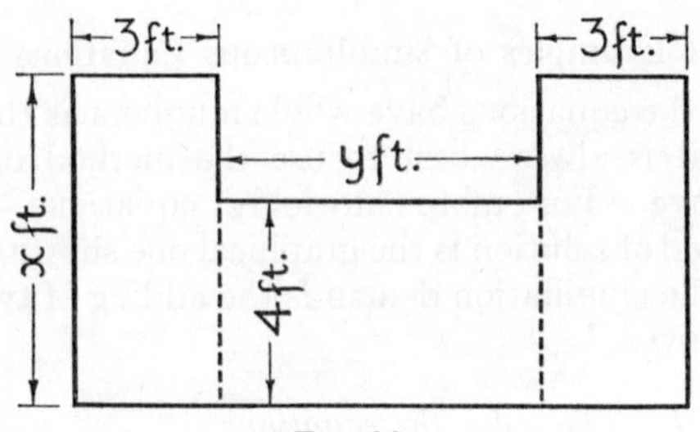

FIG. 39.

(*b*) The perimeter (the length of the boundary of the figure) is made up of two lengths of x ft., four lengths of 3 ft., two lengths of y ft., and two lengths of $(x - 4)$ ft.

so $\qquad 2x + 12 + 2y + 2(x - 4) = 48$ (the whole perimeter)

i.e., $\qquad 2x + 12 + 2y + 2x - 8 = 48$

$$\therefore \quad 4x + 2y + 4 = 48$$

$$\therefore \quad 4x + 2y = 44$$

Divide b.s. by 2, $\qquad \underline{2x + y = 22} \quad \ldots\ldots \quad (2)$

The facts given thus resolve themselves into two simple equations (1) and (2). We will eliminate y.

Multiply (2) by 2, $\qquad 4x + 2y = 44 \quad \ldots\ldots \quad (3)$

Also $\qquad 3x + 2y = 36 \quad \ldots\ldots \quad (1)$

Subtract (1) from (3) $\qquad \underline{x = 8}$

Substitute $x = 8$ in (1), $\qquad 24 + 2y = 36$

$$\therefore \quad 2y = 12$$

$$\underline{y = 6}$$

So the unknown dimensions are 8 ft. and 6 ft.

Check : In equation (2)

l.h.s. $= 16 + 6$ $\qquad$ r.h.s. $= 22$

$= 22$

Check for yourself that with these measurements the area and the perimeter are indeed as stated.

Example 5(*n*). (*Equations involving fractions.*)

To solve $x - 5 = \frac{1}{7}(y - 2)$ . . . . . . (1)

$\frac{1}{3}(x + 10) = 4y - 3$ . . . . . . (2)

Both equations need to be cleared of fractions and brackets and then "cleaned up" before eliminating x or y.

Multiply (1) by 7, $7x - 35 = y - 2$

i.e., $\underline{7x - y = 33}$ . . . . (3)

And multiply (2) by 3, $x + 10 = 12y - 9$

Whence $\underline{x - 12y = -19}$ . . . . . . (4)

The formidable equations (1) and (2) have now been reduced to neater forms as (3) and (4). We will eliminate x this time.

Then $7x - y = 33$ . . . . (3)

and multiply (4) by 7, $7x - 84y = -133$ . . . . . . (5)

Subtract (5) from (3) $+ 83y = 166$

$\therefore \underline{y = 2}$

Note : $-(-84y)$ is the same as $+84y$, and $-(-133)$ is the same as $+133$.

Substitute $y = 2$ in equation (3) [which comes from ORIGINAL equation (1)]

and $7x - 2 = 33$

$\therefore 7x = 35$

$\therefore \underline{x = 5}$

Check : In ORIGINAL equation (2)

l.h.s. $= \frac{1}{3}(5 + 10)$ r.h. . $= 8 - 3$

$= 5$ $= 5$

5.11. Awkward Subtractions

A useful verbal rule in subtracting one equation from another when negative signs are much in evidence is "change the sign of the bottom line and add".

E.g., in solving

$6p - 5q = 24$ . . . . (1)

and $9p - 4q = 22$ . . . . (2)

If we wish to eliminate q we multiply (1) by 4,

$24p - 20q = 96$ . . . . (3)

and multiply (2) by 5,

$$45p - 20q = 110 \quad . \quad . \quad . \quad . \quad . \quad (4)$$

We must clearly SUBTRACT (4) from (3) to eliminate q.

Mentally we " change the sign of the bottom line and add ".

I.e., we ADD $-45p$ to $24p$ (giving $-21p$)

we ADD $+20q$ to $-20q$ (giving zero)

we ADD -110 to 96 (giving -14)

and thus obtain $\quad -21p = -14$

or $\quad 21p = 14$

giving $\quad \underline{p = \frac{2}{3}}$

(Verify that $q = -4$, and check these values of p and q.)

5.12. Practical Difficulties

The examples above have been deliberately simplified by choosing values which lead to easy arithmetic, most of which can be performed mentally. The numbers which occur in practice are seldom so easy, but the same procedure can be applied as in the previous examples, making full use of the slide-rule to perform the necessary multiplication and division.

Example 5(*k*) might occur in a practical situation in the following modified form:

" A 24-volt battery B with an internal resistance of 0·4 ohms is being charged from 220-volt D.C. mains A, through a resistance of 60 ohms. Calculate the current supplied by the mains and by the battery when a load *L* equivalent to 11·8 ohms is connected across the battery."

As before we suppose the battery supplies x amperes and the mains y amperes, so that a total of $(x + y)$ amperes flows in the load. (Fig. 38 refers.)

For the circuit B–L we now have

$$0{\cdot}4x + 11{\cdot}8(x + y) = 24$$

which reduces to

$$12{\cdot}2x + 11{\cdot}8y = 24$$

or

$$\underline{6{\cdot}1x + 5{\cdot}9y = 12} \quad . \quad . \quad . \quad . \quad (1)$$

And for the circuit A–L

$$60y + 11{\cdot}8(x + y) = 220$$

which similarly reduces to

$$\underline{5{\cdot}9x + 35{\cdot}9y = 110} \quad . \quad . \quad . \quad . \quad (2)$$

In this case it is simpler to eliminate x rather than y. This we can do by multiplying (1) by 5·9 and (2) by 6·1. In a practical case such as this, accuracy to 2 significant figures will be adequate. So we have

Multiply (1) by 5·9,

$$36{\cdot}0x + 34{\cdot}8y = 70{\cdot}8 \quad . \quad . \quad . \quad . \quad (3)$$

Multiply (2) by 6·1,

$$36{\cdot}0x + 219y = 671 \quad . \quad . \quad . \quad . \quad (4)$$

Subtracting (3) from (4), we get

$$184{\cdot}2y = 600{\cdot}2$$

$$\therefore \quad y = \frac{600{\cdot}2}{184{\cdot}2}$$

i.e.,

$$\underline{y = 3{\cdot}26}$$

Substitute for $y = 3{\cdot}26$ in equation (1), and

$$6{\cdot}1x + 5{\cdot}9 \times 3{\cdot}26 = 12$$

i.e.,

$$6{\cdot}1x + 19{\cdot}2 = 12$$

so

$$6{\cdot}1x = -\ 7{\cdot}2$$

$$\therefore \quad x = -\frac{7{\cdot}2}{6{\cdot}1}$$

i.e.,

$$\underline{x = -\ 1{\cdot}18}$$

We see that in this situation x *has a negative value,* which can only mean that the current is flowing through the battery in the *opposite direction* to that we assumed (as shown by the arrow in Fig. 38). That is, the battery is not supplying power to the load, but on the contrary is receiving it from the mains: in other words *the battery is being charged.*

To sum up, there is a charging current through the battery of 1·2 amperes; while the mains supply a total current of 3·3 amperes, of which only 2·1 amperes is taken by the load. All the currents are given to 2 significant figures.

All that remains is to check our results $x = -1{\cdot}18$ and $y = 3{\cdot}26$ in equation (2):

$$\begin{aligned} \text{l.h.s.} &= 5{\cdot}9 \times (-1{\cdot}18) + 35{\cdot}9 \times 3{\cdot}26 \\ &= -6{\cdot}96 \quad + 117{\cdot}0 \\ &= 110{\cdot}0 \\ \text{r.h.s.} &= 110 \end{aligned}$$

When we started this problem we imagined that the battery, as well as the mains, would be supplying current to the external load. Notice how neatly the resulting mathematical argument puts us right on this point, demonstrating that the battery is being charged and not discharged under these conditions.

For those who fight shy of too much arithmetic, there is an alternative graphical technique for solving awkward simultaneous equations, to be found in the next chapter (para. 6.3).

5.13. General Summary—Simultaneous Equations (Two Unknowns)

(i) First decide which unknown is the easier to eliminate.

(ii) By suitable multiplying, obtain equations with equal amounts of this unknown.

(iii) Add or subtract to obtain the other unknown.

(iv) To find the first unknown, substitute the one thus found in the EASIEST equation above.

(v) Always CHECK in the ORIGINAL equation which was not the one used in (iv).

(vi) If in response to a problem, check results in terms of the practical data.

EXERCISE 5(*b*)

A. Solve the following simultaneous equations, making careful checks:

1. $x + 2y = 13$, $2x + y = 11$
2. $a + b = 9$, $a - b = 5$
3. $3p - q = 3$, $2p + 3q = 35$
4. $2x - 5y = 7$, $7x - 2y = 40$
5. $26x + 8y = 9$, $2y - 6x = 2$
6. $\frac{x}{4} - \frac{y}{3} = 4$, $2x + y = 10$

7. $r = 5s$
$3r - 2s = 16{\cdot}12$

8. $3x - 4y = 15$
$4x + y = 4{\cdot}8$

9. $3y - 4(x - y) = 7$
$7x = 5y + 18{\cdot}2$

10. $2x - \dfrac{4 - y}{9} = 7{\cdot}1$
$5(x + y) = 2{\cdot}4(x - y)$

B. The following examples can involve two unknown quantities, and therefore the solution of simultaneous equations :

11. A contractor shifted 15 tons of gravel on Tuesday, comprising 3 lorry-loads and 8 cart-loads. On Thursday 2 lorry-loads and 20 cart-loads shifted 21 tons of gravel. Find the average weight of a lorry load. (Suppose a lorry-load is x tons, and a cart-load y tons.)

12. In a factory employing 315 workers there are 49 more men than women. How many men are employed there?

13. A bill of £1 5*s*. 6*d*. was paid in shillings and half-crowns, 15 coins in all. How many of each coin were required?

14. Scrap aerial cable contains lead and copper worth £8 8*s*. and £1 12*s*. a cwt. respectively (fictitious prices !). If the total value per ton of the scrap is £154 8*s*., find the proportion by weight of lead to copper present. (Suppose x cwt. of lead, y cwt. of copper per ton.)

15. The currents in amperes in two wires are such that

$$5I_1 = 7I_2, \quad 8(I_1 + I_2) + 12I_2 = 104$$

Simplify the second equation, and find the two currents, I_1, I_2 amperes.

16. One short overhead route involved 7 poles carrying 10 wires : the cost was £120. A second route with 6 wires and 12 poles cost £150. Deduce the overall costs " per pole " and " per wire ".

17. From the data of Question 16 write a formula for the estimated cost £C for a route with p poles carrying W wires. Hence estimate the cost of erecting 15 poles carrying 14 wires.

18. In the problem worked out in para. 5.12 above, take the resistance of the load as 6 ohms instead of 11·8 ohms. What currents do mains and battery now supply?

For what value of the load resistance is the current supplied by the battery zero, *i.e.*, the battery " floats " (neither discharging nor being charged)? (Suppose L ohms load, $x = 0$.)

19. On a busy manual exchange an operator is found to handle on the average 120 local and 50 junction calls per hour.

On a similar exchange with much more junction traffic an operator handles 50 local and 90 junction calls hourly.

Estimate the time in seconds to be allowed for handling

(*a*) a local call (x sec.)

(*b*) a junction call (y sec.).

20. A battery is made up of nickel–iron (" NIFE ") and lead–acid cells in series. 3 NIFE cells and 5 lead–acid cells give an e.m.f. of 13·9 volts, while 12 lead–acid cells and 8 NIFE cells *reversed* (*i.e.*, opposing) give an e.m.f. of 13·6 volts. Find the e.m.f. of a single NIFE cell.

21. The annual running costs of a manual telephone exchange are estimated to be "so much a line + so much a call".

An exchange with 1000 lines, handling 300,000 calls annually, costs £7500 a year to run. For a smaller but busier exchange, with 600 lines and 450,000 calls, the figure is £5400.

Find the costs per line (£x) and per call (£y), and so derive a formula for an exchange with e lines, t calls, for the annual cost £C.

What would you estimate to be the annual cost of a 1200-line exchange handling 510,000 calls annually?

22. A storekeeper keeps a tally of scrap lead and copper, noting the total weight of each load of scrap, and the number of barrels of each metal to arrive in each load.

He receives a load of 10 barrels weighing 900 lb., but forgets to record how many contain lead.

The previous consignments were:

Total wt.	Barrels total.	Lead.	Copper.
810 lb.	8	5	3
730 lb.	9	2	7

Can the storekeeper tell from these records how many barrels of each metal are in the 10-barrel consignment, or must he unpack each barrel?

23. At 19° C. the resistance of a coil of wire is 76·3 ohms; at 45° C. it is 83·9 ohms. If a formula $R = aT + b$ connects its resistance R ohms with the temperature T° C., find a and b.

What is the resistance R_0 ohms at 0° C.? If $R = R_0\,(1 + \alpha T)$ find α (the "temperature coefficient").

24. A radio dealer cannot find the list price of a certain range of resistors, but remembers that the 100-ohm type cost him 6d. more than the 40-ohm one. An old invoice gives him the further information that 3 dozen of the former and 2 dozen of the latter cost £9 3s. Find the cost price of each type of resistor.

CHAPTER 6

THE GRAPHS OF SOME MATHEMATICAL FUNCTIONS

6.1. Straight-line Graphs—the Pattern $y = ax + b$

In Chapter 4 we found that if a straight-line graph was obtained when pairs of observations were plotted on squared paper, then a simple algebraic " law " connected the two quantities measured. Thus in Experiment C the length l cm. of a stretched spring for a given load P gm. was given by a formula

$$l = 0{\cdot}036P + 34$$

In general, for any spring we expect the values of l and P to be connected by a straight-line " law " (Hooke's Law)

$$l = aP + b$$

where a is a number depending on the design of the spring, and represents the " steepness " of the line, and b represents the length in cm. of the spring when unloaded.

Even more generally, if such a linear relation or law connects any two quantities x and y, then the graph of y against x would give a straight line of the pattern $y = ax + b$.

Let us test this with a numerical case, *e.g.*, $y = 3x + 7$, plotting y for values of x from -3 to $+3$.

Tabulating our values of x and y:

$x =$	-3	-2	-1	0	1	2	3
$3x =$ $7 =$	-9 7	-6 7	-3 7	0 7	3 7	6 7	9 7
$y = 3x + 7 =$	-2	1	4	7	10	13	16

Notice how the value of y goes up in equal steps (3 units) for equal (unit) increases in x.

The resulting graph is shown in Fig. 40, and is clearly a straight line.

We see that in the " equation " of this straight line $y = 3x + 7$, the number term 7 is just the value of y when $x = 0$. On the graph it is the height at which the straight line cuts the y-axis. This is called the " intercept " on the y-axis. The number 3

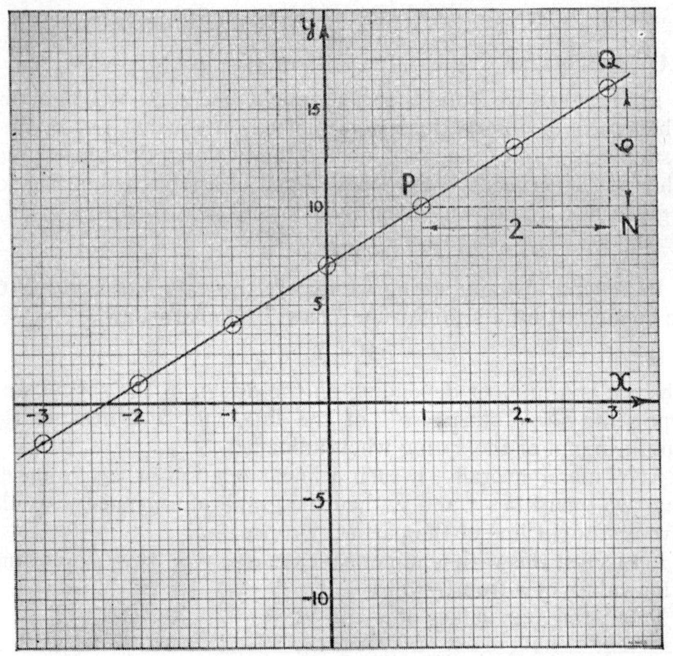

FIG. 40.—THE GRAPH OF THE EQUATION $y = 3x + 7$.

which multiplies x (we call 3 the *coefficient* of x) gives the steepness of the straight line expressed as the proportion $\dfrac{\text{vertical rise}}{\text{horizontal increase}}$

e.g., $$\frac{QN}{PN} = \frac{6}{2} = 3$$

It gives indeed the increase in y per unit increase in x. The mathematical name for this proportional steepness is " slope " or " gradient ". (It has another name in trigonometry—see Chapter 8.)

Generally then, the numbers a and b in the straight-line equation $y = ax + b$ represent the slope of the straight line and its intercept on the y-axis.

To graph such an equation only TWO points are necessary. We generally plot THREE as a check of our accuracy.

The x and y values at any point are called its " co-ordinates ". The point where $x = 2$ and $y = 13$ is often referred to as (2, 13), or as the point whose co-ordinates are 2 and 13.

Verify from the graph that the co-ordinates of any point on our straight line " satisfy " the equation $y = 3x + 7$.

6.2. Some Variations of the Straight-line Graph (Fig. 41)

(*a*) The intercept may be negative (*i.e.*, " below the x-axis " *e.g.*, $y = 2x - 5$.

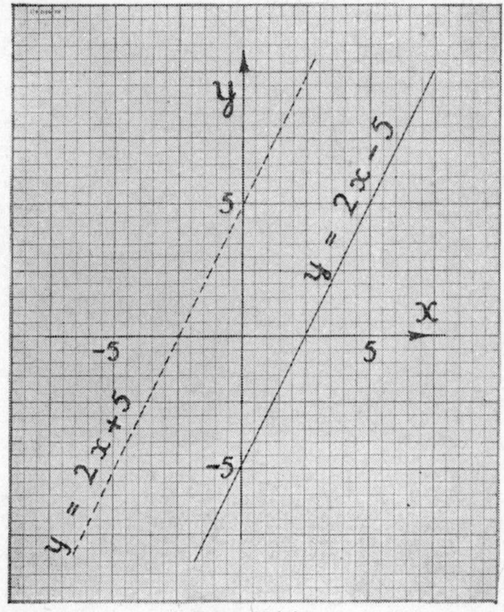

FIG. 41(*a*).

The graph $y = 2x + 5$ is shown dotted on the same diagram to bring out the difference here.

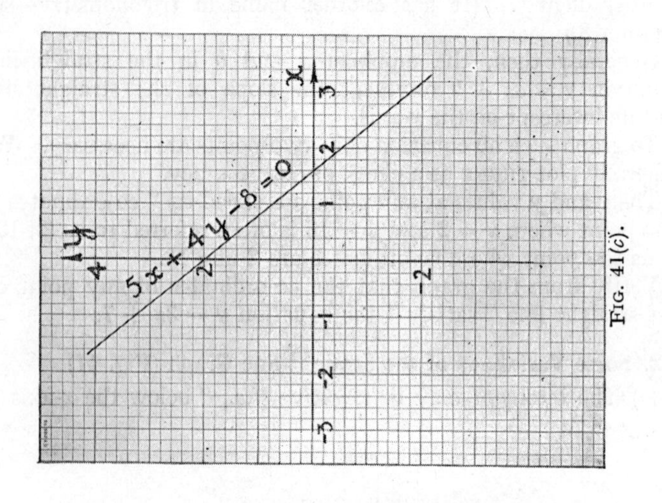

FIG. 41(c).

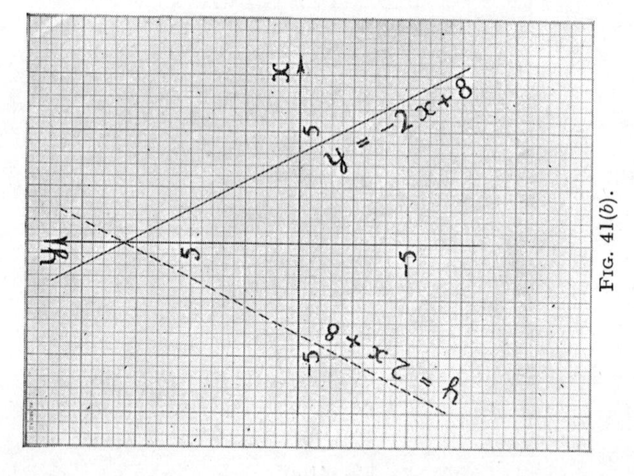

FIG. 41(b).

(*b*) The slope may be negative (*i.e.*, "downhill"), *e.g.*, $y = -2x + 8$.

The graph $y = 2x + 8$ is shown dotted to show the contrast in this case.

(*c*) The "equation" of a straight line may be disguised, *e.g.*, $5x + 4y - 8 = 0$ is an equation of the "first degree" in x and y (*i.e.*, it does not contain squares or higher powers of x and y).

We can rearrange this equation to read

$$4y = -5x + 8$$

and dividing both sides by 4, we obtain

$$y = -\tfrac{5}{4}x + 2$$

which is a straight line with a "downward" slope ($a = -\frac{5}{4}$) and an intercept of 2 units ($b = 2$) on the y-axis.

(The intercept on the x-axis, when $y = 0$, is $x = \frac{8}{5}$ or 1·6.)

Generally we can say that any equation of the form

$$px + qy + r = 0,$$

where p, q, r are numbers, will be represented graphically by a straight line.

Example 6(*a*). *In testing a certain lifting machine the effort* E *lb. wt. required just to raise a load* L *lb. was recorded as shown in the following table.*

L . .	40	60	80	100	120	140	160
E . .	8·3	10·3	12·1	14·2	16·0	17·9	19·8

Verify that a relation E = aL + b *holds between* E *and* L *and find from your graph the values of* a *and* b.

Comparing the required formula $E = aL + b$ with $y = ax + b$, it is sensible to take the L-axis as "base-line". Note that we have "measured up" the effort E lb. wt. for selected values of the load, so that this is the more logical arrangement. The resulting graph is shown in Fig. 42, and is clearly a straight line.

From the graph the intercept b on the E-axis is approximately 4·5 lb. wt.

This represents the minimum effort to operate the machine when unloaded. The " slope " is conveniently found from the triangle PQN, as

$$a = \frac{QN}{PN} = \frac{19{\cdot}8 - 10{\cdot}3}{160 - 60} = \frac{9{\cdot}5}{100} = 0{\cdot}095$$

(Note how PN is chosen to simplify the division.)

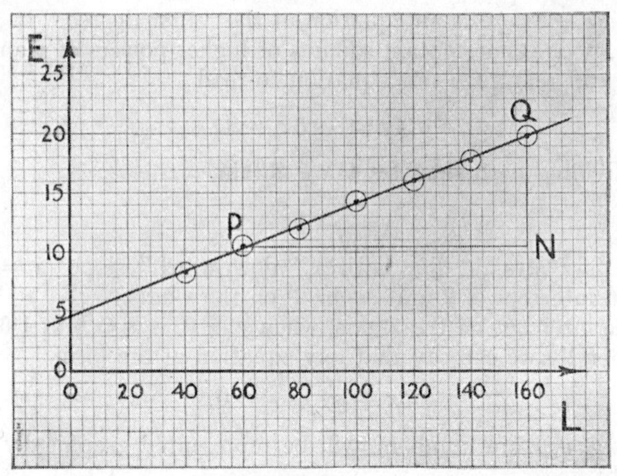

FIG. 42.

The relation between effort and load for this machine is thus given by

$$E = 0{\cdot}095L + 4{\cdot}5$$

This is called the " Law " of this machine.

6.3. Use of Graphs to Solve Simultaneous Linear Equations

As suggested in Chapter 5 (para. 5.10) simultaneous equations of the linear type may be solved graphically. This is slower and less accurate than the " elimination method " when the numbers concerned are simple, but considerably quicker where large or decimal numbers are involved (as in Example 6(*c*) below).

Example 6(*b*). *Obtain graphically the values of* x *and* y *which satisfy*

$$7x - y = 33 \quad . \quad . \quad . \quad . \quad . \quad (1)$$

and

$$x - 12y = -19 \quad . \quad . \quad . \quad . \quad . \quad (2)$$

(*These equations were solved by algebra in Example* 5(*n*).)

The easiest method of drawing each straight line (Fig. 43) is to note where it crosses the axes of x and y.

Thus for equation (1). When $y = 0$, $7x = 33$, *i.e.*, $x = 4{\cdot}71$.

So this straight-line graph (1) cuts the x-axis at the point A, where $x = 4{\cdot}71$.

Similarly, when $x = 0$, $-y = 33$ and $y = -33$, giving the point B where the straight line (1) cuts the y-axis.

The straight line whose equation is $7x - y = 33$ is then obtained by joining AB. The co-ordinates of every point on this line satisfy the single equation $7x - y = 33$.

Similarly for equation (2), a straight line graph CD can be drawn, where C is the point $(-19, 0)$, and D is the point $(0, 1{\cdot}58)$. So the co-ordinates of every point on this line satisfy the equation $x - 12y = -19$.

From the graph (Fig. 43)it will be seen that the straight lines AB and

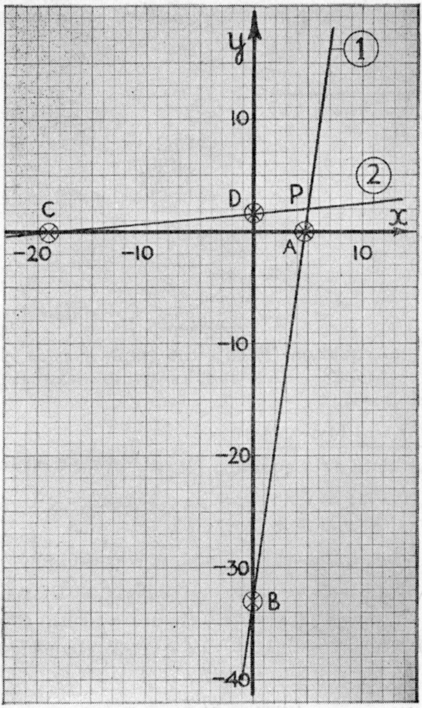

FIG. 43.

CD meet at the point P where $x = 5$ and $y = 2$. The co-ordinates $x = 5$, $y = 2$ of this point P, which alone lies on BOTH lines, must therefore satisfy both equations.

The solution of these equations is thus $x = 5$, $y = 2$.

Note how much quicker is the algebraic solution in this case, where only whole numbers are involved in the equations and their solution. But if we are faced with simultaneous linear equations in which the coefficients are awkward decimals or large numbers, it is generally more efficient to solve them by drawing graphs.

Example 6(*c*). *Solve the equations*

$$2{\cdot}13x + 4{\cdot}8y = 27{\cdot}2 \quad . \quad . \quad . \quad . \quad (1)$$

and

$$5{\cdot}93x - 3{\cdot}14y = 14{\cdot}37 \quad . \quad . \quad . \quad . \quad (2)$$

The method of elimination (para. 5.9) would be somewhat laborious.

But if we draw a straight-line graph for each equation we can solve the problem quite speedily.

<u>To graph $2{\cdot}13x + 4{\cdot}8y = 27{\cdot}2$</u>

When $y = 0$,

$$2{\cdot}13x = 27{\cdot}2$$

$$\therefore \quad x = \frac{27{\cdot}2}{2{\cdot}13}$$

$$i.e., \quad \underline{x = 12{\cdot}81} \quad \text{(by slide-rule}$$

When $x = 0$,

$$4{\cdot}8y = 27{\cdot}2$$

$$\therefore \quad y = \frac{27{\cdot}2}{4{\cdot}8}$$

$$i.e., \quad \underline{y = 5{\cdot}68} \quad \text{(by slide-rule)}$$

$\therefore$ The graph of equation (1) cuts the graph axes at A and B, where $x = 12{\cdot}81$ and $y = 5{\cdot}68$. This enables us to draw the straight line (marked (1) in Fig. 44).

<u>To graph $5{\cdot}93x - 3{\cdot}14y = 14{\cdot}37$</u>

By similar calculation for equation (2)

$$y = 0 \text{ at } C, \text{ where } x = 2{\cdot}49$$

$$x = 0 \text{ at } D, \text{ where } y = -4{\cdot}56$$

i.e., CD gives the graph of equation (2).

Where the straight lines intersect on the graph at the point P,

$$x = 4{\cdot}45, \ y = 3{\cdot}70 \text{ (to nearest 20th)}$$

Check: In (1),

$$\text{l.h.s.} \simeq 9{\cdot}4 + 17{\cdot}8 \quad \text{(by slide-rule)}$$

$$\simeq 27{\cdot}2$$

$$\text{r.h.s.} = 27{\cdot}2$$

also in (2), l.h.s. $\simeq 26{\cdot}3 - 11{\cdot}6$

$$\simeq 14{\cdot}7$$

r.h.s. $= 14{\cdot}37$

$\therefore\ x = 4{\cdot}45,\ y = 3{\cdot}70$ is the solution of these equations.

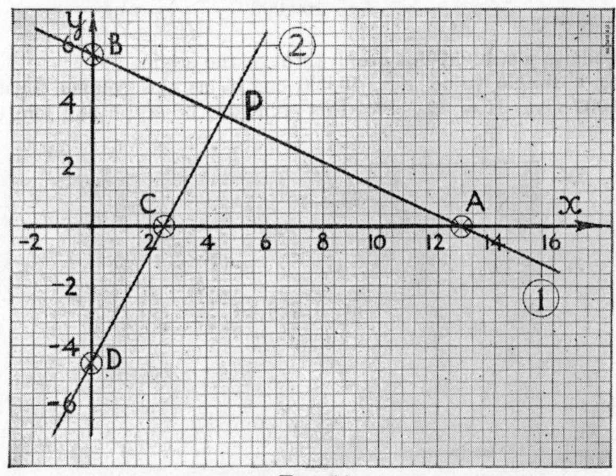

FIG. 44.

Note the slight discrepancy (about 2% error) in the second check. We cannot expect more than 2 significant figures accuracy unless the graph is enlarged. In most cases no greater accuracy is needed.

6.4. " Families " of Straight Lines

In many technical charts a series or " family " of lines may be displayed—giving " at-a-glance " information which would otherwise involve printed tables.

Example 6(*d*). *The calorific value of gas for domestic use was* 560 *B.Th.U. per cu. ft. before the war, but since* 1940 *has been fixed at* 530 *B.Th.U. If* 100,000 *B.Th.U.* = 1 *therm, draw graphs to read off the heating value* H *therms against the volume* V *cu. ft. of gas consumed (as shown on the gas meter). The graphs should include up to* 12,000 *cu. ft. of gas per quarter, and allow for differing qualities* 500, 530, *and* 560 *B.Th.U per cu. ft.*

For any one quality of gas the consumption in therms will be directly proportional to the volume of gas in cu. ft. passing through the meter.

Thus, at 530 B.Th.U. per cu. ft. the heating value of V cu. ft. would be $530V$ B.Th.U. or $\frac{530V}{100,000}$ therms. This formula

$$H = \frac{530V}{100,000}$$

corresponds to a straight-line graph through the origin.

Taking a simple value, *e.g.*, 10,000 cu. ft. as the volume consumed, $V = 10,000$ gives $H = 53$. Plot this point on the graph and join to the

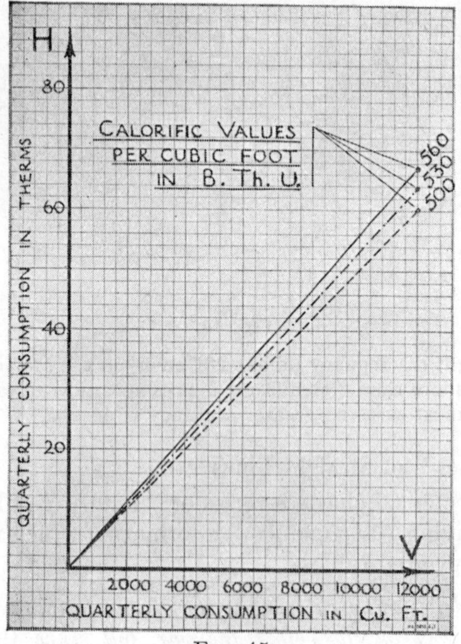

FIG. 45.

origin, and we have a graph from which we can read directly the number of therms for any meter consumption in cu. ft. when the quality of the gas is 530 B.Th.U.

Similarly, two other straight lines can be drawn for 500 and 560 B.Th.U. (see Fig. 45), and we have a " family " of straight lines through the origin which replaces a 3-column conversion table.

EXERCISE 6(*a*)

A. Express the following linear equations in the form $y = ax + b$ (a is the " slope ", b the intercept on the y-axis).

1. $3x + y = 16.$ 2. $y - 7x = 4.$ 3. $x + y = 72.$

4. $2x + 5y = 11.$ 5. $x - 4y = 8.$ 6. $\frac{x}{2} + \frac{y}{3} = 1.$

7. $3x - 4y - 7 = 0.$ 8. $px + qy + r = 0.$ 9. $\frac{x}{m} + \frac{y}{n} = 1.$

10. $3{\cdot}8x + 7{\cdot}2y + 8{\cdot}3 = 0.$

B. Plot the following sets of data, and obtain the best estimate of the slope and intercept of the straight-line law which best fits each graph.

11. The load L lb. lifted by a pulley system with effort E lb. wt.

E . . .	6·6	8·2	9·8	11·4	12·9
L . . .	100	200	300	400	500

Law of Machine : $L = aE + b.$

What load will an effort of 10 lb. wt. raise ?

12. Maximum safe current y amperes for V.I.R. insulated copper wire of cross-section x sq. in.

x .	0·25	0·4	0·5	0·6	0·7	0·8
y .	236	350	422	492	561	625

Law: $y = ax + b.$

Find the least diameter of such a wire to carry 300 amperes.

13. A weight W lb. is hung from the end of a helical spring and the length l in., of the spring, is observed as follows :

W (lb.) . .	50	100	150	200	250
l (in.) . .	28	31	34	37	40

Law : $l = aW + b$

What is the *unstretched* length of the spring? What weight would stretch the spring to twice its unloaded length ?

14. R ohms per mile is the resistance of copper wire weighing l yd. per lb.

S.W.G. . .	18	20	22	23	24
l . . .	47·8	85	140·4	191·2	227·8
R . . .	23·1	40·9	67·7	93·6	111·3

Law : $R = al + b$.

Estimate the resistance per mile of No. 21 Gauge (S.W.G.) equivalent to 107·6 yd. per lb.

15. The resistance R ohms of a specimen of wire at temperature $T°$ C. :

T .	20	40	60	80	100	120
R .	100·9	101·5	102·5	103·1	104	104·8

Law: $R = R_0(1 + aT)$.

What is the meaning of R_0 ?

C. Solve the following simultaneous equations by drawing graphs. Check your results in each case. (Answers to 2 significant figures.)

16. $5x + 2y = 16$
$7x - 2y = 32$

17. $7x + 11y = -16·5$
$4x - 5y = 7·5$

18. $44x + 3y = 62$
$20x - 9y = 4$

19. $1·5x + 2·4y = 1·4$
$1·8x - 14y = 5·9$

20. $3x + 2y + 1 = 0$
$4y = 19 + x$

21. $\frac{x}{3} + \frac{y}{4} + 2 = 0, \ y = 3x + 10$

D. Construct straight-line graphs for converting the following :

22. Resistance in ohms per mile to ohms per kilometre (to read up to 250 ohms per kilometre).

23. Tyre pressures in pounds per sq. in. to kg. per sq. cm. (to 120 lb. per sq. in.)

24. Kilowatts to horse-power, given that 1 h.p. = 1·34 kW. (up to 750 kW.)

25. Quarterly gas consumption in cu. ft. to cost in shillings, if a therm costs 1*s.* 5½*d.* and the meter rent is 1*s.* per qtr. (Up to 6000 cu. ft. per qtr. Calorific value : 1 cu. ft. = 530 B.Th.U.) (See Example 6(*d*).)

E. Miscellaneous Problems.

26. The deterioration in candle-power of a carbon- and a metal-filament electric lamp is compared below :

Time t hr. .		0	100	200	400	600	800	1000	1200	1400
Percentage P of initial candle-power . .	Carbon	100	112	101	92	85	78	71	63	—
	Metal	100	102	100	98	95·5	93	90	86	81·5

It is customary to replace a lamp when its candle-power falls below 80% of its initial value. From a graph find by how much per cent. the metal-filament lamp outlives the carbon-filament type.

27. Using the graphs obtained in Question 26, express P in terms of t in the form $P = at + b$ for the straight-line portion of each graph.

28. Construct a chart for reading directly the loop resistance of an overhead line of any length up to 20 miles, for the following copper conductors—100 lb., 200 lb., 400 lb., 600 lb. per mile.

It helps to use the " Ohm-Mile " constant, *i.e.*, that

Loop Resistance per Mile in Ohms × Conductor Weight per Mile in lb. = 1760.

29. Using the chart constructed in Question 28: (*a*) What is the loop resistance of 15½ miles of " 100-lb." conductor? (*b*) Which gauge of the four quoted should be used for an 18-mile route if the loop resistance must not exceed 90 ohms?

30. An accumulator is being charged in 10 hr., and the specific gravity S of the electrolyte is measured hourly :

Time t hr .	0	1	2	3	4	5
Sp. Gr. S .	1·174	1·175	1·178	1·182	1·187	1·192
Time t hr. .	6	7	8	9	10	
Sp. Gr. S .	1·197	1·202	1·207	1·210	1·212	

From a graph of these results, obtain a formula for S in terms of t, and state for what range of t this is valid. (Use a large scale for S.)

6.5. Solving a Quadratic Equation by a Graph

Sometimes equations contain the square of an unknown quantity. These equations are called quadratic, and cannot be solved by the methods of Chapter 5.

In Example 5(*e*) (p. 116) we obtain such an equation,

$$r^2 + 4r = 14$$

for finding the radius of a cylinder 4 in. long, the total surface area of which is 88 sq. in.

We will now solve this equation graphically.

By trial we note that when $r = 2$ (in.) the expression $r^2 + 4r$ is $(2 \times 2 + 4 \times 2)$, *i.e.*, 12, and $r = 3$ gives a value $(3 \times 3 + 4 \times 3)$, *i.e.*, 21. So common-sense reasoning leads us to see that the required radius must be between 2 and 3 in.—and probably nearer the former measurement.

By substituting 2·1, 2·2 . . . etc., for r it is possible to narrow down the answer to a radius between 2·2 and 2·3 in. and so compile a short table of values, draw a graph of the " function " $r^2 + 4r$ and find graphically when this function is exactly 14.

$r =$	2·0	2·1	2·2	2·3	2·4
$r^2 =$	4·0	4·41	4·84	5·29	5·76
$4r =$	8·0	8·4	8·8	9·2	9·6
$r^2 + 4r =$	12·0	12·81	13·64	14·49	15·36

Clearly for any value of r in the above table the sum of the figures in the 2nd and 3rd row will give a value for $r^2 + 4r$. It is not necessary to go beyond $r = 2·3$, as $r^2 + 4r$ then exceeds 14.

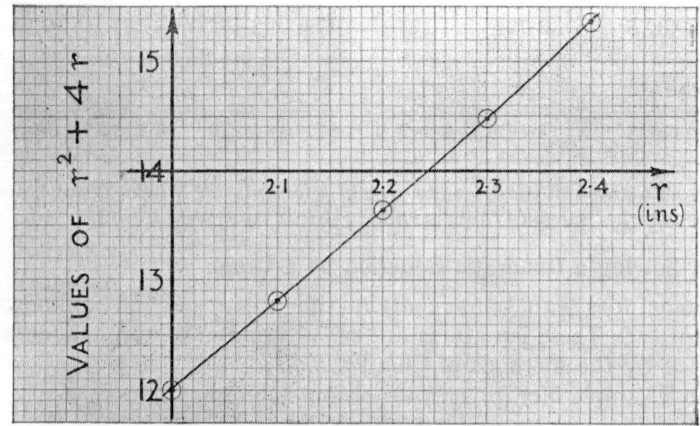

FIG. 46.—TO SOLVE THE EQUATION $r^2 + 4r = 14$.

To obtain our value of r accurately (*e.g.*, to nearest " thou ") we must use as large a scale as possible for both axes. Thus in Fig. 46 we have used 1 in. to represent $\frac{1}{10}$ in. on the r-axis and 1 in. for each " $r^2 + 4r$ " unit. As we started with values of r and calculated values of $r^2 + 4r$ from them, we take r as our " base-axis ". In mathematical language, r is our independent variable, and $r^2 + 4r$ our dependent variable.

When we plot the above values we observe that the five points do not lie on a straight line but on a gentle curve.

From the graph $r^2 + 4r$ will be exactly 14 only when $r = 2{\cdot}243$ (the third decimal place is uncertain). So this is our required radius in inches.

So the solution we require of the equation $r^2 + 4r = 14$ is $r = 2{\cdot}24$ (3 significant figures).

An algebraic way of finding this value is shown as an Appendix to the present chapter (p. 159).

6.6. Graphs of Some Typical Quadratic Functions

In the above discussion we found that the graph of a quadratic function $(r^2 + 4r)$ was a curve. Let us consider the curves obtained by graphing other quadratic functions.

(a) *Graph of* $y = x^2$

This is the simplest possible quadratic function—it gives a curve from which we can read off the squares of numbers (and conversely, square roots too).

To make our picture more complete, we will include negative values of x in our table of values:

Independent x	-3	-2	-1	0	1	2	3
Dependent . $y = x^2$	9	4	1	0	1	4	9

The graph of the function $y = x^2$ is shown in Fig. 47.

(b) *Graph of* $y = Kx^2$, *where* K *is a number*

On the same graph (Fig. 47) we have drawn the graphs of $y = 3x^2$, $y = \frac{1}{2}x^2$, $y = -2x^2$ (giving K values 3, $\frac{1}{2}$, -2). Notice that $y = x^2$ belongs to the same " family " of curves; it is indeed the special case of $K = 1$.

Note the strong " family likeness " between all the curves drawn: all are " balanced " about the y-axis—we say they are all " symmetrical " about this axis, which is the " axis of symmetry " of each curve. The shape of each curve is that of a parabola—and is important in scientific and engineering work; typical examples of the occurrence of a parabola are the cross-section of a parabolic-mirror for an efficient spot-light, and the shape of a normally tensioned telegraph wire, or of the supporting hawser for a suspension bridge with a uniformly loaded roadway.

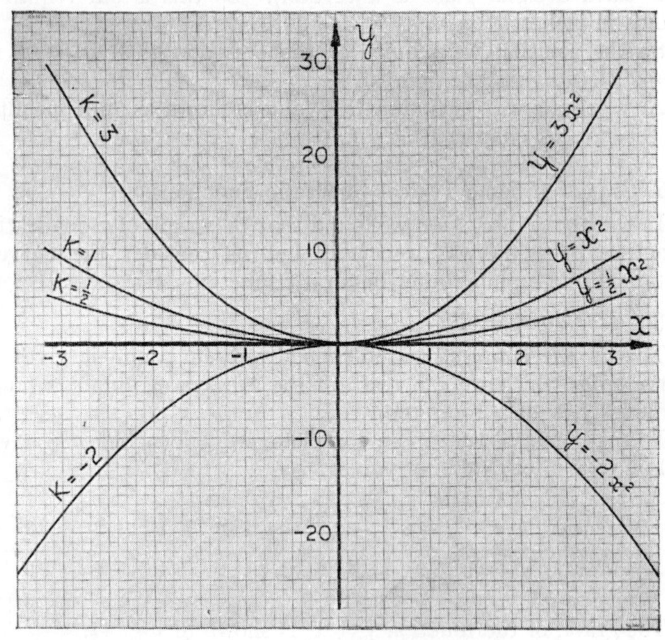

FIG. 47.—THE FAMILY $y = Kx^2$.

For example the " sag " y ft. of a copper wire in terms of the semi-span x ft. (Fig. 48) is given by the formula

$$y = \frac{x^2}{2c}$$

where c is a constant depending on the weight of the wire * and

* It is in fact the length of wire in feet whose weight is equal to the tension in the wire at the mid-point of the span.

the tension in it. For " 200-lb." copper conductors under normal tensioning c is approximately 5000.

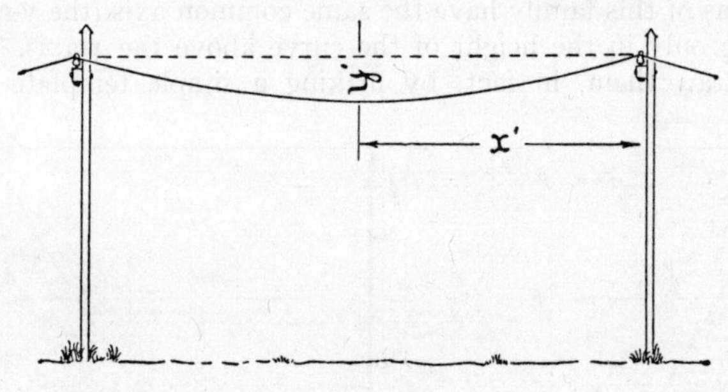

FIG. 48.

Thus for a 200-ft. span, $x = 100$ (ft.)

and
$$\begin{aligned} y &= \frac{x^2}{2c} \\ &= \frac{100^2}{2 \times 5000} \\ &= \frac{10^4}{10^4} \\ &= 1 \end{aligned}$$

So the normal sag of copper conductor in the middle of a 200-ft. bay is about 1 ft.

(c) *Graphs of the type* $y = ax^2 + c$

Let us consider a numerical example $y = 3x^2 + 7$. We have already drawn $y = 3x^2$: to derive from this graph the curve $y = 3x^2 + 7$ we simply increase by 7 the " height ", or " ordinate " y for each value of x.

In Fig. 49 we have drawn $y = 3x^2$ (chain dotted) and $y = 3x^2 + 7$ (dotted line) with the same graph-axes.

Similarly, $y = 3x^2 - 8$ (shown with a continuous line in Fig. 49) provides a graph which is everywhere 8 units below the corresponding ordinate in $y = 3x^2$.

We can imagine a large number of similar graphs, of the pattern $y = 3x^2 + c$, where c is any number we choose. All the parabolas of this family have the same common axis (the y-axis), differing only in the height of the curve above the x-axis. We could draw them, in fact, by making a simple template (the

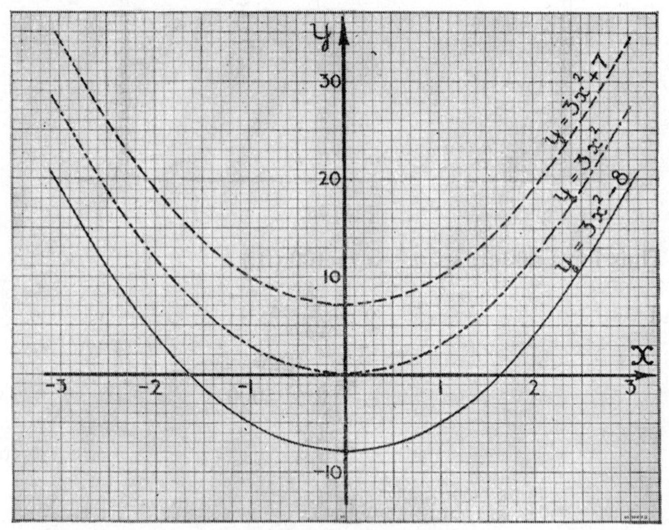

FIG. 49.—THE FAMILY $y = 3x^2 + c$.

curve $y = 3x^2$) and moving it up and down the y-axis, drawing different positions of the parabola as we go.

Such families of curves, whose equations differ only by a constant number, are very common in telecommunications practice, particularly in recording valve-test data. Few of them are so simple mathematically as the parabolas in our illustration here.

In general, $y = ax^2 + c$ represents a parabola with the y-axis as its axis of symmetry, which meets the y-axis at the point where $y = c$. This point is called the vertex of the parabola. Either a or c, or both, may be negative.

Example 6(*e*). *The form of the supporting chain of the main span of a suspension bridge (Fig.* 50(*a*)) *referred to axes through the mid-point of the roadway can be represented mathematically by the formula* $y = \frac{x^2}{400} + 12$ (x *and* y *both measurements in feet). Find the length of vertical supporting cable at a distance of* 20 *yd. from the centre of the bridge. How high must each supporting tower be above the roadway if the span is* 320 *ft.?*

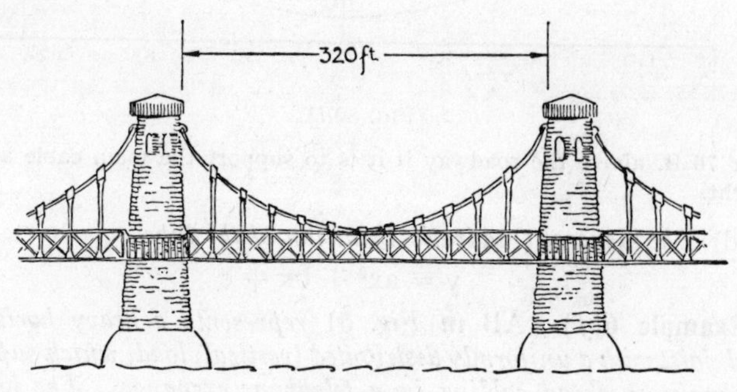

FIG. 50(*a*).

Calculating y for values of x from -160 to $+160$ ft. at intervals of 40 ft., we obtain the following table:

x . .	-160	-120	-80	-40	0	40	80	120	160
$\frac{x^2}{400}$ . .	64	36	16	4	0	4	16	36	64
12 . .	12	12	12	12	12	12	12	12	12
$y = \frac{x^2}{400} + 12$	76	48	28	16	12	16	28	48	76

These are the values of x and y graphed in Fig. 50(*b*).

Now y ft. represents the vertical height of the cable above the roadway at any distance x ft. from the centre of the span. Thus when $x = 60$ (*i.e.*, 20 yd.) we read off from the graph the height $y = 21$. So the vertical supporting cable at this point of the roadway must be 21 ft. long. Similarly, we could find from the graph the length of supporting cable needed at any point of the roadway.

The supporting tower at either end of the bridge must clearly be well

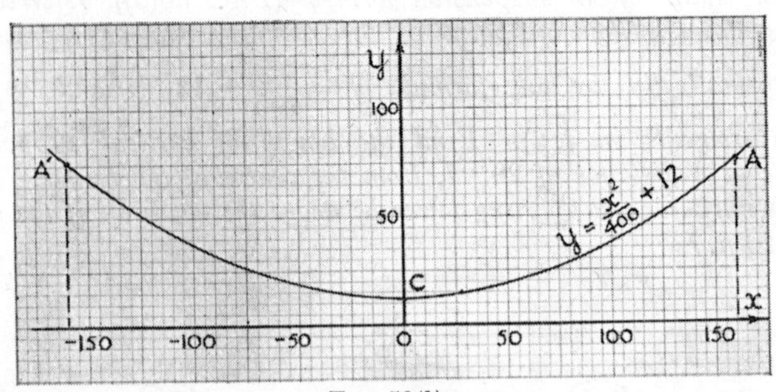

FIG. 50(*b*).

over 76 ft. above the roadway if it is to support the main cable at this height.

(d) *The General Quadratic Function of the type*

$$y = ax^2 + bx + c$$

Example 6(*f*). AB *in Fig.* 51 *represents a heavy horizontal steel joist with a uniformly distributed (vertical) load, which supports the main overhead cabling in a telephone exchange. The joist is* 10 *ft. long and is firmly embedded in masonry at each end. Under these conditions the tendency to* bend *at any point* P *of the beam,* x *ft. from* A, *can be shown to be proportional to* $3x^2 - 30x + 50$. *In order to gain an impression of how this tendency to bend (the "bending moment" of the mechanical engineer) varies at different points of the joist we may graph the function* $y = 3x^2 - 30x + 50$ *from* $x = 0$ *to* $x = 10$. *This will "cover" the whole joist, and help us to appreciate where the bending stresses are greatest.*

$x =$	0	1	2	3	4	5	6	7	8	9	10
$3x^2 =$ $-30x =$ $+50 =$	0 0 50	3 −30 50	12 −60 50	27 −90 50	48 −120 50	75 −150 50	108 −180 50	147 −210 50	192 −240 50	243 −270 50	300 −300 50
$y =$	50	23	2	−13	−22	−25	−22	−13	2	23	50

Note how values of $y = 3x^2 - 30x + 50$ have been obtained in the table above. This is far more accurate than working separate "sums" on scraps of paper!

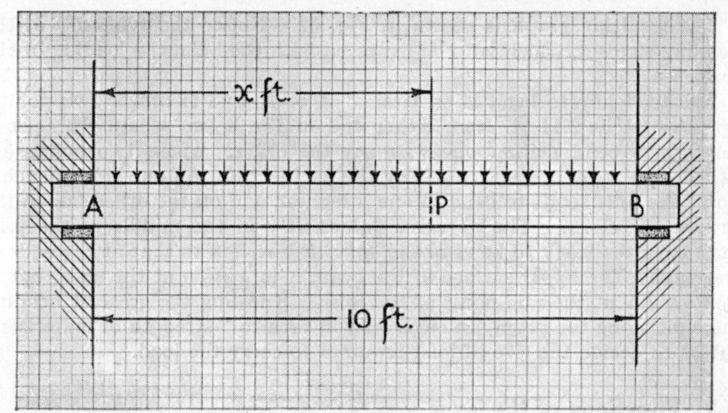

FIG. 51.—A UNIFORMLY LOADED BEAM, CLAMPED AT BOTH ENDS.

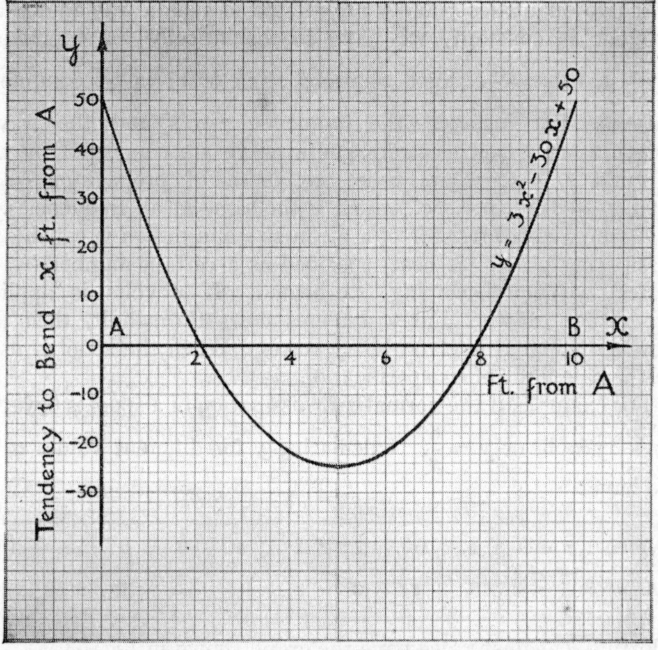

FIG. 52.—THE TENDENCY TO BEND IN THE BEAM AB.

The resulting graph is shown in Fig. 52. Notice the *symmetry* of the bending stresses, which is to be expected with the balanced loading of the joist.

The tendency to bend is greatest at A or B, and is NIL at two points, approximately 2·11 ft. from either end of the beam, where the joist is tending to bend neither one way nor the other. These two points are called "*points of contraflexure*" by the engineer, for on one side of such a point the joist tends to bend in one direction, and on the other side in the reverse direction. The negative values of y in the middle of the joist indicate a tendency to bend in the *opposite direction* to that at the ends of the joist. The *upper* surface of the beam is in *tension* at the ends, but in *compression* at the mid-point of the joist. Knowing the greatest bending moment along the beam, an engineer will turn to "strength of materials" tables to decide the size of joist required to carry the load.

6.7. Functions of the Type $y = \frac{a}{x}$

If two quantities x and y are related in such a way that their product is constant, we have mathematically an expression $xy = a$, or $y = \frac{a}{x}$. Such relations often occur in telecommunications. For instance, the frequency f c/s and the wavelength λ cm. of an electromagnetic wave are connected by the formula $\lambda f = c$, where c is the speed of propagation, *i.e.*, $2{\cdot}99 \times 10^{10}$ cm. per sec., which is the same at all frequencies.

Again, a spool resistor of a given size and design is capable of dissipating power (as heat) specified as a certain number (w) of watts, without becoming overheated. If x volts is applied across the resistor, and y amperes flows through it, then the power dissipated is given by xy watts. When the spool is used to the safe limit of its specification, $xy = w$.

For example, if the normal safe dissipation of a spool is 6 watts, then $xy = 6$, or $y = \frac{6}{x}$. Calculating y for different values of x, we have:

(Volts) $x = 0$	1	1·5	2	3	4	5	6	8	10
(Amperes) $y = \infty$	6	4	3	2	1·5	1·2	1	0·75	0·6

Plotting y against x, we have the graph shown in Fig. 53.

If we give x negative values, we obtain the same series of values for y, but each one is negative. This may be interpreted as a reversal of voltage and therefore of the current passing through the spool. Thus an adequate representation of the formula

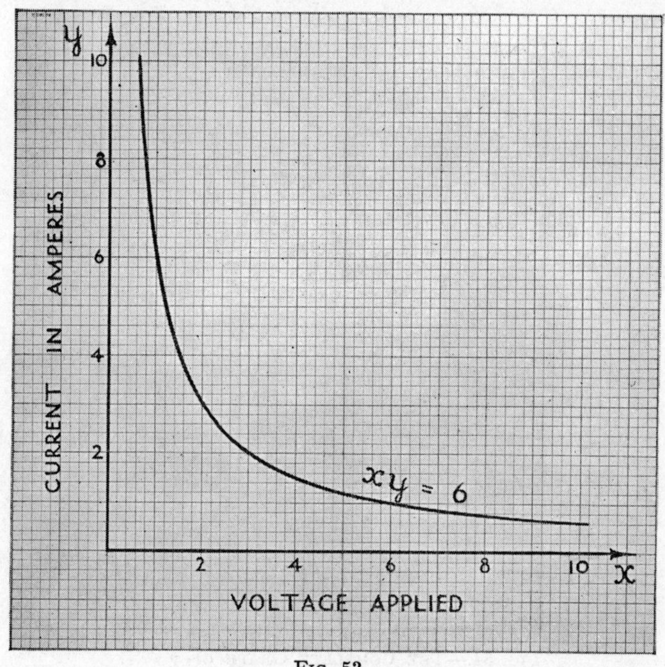

FIG. 53.

$y = \frac{a}{x}$ requires the complete graph of Fig. 54, with two distinct symmetrical " branches ". (A mathematical curve known as a " *rectangular hyperbola.*")

The spool with a 6-watt rating can be used in a circuit, so long as the voltage and current concerned give a point on the graph on the origin side of the curve. For example, if its resistance were 40 ohms, and it were to have 12 volts across it in a certain circuit, the current passing would be $\frac{12}{40}$ or $\frac{3}{10}$ amperes (by Ohm's Law). The point $x = 12$, $y = \frac{3}{10}$ lies below the graph $y = \frac{6}{x}$ (in fact, $xy < 6$), so this spool would be appropriate to the circuit concerned.

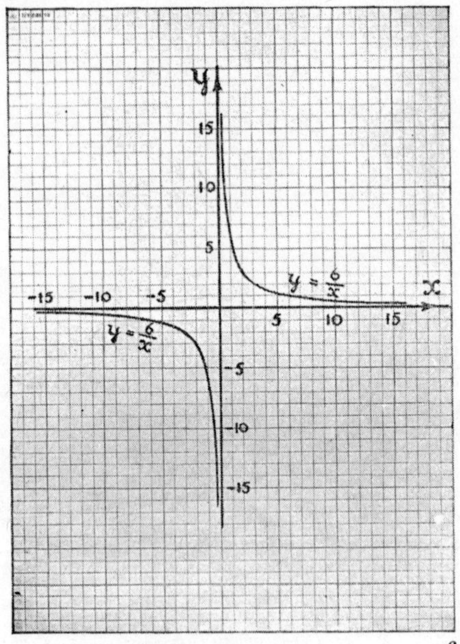

FIG. 54.—THE COMPLETE GRAPH OF $y = \frac{6}{x}$.

In Fig. 55 we have drawn a " family " of curves for spools of different sizes, each capable of a specified power dissipation. We might use such a chart to determine quickly which size spool to use, where we know the current and voltage involved in a particular circuit. (See Exercise 6(*b*), Question 30.)

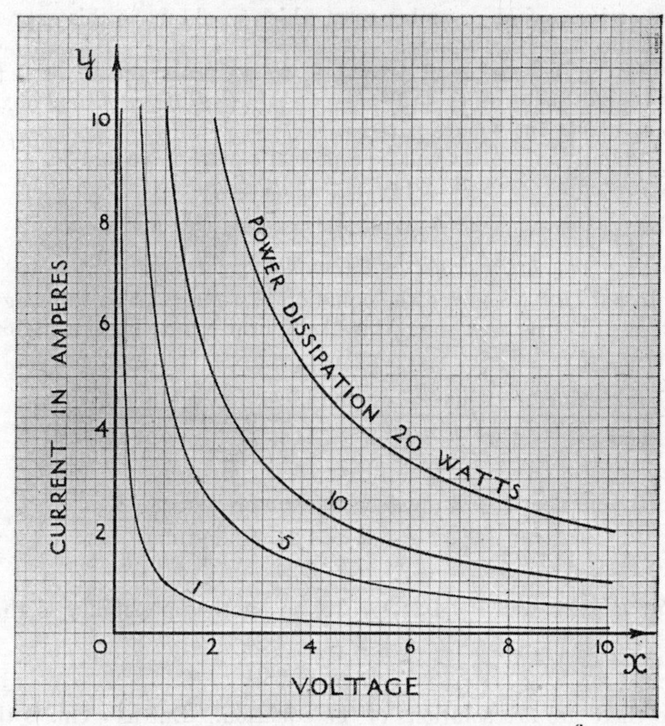

FIG. 55.—A FAMILY OF CURVES OF THE TYPE $y = \frac{a}{x}$.

* APPENDIX TO CHAPTER 6

An Algebraic Method of Solving Quadratic Equations by Completing the Square

(a) *The Perfect Square: Completing the Square*

In Chapter 4, para. 4·4(c), and in Example 4(e), we noted the pattern of a perfect square:

Either $\qquad (x + a)^2 = x^2 + 2ax + a^2$

or $\qquad (x - a)^2 = x^2 - 2ax + a^2$

Sometimes we begin with a quadratic expression like $x^2 + 6x$

and strive to find a number term which would make this expression into a perfect square.

Clearly $(x + 3)^2 = x^2 + 6x + 9$, so we must add 3^2 or 9 to $x^2 + 6x$ to make up $(x + 3)^2$, *i.e.*, we add (*half the coefficient* of $x)^2$ to complete the square pattern.

Similarly, $x^2 - 14x$ will be a perfect square if we add 7^2,

i.e., $$x^2 - 14x + 7^2 = (x - 7)^2$$

Also $$y^2 + 5y + (\tfrac{5}{2})^2 = (y + \tfrac{5}{2})^2, \quad \text{and so on.}$$

(b) *To Solve a Quadratic by "Completing the Square"*

In para 6.5 we solved the equation $r^2 + 4r = 14$ by drawing a graph. Let us now solve this algebraically, *i.e.*, by calculation.

Then if $$r^2 + 4r = 14$$

"completing the square" on the l.h.s., *i.e.*, adding 2^2 to b.s.

then $$r^2 + 4r + 2^2 = 14 + 2^2$$

i.e., $$(r + 2)^2 = 18$$

So the number $r + 2$ must be the square root of 18.

So $$r + 2 = 4{\cdot}24 \quad \text{(slide-rule)}$$

$$\therefore \quad \underline{r = 2{\cdot}24}$$

But equally well $(-4{\cdot}24)^2 = 18$, so another possibility is that

$$r + 2 = -4{\cdot}24$$

$$\therefore \quad \underline{r = -6{\cdot}24}$$

So there are *two* solutions of the equation. But as r cm. is the radius of a cylinder, the only solution which has a physical meaning is $r = 2{\cdot}24$.

Example 6(g). *Solve the equation* $x^2 - 8x + 7 = 0$

Now $$x^2 - 8x = -7$$

Completing the square (add 4^2 to both sides),

$$x^2 - 8x + 4^2 = -7 + 4^2$$
$$= -7 + 16$$

i.e., $$(x - 4)^2 = 9$$

Taking the square root of b.s. $$x - 4 = \pm 3$$

(*i.e.*, the square root of 9 is either $+3$ or -3)

$$\therefore \quad \underline{x = 7 \text{ or } 1}$$

Both solutions in this case are *positive*.

Example 6(*h*). *Solve the equation* $2x^2 - 7x + 3 = 0$.

We first make the coefficient of x^2 unity.

So dividing b.s. by 2, $x^2 - \frac{7}{2}x + \frac{3}{2} = 0$

Take $\frac{3}{2}$ from b.s., $x^2 - \frac{7}{2}x = -\frac{3}{2}$

Complete the square: add $(\frac{7}{4})^2$ to b.s.,

$$x^2 - \frac{7}{2}x + (\frac{7}{4})^2 = -\frac{3}{2} + (\frac{7}{4})^2$$
$$= -\frac{3}{2} + \frac{49}{16}$$
$$\therefore \quad (x - \frac{7}{4})^2 = \frac{25}{16}$$

Taking the square root of b.s., $x - \frac{7}{4} = \pm\frac{5}{4}$

$$\therefore \quad x = \frac{7}{4} \pm \frac{5}{4}$$

Taking the + sign first, $x = \dfrac{7+5}{4}$ or $\dfrac{7-5}{4}$

So the solutions are $x = 3$ *or* $\frac{1}{2}$.

As a general rule, if the numbers in the original equation are whole numbers, keep all subsequent working in improper fraction form. *Always* make sure the *denominator* in the r.h.s. is a perfect square before taking the square root.

Example 6(*i*). *Solve* $3x^2 + 10x - 2 = 0$

Divide b.s. by 3, $x^2 + \frac{10}{3}x - \frac{2}{3} = 0$

i.e., $x^2 + \frac{10}{3}x = \frac{2}{3}$

Complete the square: add $(\frac{5}{3})^2$ to b.s.,

$$x^2 + \frac{10}{3}x + (\frac{5}{3})^2 = \frac{2}{3} + (\frac{5}{3})^2$$
$$= \frac{2}{3} + \frac{25}{9}$$
$$\therefore \quad (x + \frac{5}{3})^2 = \frac{31}{9}$$

Taking the square root of b.s.,

$$x + \frac{5}{3} = \frac{\pm 5{\cdot}57}{3} \text{ (slide-rule)}$$
$$\therefore \quad x = -\frac{5}{3} \pm \frac{5{\cdot}57}{3}$$
$$= \frac{0{\cdot}57}{3} \text{ or } \frac{-10{\cdot}57}{3}$$

i.e., $x = 0{\cdot}19$ or $-3{\cdot}52$

(c) *Note on the Use of Factors to Solve Quadratic Equations*

In Examples 6(*g*) and 6(*h*) above the solutions were rational, *i.e.*, 7, 1 and 3, $\frac{1}{2}$. In other words, the square roots involved in the process were exact.

Such cases are exceptional, but when they occur the quadratic equation may be solved by *factorising* (ref. p. 91).

Thus, in Example 6(*g*) $x^2 - 8x + 7 = 0$.

Factorising the l.h.s., $(x - 7)(x - 1) = 0$ which means that either $x - 7$ or $x - 1$ must be zero, giving the alternatives

$$x - 7 = 0 \text{ or } x - 1 = 0$$

i.e.,
$$x = 7 \text{ or } \quad x = 1$$

which was the result obtained by completing the square.

It must be stressed that this method is only operative if the quadratic expression has exact factors, *i.e.*, rational factors involving whole numbers. It is therefore of very limited application, although sometimes a useful concept in mathematical argument.

EXERCISE 6(*b*)

A. Solve graphically the following quadratic equations (2 significant figures are sufficient):

1. $x^2 - 5x + 2 = 0$ (plot from $x = 0$ to 6).
2. $x^2 + 8x = 40$ (rom $x = -15$ to 5).
3. $3x^2 + x - 4 = 0$ (from -3 to $+3$).
4. $2x^2 - 5x + 1 = 0$ (from 0 to 3).
5. $5x^2 + 2x = 3$ (from -2 to $+1$).
6. $3x^2 = 5 - 18x$ (from -8 to 2).

In the following examples, solve by "completing the square" (3 significant figures) and verify by sketched graphs, or by factorising where this is possible.

7. $x^2 - 6x = 14$.
8. $x^2 + 5x + 3 = 0$.
9. $3x^2 - 10x + 3 = 0$.
10. $2x^2 + 7x - 6 = 0$.

B. *Quadratic Functions*

11. The weight W lb. per 100 yd. of steel wire in terms of its diameter d in. is given by a formula $W = ad^2$. If S.W.G. No. 8 gauge is 0·16 in. in diameter and 100 yd. of it weighs 19·8 lb. calculate the constant a.

Plot W against d for values up to 0·5 in. Use your graph to find the weight of 5 miles of No. 3 gauge (0·252 in. in diameter) to 3 significant figures.

12. Construct a chart to give the cross-sectional area A sq. cm. of a cylindrical tube d INCHES internal diameter (d from 1 to 6 in.). What is the diameter of a tube whose cross-sectional area is to be 80 sq. cm.? Check by calculation.

13. The height y ft. of a ship's radar "aerial" above the water for a scanning range or "horizon" of x miles is given by an approximate formula $y = \dfrac{2x^2}{3}$.

Plot a graph for horizons up to 60 miles. Hence find: (*a*) the height required to "scan" over a radius of 32 miles; (*b*) the range if the "aerial" is 185 ft. above the water level. Check by the formula.

14. For a certain grade of Swedish Redwood the maximum permissible compression load (P lb.) parallel to the grain for a vertical strut of cross-section l in. by l in. is given by $P = 940l^2$.

Plot P against l (up to $l = 6$); from the graph give the minimum square cross-section of a strut which has to withstand an estimated maximum compression load of 2 tons.

15. A train travelling at 60 m.p.h. can pull up in 120 yd. in an emergency. At 20 m.p.h. the stopping distance is 24 yd.

Assume a formula $y = ax^2 + b$ connecting the speed x m.p.h. and the distance y yd. Find a and b and plot a graph for speeds up to 80 m.p.h. to show the emergency stopping distance. At what speed can the train pull up in 40 yd.?

16. The form of a certain suspended cable is given by the formula

$$y = \frac{x^2}{160} + 30$$

where y (ft.) is the height of a point P of the cable above the level ground and x (ft.) the horizontal distance of P from the lowest point of the cable. What is the clearance of the cable, and how high must the supports be at each end of an 80-ft. bay? Sketch the side elevation of the cable and its supports. If the sag must not exceed 6 ft. what must be the maximum length of a bay?

17. A beam 12 ft. long, uniformly loaded with 60 lb. per foot run, is supported at each end (but not rigidly fixed). The "bending moment" M (tendency to bend) at a point x ft. from one end of the beam is $360x - 30x^2$ lb. ft. units. Plot M against x. Where is this beam likely to bend most?

18. A spring-board (weighing 6 lb. per ft.) is clamped at a point A and its free end is at B, 10 ft. from A. If a man weighing 11 st. 6 lb. stands at C, 8 ft. from A, the bending moment in the beam x ft. from A is given by $M = 3x^2 - 220x + 1580$. Plot M from $x = 0$ to $x = 8$ (*i.e.*, from A to C). Where is the bending stress greatest?

19. The Menai Suspension Bridge has a total span of 560 ft., and the "equation" of its suspension cable referred to horizontal and vertical axes through the mid-point of the roadway is approximately

$$y = 0{\cdot}000204x^2 + 15.$$

Plot this on graph-paper to obtain an impression of the side elevation of the bridge.

20. The Brooklyn Suspension Bridge in New York has a road span of $1595\frac{1}{2}$ ft., and the ends of its suspension chains are 237 ft. above the roadway, while the mid-point of each chain is 10 ft. above it. Obtain an equation of the form $y = ax^2 + b$ (x, y same meaning as in Question 19) and sketch the side elevation of this bridge.

C. *Problems involving Quadratic Equations.* Solve these graphically.

21. A room is 12 ft. by 10 ft. By what equal amount (x ft). must length and breadth be increased, to increase the floor space by 50%. (A sketch diagram will help to obtain your quadratic equation.)

22. A certain cylindrical coil requires at least 60 sq. in. of surface area to allow for heat dissipation. If its length be 10 in., show that its minimum radius r in. is a solution of the equation $r^2 + 10r = \frac{30}{\pi}$. Find this radius by drawing a graph.

23. A length of metal, 14 in. wide, is bent to form an open rectangular gutter. If the cross-sectional area of this gutter is to be 24 sq. in. find its dimensions. (Suppose x in. bent up on either side.)

24. Write down a concise formula for the total surface area A sq. cm. of a solid cone, of base radius r cm. and " slant height " s cm. Find graphically the radius of the base of such a cone, if its surface area is 48 sq. cm. and its slant height 5·2 cm. (result to 2 significant figures).

25. Graph the function $y = 3x^2 - 7x + 2$ and use it to solve the equations $3x^2 - 7x + 2 = 0$, $7x + 1 = 3x^2$.

Check your solutions to the second of these equations by " completing the square ".

D. *Examples involving* $y = \frac{a}{x}$ *(Inverse Proportion)*

26. The loop resistance R ohms per mile and the weight W lb. per mile of conductor of hard-drawn copper wire are connected by the approximate formula $RW = 1760$ (the so-called " Ohm-Mile " constant). Construct a graph to show the resistance per mile loop of such conductors from 100 to 600 lb. per mile. From your graph: (*a*) What would be the weight per mile of S.W.G. 12 (9·78 ohms per loop mile)? (*b*) Estimate the resistance of 150-lb. copper conductors per mile of route.

27. If a car does x miles to the gallon, and petrol costs p pence per gallon, find a formula expressing the cost y shillings of the petrol per 1000 miles in terms of x and p.

Draw three graphs with the same axes, plotting y against x, for petrol costing 2*s*. 6*d*., 3*s*., 3*s*. 6*d*. a gallon.

(*a*) If petrol costs 3*s*. a gallon, find the saving per 1000 miles on a 24-m.p.g. car, when its performance (in m.p.g.) is improved by 20%.

(*b*) For a car which averages 32 m.p.g., find the increased cost of a 500-mile journey if the price of petrol rises from 2*s*. 6*d*. to 3*s*. 6*d*. a gallon.

28. The voltage V volts which must be applied across a parallel-plate capacitor, of capacitance C farads, to obtain a charge of +20 microcoulombs on the positive plates of the capacitor is given by $V = \frac{20 \times 10^{-6}}{C}$.

Plot V against C, for values of the capacitance from 2 to 10 μF. Estimate the capacitance of an unmarked capacitor, which requires 6·4 volts to give it a charge of 20 microcoulombs.

29. Construct a chart from which you can read off the wave-length (λ cm.) of an electromagnetic wave of frequency f c/s, to cover the frequency band 0·5 to 6 Mc/s. ($\lambda f = 3 \times 10^{10}$ approx.) Use your chart to convert to frequencies the wavelenths of the following programmes: Midland Regional 276 m., Lyons 498 m., Hilversum 402 m. Find the wavelength which is numerically the same in cm. as the frequency in c/s.

30. In a certain trunk exchange a sleeve circuit relay (85 ohms) has to carry 0·28 amperes during a call. Find the voltage across the relay coil, and test whether the power dissipated exceeds the safe power rating of 7 watts for the winding of this relay.

REVISION PAPERS B

B.1

1. Solve the simultaneous equations:

$$3x - 2y = 0$$
$$\frac{y}{4} - \frac{x}{3} = 3$$

Verify your algebraic solution by drawing two straight-line graphs.

2. The total external surface of an open rectangular tank is 812 sq. ft. The base of the tank is a square, the side of which is 3 ft. greater than the height of the tank. Find the dimensions of the tank (neglecting the thickness of the material), and calculate its capacity in gallons. Solve graphically any equation you obtain.

3. Solve the equations:

$$\text{(i)} \quad \frac{7y}{10} - \frac{9}{5} = \frac{y}{5} + \frac{7}{10}$$

$$\text{(ii)} \quad \frac{2x+3}{2} - \frac{2x-2}{3} = \frac{x-1}{12}$$

4. A sphere of lead of diameter 4·8 in. is beaten into a rectangular sheet 0·04 in. in thickness, with sides in the ratio 2 : 1. Find the dimensions of the sheet. If the same quantity of lead were rolled into foil 0·002 in. in thickness and 8 in. wide, how long would the roll be?

5. Use logarithms to evaluate:

$$\text{(i)} \quad \sqrt[3]{\frac{4 \cdot 5}{12 \cdot 46 \times 3 \cdot 146}}$$

$$\text{(ii)} \quad G = 29 \cdot 4 \times \sqrt{\frac{HD^5}{L}},$$

where $D = 0 \cdot 894,\ H = 31,\ L = 189.$

6. A lead–acid accumulator is being charged at the normal "10-hr." rate, and the specific gravity of the acid is checked hourly by means of a hydrometer.

Time t hr. from commencement of charge	0	1	2	3	4	5
Specific gravity, S .	1·175	1·176	1·179	1·183	1·187	1·192
Time t hr. from commencement of charge	6	7	8	9	10	11
Specific gravity, S .	1·197	1·201	1·205	1·209	1·212	1·214

Plot S against t. For what range of t is the formula $S = at + b$ applicable (within limits of observational error)? Estimate the constants a and b from your graph.

B.2

1. A quarterly electricity bill consists of a fixed charge of $1\frac{1}{4}\%$ of the rateable value of the house together with " so much a unit ".

In one quarter 324 units are consumed, and the bill comes to £1 6*s*. Next quarter 216 units are used, and the bill is 19*s*. 3*d*. Find the cost per unit and the rateable value of the house.

2. Solve the equations:

$$(a)\ 3(x + 4) - 5(x - 1) = 19$$

$$(b)\ \frac{5}{8y} - \frac{7}{11y} = 4 \qquad \text{(2 significant figures)}$$

3. In replacing certain old underground cable $4\frac{1}{2}$ tons of lead and 7 cwt. of copper were recovered and sold as scrap for £360 10*s*. A previous consignment of $1\frac{1}{2}$ tons of lead and 17 cwt. of copper realised £215 10*s*. Find the prices of scrap lead per ton and scrap copper per cwt.

4. Plot the graph of $y = 4x^2 + 7x - 3$ from $x = -3$ to $x = +3$. From the graph: (*a*) What are the approximate solutions of the equation $4x^2 + 7x - 3 = 0$? Check them. (*b*) What values of x make the function $4x^2 + 7x - 3$ equal to $+8$? State the quadratic equation of which these are the roots.

5. Find the diameter of a cylindrical oil-storage tank which is 20 ft. high and holds 120,000 gall. of oil. (1 cu. ft. $= 6\frac{1}{4}$ galls.)

6. The anode (plate) voltage of a triode valve under test is maintained constant at 150 volts, while the control grid voltage V_g is varied. The current I passing through the valve is recorded as follows:

V_g (volts) .	−10	−9	−8	−7	−6	−5	−4	−3
I (mA.) .	0	0·2	0·8	1·8	2·6	3·7	4·8	6·0

V_g (volts) .	−2	−1	0	1	2	3	4	
I (mA.) .	7·1	8·2	9·3	10·4	11·5	12·3	12·9	

Plot I against V_g. For what range of grid voltage may a straight-line law $I = aV_g + b$ be assumed? Estimate a and b from your graph.

B.3

1. A bright object is placed at a distance d cm. from a lens, and the position of a screen is adjusted until a sharp image is obtained. The screen is then S cm. from the object. The following results are obtained:

d . .	26	22	18	14	10	8
S . .	34·0	30·2	27·0	24·6	25·1	32·2

Find graphically the least distance between object and image.

2. In the U.K. television system 25 pictures each made up of 405 lines are transmitted every second. If the carrier frequency for the London region is 45 megacycles per second, what is the ratio between the carrier-frequency and the line-frequency?

3. Solve the equations:

$$(a)\ \frac{3y-1}{4} - \frac{y+1}{6} = 1 + \frac{y}{2}$$

$$(b)\ (2x-3)^2 = 4x^2 - 27$$

4. A man buys a television set for £x and sells it for £12 15s. at a loss of x % (of the buying price). What did he pay for it?

5. An old rule says: " to convert pence per day into pounds per annum, add to the number of pence its half ". Find the percentage error in using this rule. Express the rule as a concise algebraic formula.

6. The following data apply to the large lead–acid accumulators used in telephone exchanges:

Capacity in amp.-hr. when discharging in 10 hr. . .	1000	1400	2000	2400	3000	3400	4000
Wt. per cell in lb. .	630	820	1110	1300	1620	1810	2100
Gallons of acid per cell . . .	14	18	24	28	34	38	44

On the same graph display the weight W lb. of the cell and the quantity of acid Q gall. required for any given capacity x amp.-hr. Derive approximate formulæ giving W and Q in terms of x.

B.4

1. (a) A man can lift a field of potatoes in x days. His son could do it in y days. How long should they take working together?

(b) If $T = \frac{R}{\pi}\left(\frac{1}{M} - \frac{1}{S}\right)$, express S in terms of T, R, M.

2. Solve graphically the equations:

$$\text{(i)}\ \begin{cases} 1{\cdot}68x - 2{\cdot}71y = 4{\cdot}8 \\ 3{\cdot}42x + 1{\cdot}8y = 7{\cdot}4 \end{cases}$$

$$\text{(ii)}\ 3x^2 - 7x - 2 = 0$$

Check (ii) by " completing the square ".

3. A body has an initial speed of u f.p.s. and moves with an acceleration of a ft./sec./sec. At the end of t sec. it has travelled s ft. and attained a speed of v f.p.s. Then $v^2 = u^2 + 2as$, and $v = u + at$. Obtain a formula for s in terms of u, v, and t only (*i.e.*, " eliminate " a) and explain your result in words.

4. The cost of a book including postage used to be 4*s*. 10*d*. The book has now risen in price by 30%, and the postage on it is up 50%. So now it costs 6*s*. 4*d*. including postage. Calculate the original price of the book.

5. In electrolytic action 1 ampere will remove approximately $0{\cdot}66 \times 10^{-3}$ gm. per sec. of a certain metal from the positive electrode. If an earth plate of this metal weighs 5 kg., and acts as the positive electrode with an electrolytic leakage current of 5 mA., how many years would you expect the plate to last? (Assume still effective until 75% of original weight lost.)

6. The effort E lb. wt. required to lift a load of W cwt. was measured in testing a crane with the following results :

$W =$	3·6	5·4	7·6	11·2	13·8	16·5
$E =$	21	28	36	50	59	70

Show graphically that a law for this machine in the form $E = aW + b$ is reasonable. Estimate from your graph the values of a and b.

B.5

1. At 30 m.p.h. the frictional resistance of a train is approximately 1380 lb. wt., while at 60 m.p.h. it has risen to 3800 lb. wt. Assuming $F = a + bv^2$ gives the frictional resistance F lb. wt. at any speed v m.p.h., find the numbers a and b. Plot F against v, and so find the speed at which F first exceeds a ton wt.

2. The capacitance of a capacitor having air as its dielectric is given in electrostatic units (e.s.u.) by $C = A/4\pi d$, where A sq. cm. is the effective area of each plate, and d cm. the distance between the plates. If 1 farad $= 9 \times 10^{11}$ e.s.u., find the effective plate area required to give a capacitance of 20 $\mu\mu$F., assuming a separation of 1·5 mm. between the plates.

3. A closed (cylindrical) tin container is to be $4\frac{1}{2}$ in. high and is to have a total surface area of 84 sq. in. What will its radius be? (Solve the resulting quadratic by drawing a graph, and check by " completing the square ".)

4. Because of increased prices a manufacturer reduces his consumption of solder by 25%, and thereby reduces the cost of this item by 12%. What was the percentage rise in the price of solder?

5. Assuming the formula $t = 2\pi\sqrt{\frac{l}{g}}$ for the period t sec. (time of swing)

of a simple pendulum of length l cm., where $g = 981$ cm./sec./sec., express l in terms of t in the form $l = Kt^2$ and plot l against t for periods up to 4 sec. From the graph find the period of a simple pendulum 1 metre long.

6. A storekeeper has trays which are of three widths, 4·5 in., 8 in., and 12 in.

(*a*) What must be the width of his racks if the shelves have to be just filled with trays of any one of the three sizes?

(*b*) To take the trays the shelves must be at vertical centres of $3\frac{1}{2}$ in., $5\frac{1}{4}$ in., and 7 in. respectively. The shelves rest on stops which can be inserted into holes drilled in the upright members. At what centres must the holes be drilled if any rack is to be capable of being filled with any one kind of tray?

B.6

1. Cars A and B make a journey of 112 miles in the same direction. A is 10 m.p.h. faster than B but starts 3 hr. 20 min. later. They arrive together.

Complete the following table:

	Car B	Car A
Speed in m.p.h. . .	x	
Time taken in hrs. .		

Build up a quadratic equation to find x.
Solve this graphically.
What are the speeds of the two cars?

2. The pull P gm. wt. of an electromagnet is given by $P = \dfrac{B^2A}{8\pi g}$, where B lines per sq. cm. is the flux density across the air-gap, A sq. cm. is the area of cross-section of the air-gap, $\pi = 3{\cdot}142$, $g = 981$ cm./sec./sec. Express B in terms of A and P, and hence determine the flux-density across the air-gap of an electromagnet 6·4 sq. cm. in cross-section, if it exerts a pull equal to a weight of 3·6 kg.

3. A man takes 5 min. longer to swim 1000 yd. against a current than with it. Normally he swims 45 yd. per minute in still water; estimate the speed of the current. (Assume it is x yd. per min.)

4. A length of 50 yd. of 60 pr./20 lb. aerial cable is recovered as scrap. If the external diameter of the cable is 1·02 in., the lead sheath is 0·08 in. thick, and each copper conductor is 0·0355 in. in diameter, calculate the weight of lead and copper recoverable. (1 cu. ft. of water weighs $62\frac{1}{2}$ lb. Specific gravity of lead is 11·36, of copper 8·89.)

5. The charge on a single electron is approximately $1{\cdot}6 \times 10^{-19}$ coulombs. The average current flowing from the battery of a large telephone exchange

during the busy hour is 7500 amperes. How many electrons leave the battery during the busy hour? (1 ampere = 1 coulomb per sec.)

6. In an experiment a quantity of gas was heated, and its volume V c.c. was noted for different values of the temperature $T°$ C., the gas being maintained at constant pressure throughout.

T . .	13·5	25·5	35·0	45·5	58·0	68	83·5
V . .	30·4	31·7	32·7	33·7	35·1	36·2	37·8

Choosing scales carefully, plot T against V and verify Charles' Law that $V = aT + b$, where a and b are constants. Estimate a and b from your "line of best fit". At what temperature would the initial volume be doubled?

CHAPTER 7

"TRIGEOMETRY"—THE SPATIAL ASPECT OF MATHEMATICS

An Introduction to Geometry and Trigonometry

7.1. Practical Problems which Involve Thinking in Space

Many problems which beset the engineer cannot be solved by the ordinary methods of arithmetic and algebra. They contain situations involving *space* relationships—not just pure numbers or quantities. Sample situations are suggested in Problems A to C below, which introduce us to the twin techniques of Geometry and Trigonometry.

Problem A. *What is the shortest length of flexible cabling which can be attached to the walls of a rectangular chamber* 10 *ft. long,* 8 *ft. wide, and* 6 *ft. deep if it is to enter the chamber at a bottom corner and leave by the opposite top corner? At what angle does it ascend each wall?*

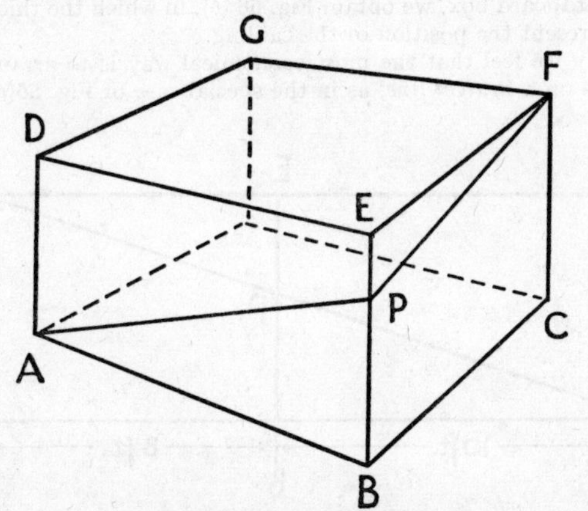

Fig. 56(*a*).—Perspective Sketch of Cable Chamber.

Arguing it Out

In Fig. 56 (*a*) we have drawn a perspective sketch of the cable chamber. The thick line A to P, P to F indicates a possible run for the cable. From experience we know we must have the cable straight across each of the two walls it has to cross. (In geometrical language " a straight line is the shortest distance between two points ".) *But how far up the wall must* P

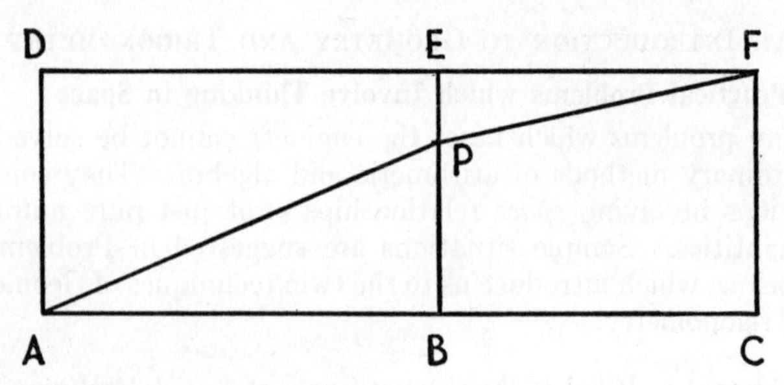

FIG. 56(*b*).—TWO WALLS OPENED OUT : ANY POSITION OF *P*.

be to make the complete cabling as short as possible ? That is our problem, for once P is fixed we can easily mark the run of the cable on each wall with a chalked string (stretched tightly from A to P, then plucked).

If we imagine the walls of the chamber to be " opened out " flat like the sides of a cardboard box, we obtain Fig. 56 (*b*), in which the thick lines AP and PF represent the position of the cabling.

Intuitively we feel that the most economical way is to arrange that A, P, and F lie on a *straight line*, as in the special case of Fig. 56(*c*).

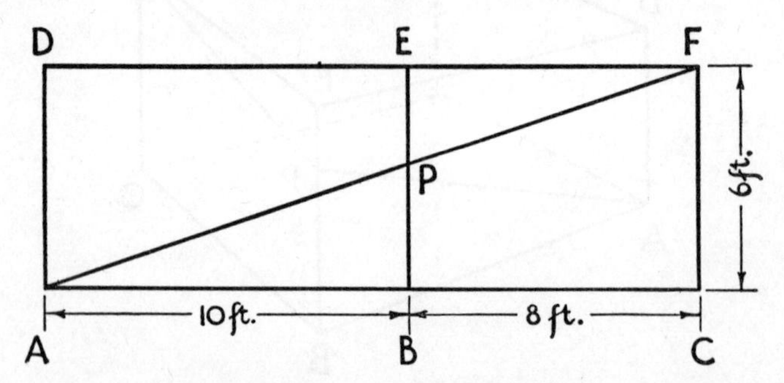

FIG. 56(*c*).—THE BEST POSITION OF *P*.

Solving by Scale Drawing and Measuring

As we know the essential measurements of Fig. 56(*c*), we could *draw* it accurately to scale and find the height PB and the total length AF by *measurement.* From our scale drawing, too, we could measure with a protractor the angle α which the cabling makes with the horizontal line AC (the " angle of ascent ").

Solving by Calculation

If drawing-instruments are not available, we use *geometrical* reasoning to *calculate* the position of P.

The triangles APB and AFC are " in proportion " (see " Similar Triangles ", para. 7.10 below) so that

$$\frac{PB}{10} = \frac{6}{18}$$

i.e.,

$$PB = \frac{10 \times 6}{18}$$

$$= \underline{\underline{3\tfrac{1}{3}}}$$

and as the measurements are in feet, the height of P above B is $3\frac{1}{3}$ ft., or 3 ft. 4 in.

The length of cabling used in this case corresponds to the length AF in Fig. 56(*c*), AFC is a *right-angled* triangle (for AC is horizontal and FC vertical), and an important geometrical property of such a triangle (see " Theorem of Pythagoras " in para. 7.11 below) is that

$$AF^2 = AC^2 + FC^2$$

$$= 18^2 + 6^2$$

$$= 324 + 36$$

$$= 360$$

so, taking the square root,

$$AF = 19{\cdot}0 \text{ (slide-rule, 3 significant figures)}$$

Thus 19 ft. of cabling is the least required for the job.

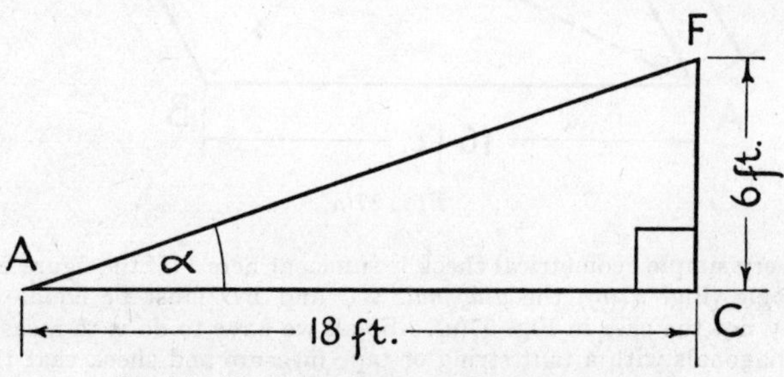

FIG. 56(*d*).—INTRODUCING TRIGONOMETRY.

The precise angle of ascent α may be calculated by using trigonometrical tables. In Fig. 56(*d*) the "slope" of the line AF, which is called the "tangent" of angle α (tan α for short) is $\frac{6}{18}$ or $\frac{1}{3}$.

$$\left(\text{The proportion } \frac{\text{vertical rise}}{\text{horizontal distance}}\right)$$

We write $\tan \alpha = \frac{1}{3} = 0{\cdot}3333$ and from a *table of tangents* we find that the angle which has this "tangent" of 0·3333 is nearly $18\frac{1}{2}°$ (more accurately 18° 26′). You can verify this from the tables at the end of this book.

The basic notions of trigonometry are discussed in detail in Chapter 8. You are advised to confirm the above calculations by measurement in a scale drawing.

Problem B. *The site for a small rural automatic exchange has been marked out on the ground, with a base rectangle* 16 *ft. by* 12 *ft. How would you check that this measured rectangle is "square"?*

It is easy to measure *lengths* accurately, but *angles* are more difficult, particularly in marking out on the ground. How are we to prevent the effect shown in Fig. 57(*a*)? (Where the figure marked out is a parallelogram.)

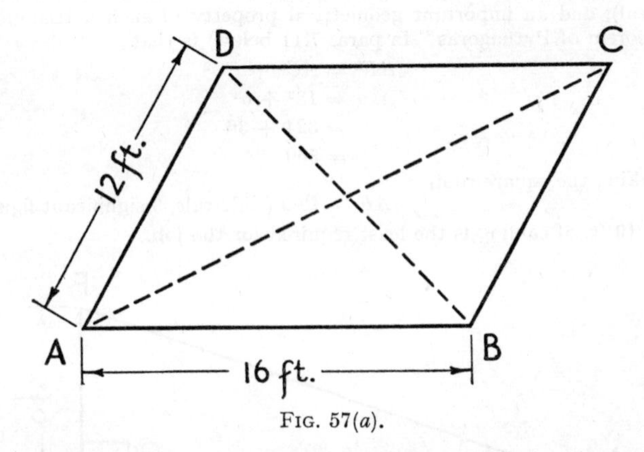

FIG. 57(*a*).

A very simple geometrical check is sufficient here. If the figure is a true rectangle (Fig. 57(*b*)) the *diagonals* AC and BD must be equal—this is clearly not the case in Fig. 57(*a*). So all we have to do is to measure the two diagonals with a taut string or tape measure and check that they are equal.

We could use *Pythagoras' theorem* here to *calculate* the diagonal BD

$$\begin{aligned} \textit{i.e.,} \qquad BD^2 &= 16^2 + 12^2 \\ &= 256 + 144 \\ &= 400 \\ \text{Whence } BD &= 20 \text{ (ft.)} \end{aligned}$$

And this is often done in practice—to check the alignment of AB and AD before marking out C.

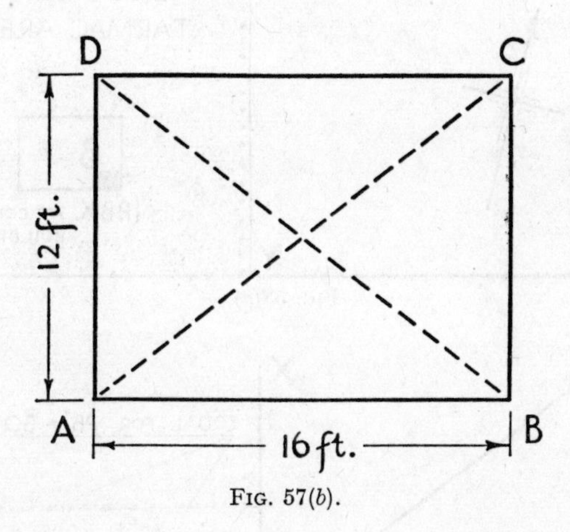

FIG. 57(*b*).

Note that in this particular case the diagonal is an exact number of feet—the sides of the triangle ABC are in the proportion 3 : 4 : 5. See para. 7.11 below (p. 224).

Problem C. *A contractor has to lay an underground telephone cable from the distribution point at* A *(see sketch map in Fig.* 58(a)) *across an open field and a stretch of tarmac to the P.B.X. of an airfield control room. If it costs twice as much per foot (including cable, excavations, and re-instatement) to bury the cable across the tarmac, how should he route the cable to make the cost of installation as low as possible?*

This problem is an example of a geometrical situation. Any attempt to solve it by algebraic calculation would result in a frightening " fourth-degree " equation like

$$2x^4 - 24x^3 + 86x^2 - 243x + 729 = 0$$

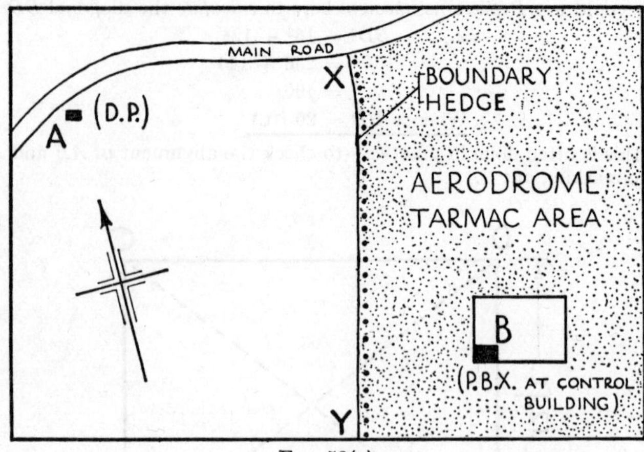

Fig. 58(*a*).

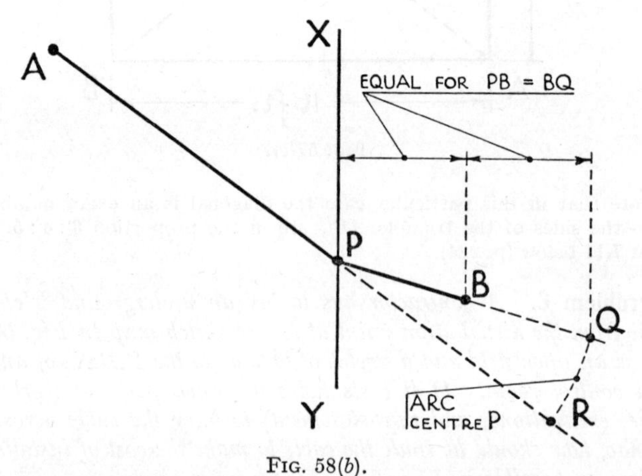

Fig. 58(*b*).

The only workmanlike solution is by practical geometrical construction.

Let us consider one possible routing for the cable A to P to B, where P is any point on the boundary hedge XY. (Fig. 58(*b*).)

As the work across the tarmac is twice as expensive as across the field, the total cost is proportional to the combined lengths of AP and *twice* PB.

If we produce the line PB to Q, making PQ twice PB, then $AP + PQ$ is proportional to the total cost of the job. (Use compasses or dividers to make $BQ = PB$.)

Produce AP and mark off along AP produced a length PR equal to PQ. (Use compasses or dividers.) Then the total cost will be least when the total length AR (which equals $AP + PQ$) is shortest.

In Fig. 58(c) different positions of P (P_1, P_2, etc.) have been chosen, and in each case the point R has been found by repeating the same construction.

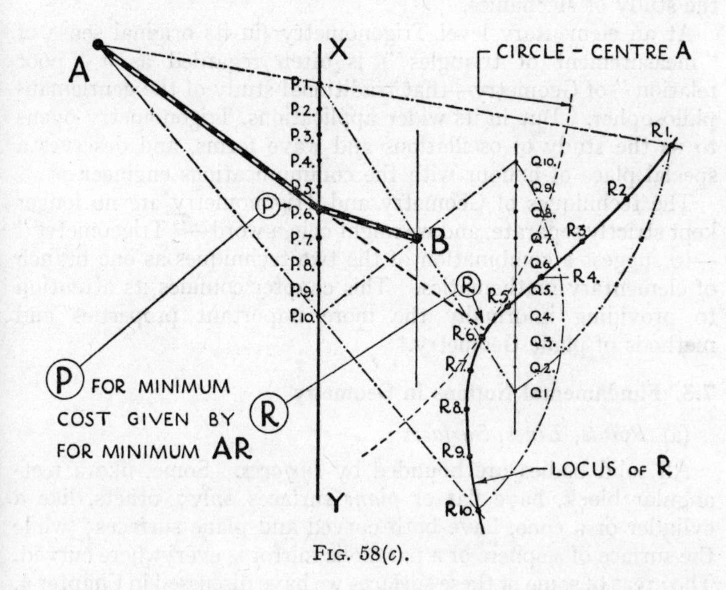

FIG. 58(c).

Notice that we can draw a curved path on which lie all possible positions (R_1, R_2, etc.) of the point R. This we call the *locus* of R.

The particular point R we want is the one *nearest to A*. So with compasses centred at A find this position of R; *i.e.*, where a circle with centre A just touches the R-locus. Join this point to A, and this will give the direction of the cable across the field.

The remainder of this chapter is devoted to the more important geometrical facts and techniques which will on occasion be helpful to the practical engineer. In Chapter 8 the rudiments of plane trigonometry are introduced.

7.2. What " Trigeometry " is About

In Arithmetic and Algebra we deal with relations between numbers and physical quantities. But the language of mathematics has also to interpret relationships of space and movement. These are the provinces of Geometry and Trigonometry—and in their application to time-and-motion problems they lead us to the study of Mechanics.

At an elementary level Trigonometry (in its original sense of " measurement of triangles ") is often regarded as a " poor relation " of Geometry—that traditional study of the gentleman-philosopher. But in its wider applications, Trigonometry opens to us the study of oscillations and wave forms, and deserves a special place of honour with the communications engineer.

The techniques of Geometry and Trigonometry are no longer kept strictly separate, and we might coin a word—" Trigeometry " —to suggest a combination of the two techniques as one branch of elementary mathematics. This chapter confines its attention to providing informally the more important properties and methods of plane Geometry.

7.3. Fundamental Notions in Geometry

(a) *Points, Lines, Surfaces*

All solid bodies are bounded by *surfaces*. Some, like a rectangular block, have flat or *plane* surfaces only; others, like a cylinder or a cone, have both curved and plane surfaces; while the surface of a sphere or a parabolic mirror is everywhere curved. The areas of some of these surfaces we have discussed in Chapter 4.

Any two surfaces usually meet (or " *intersect* ") in a *line* which may be curved or straight. The boundary between two intersecting plane surfaces (*e.g.*, wall and ceiling) is always a straight line.

Two straight lines will intersect only if they lie in the same plane and are *not* parallel. The intersection of two lines we call a *point*.

We shall confine our attention to figures lying in one plane bounded by straight lines or circles. We call this study " Plane Geometry " to distinguish it from " Solid Geometry " (a study in three dimensions) and "Spherical Geometry" (the study of figures bounded by circles on a sphere—the geometry of navigation).

(b) *Angles*

We sometimes think of an angle as a *fixed shape* or a corner—thus a sharp hair-pin bend on a road-map or the cutting-angle of a tool are essentially "shapes". Such shapes we can readily measure with a protractor.

But it is often useful to think of an angle as an *amount of turning*, as in changing the course of an aircraft or in bending a sheet of tinplate to a given shape. This dynamic idea of an

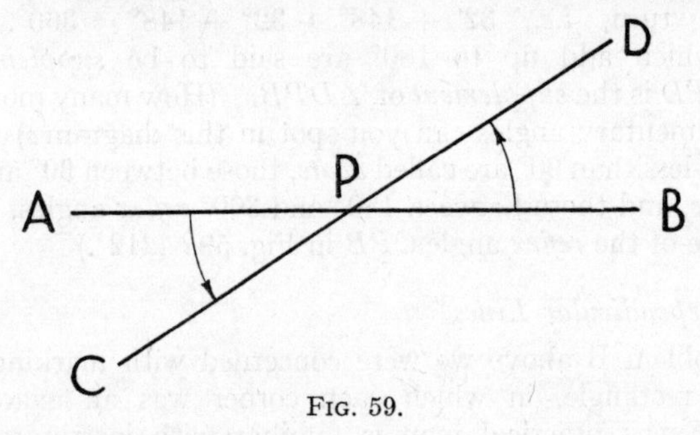

FIG. 59.

angle is in fact more generally useful, as it gives us a meaning to angles greater than 360°.

One complete turn represents 360° on modern scales, with a quarter-turn (or "right angle") as 90°. Recent attempts to replace this scale by one having 100° in a right angle have failed. And fortunately so, for such important angles as 30° and 60° would then have become $33\frac{1}{3}$° and $66\frac{2}{3}$°! For accurate work each degree is subdivided into 60 "minutes". Thus $42\frac{1}{2}$° may be written 42° 30′, and so on.

In Fig. 59, we imagine the line CD to have started in line with AB and then to have turned anti-clockwise about P as a pivot. The angle CD has turned through can be described either as "angle DPB" or as "angle CPA" (each is approximately 32° in our diagram).

The two angles DPB and CPA will clearly be equal for any position of CD. Such a pair of angles are called "*vertically opposite*" angles, and in the shorthand of Geometry we may write

$$\angle CPA = \angle DPB$$

Note also that there is another pair of vertically opposite angles, APD and BPC, which are also equal, and it is clear that in our figure

$$\angle APD = 180° - 32°$$
$$= 148°$$

We can easily check this, for all four angles at P make one complete turn, *i.e.*, $32° + 148° + 32° + 148° = 360°$. Two angles which add up to 180° are said to be *supplementary*, *i.e.*, $\angle APD$ is the *supplement* of $\angle DPB$. (How many more pairs of supplementary angles can you spot in this diagram?)

Angles less than 90° are called *acute*, those between 90° and 180° are *obtuse*, and those between 180° and 360° *reflex* angles. What is the size of the *reflex* angle CPB in Fig. 59? (212°.)

(c) *Perpendicular Lines*

In Problem B above we were concerned with marking out a "true" rectangle, in which each corner was an exact right angle. Every practical man is familiar with instruments like the T-square, the set-square, and the try-square, which help to make true right angles in technical drawing and in accurate work in wood or metal.

In ordinary speech the word "perpendicular" is often used in the sense of "vertical". But in Geometry "perpendicular" means "at right angles to", and is not restricted to vertical and horizontal lines. Thus in Fig. 60(a) a square framework $ABCD$ is kept in true shape (*i.e.*, AB at right angles to BC) by equal bracing wires AC and BD. These wires cross at right angles, and are correctly thought of as *perpendicular*.

In Fig. 60(b) a framework $PQRS$ of equal members PQ, QR, RS, SP is fixed in position by *unequal* bracing wires. There the wires are perpendicular, but the members of the framework are not.

(The figure $PQRS$ is called a *rhombus*; wire lines PR and QS are called *diagonals.* A well-known property of the rhombus is that its diagonals bisect each other *at right angles.*)

The word *horizontal* is derived from the horizon, and means " dead level ". We use a spirit-level to check a horizontal line

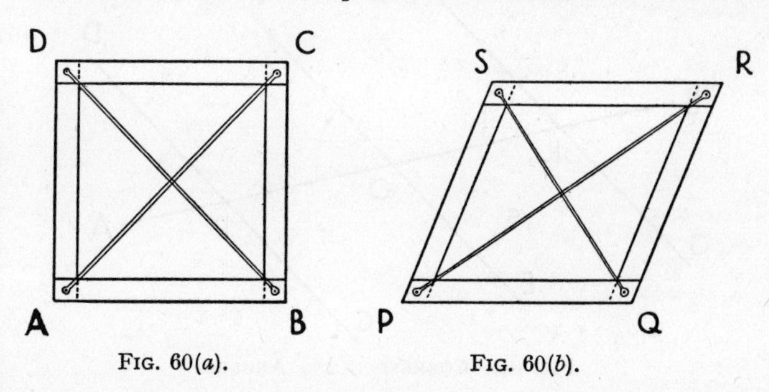

FIG. 60(*a*). FIG. 60(*b*).

or plane surface. A billiard-table top is an example of a horizontal plane. The " artificial horizon " in an aircraft is simply a device for obtaining a true horizontal, so that the navigator can plot his position accurately. A *vertical* line is simply one through the centre of the Earth, and is checked by means of a plumb-line. The upright face of a wall should be a vertical plane.

A vertical line is always perpendicular to a horizontal one, and *vice versa.*

(d) *Parallel Lines (in a Plane)*

In Fig. 61 the three lines CD, EF, GH will clearly never meet —they are all in the same direction (like the tracks of three boats sailing North-east across a lake). They are said to be *parallel* lines.

The line AB cuts the three parallel lines at P, Q, and R. (AB is called a *transversal.*)

Now look at the three angles APD, PQF, QRH marked in Fig. 61. They measure the *difference in direction* between each parallel line and the transversal. As the parallel lines are all in the same direction it follows that these three angles must be exactly equal. We call such a set of angles "*corresponding angles*". How many more sets of corresponding angles can you spot?

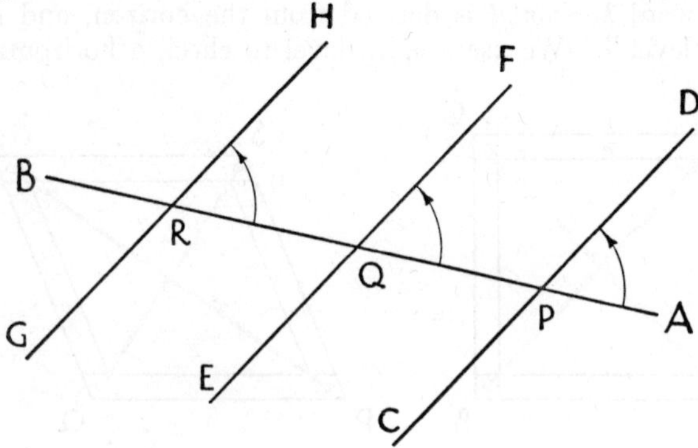

FIG. 61.—CORRESPONDING ANGLES.

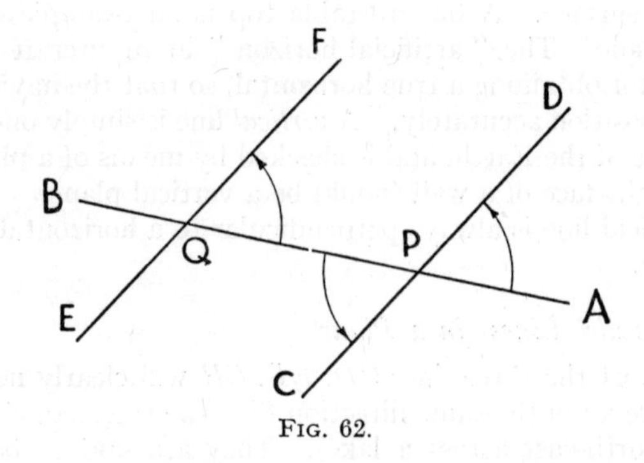

FIG. 62.

In Fig. 62 we note that the angle APD is also equal to the vertically opposite angle QPC. So that the angles QPC and PQF must also be equal. We can show this more vividly in our "geometrical shorthand", thus:

$$\angle QPC = \angle APD \text{ (vertically opposite angles)}$$
$$= \angle PQF \text{ (corresponding angles)}$$

We often use this " chain method " in geometrical arguments, *i.e.*, using the fact that two quantities are each equal to a third, to prove they are equal to each other. In this particular " chain " $\angle APD$ acts as the link between $\angle QPC$ and $\angle PQF$.

Angles like QPC and PQF, which lie between parallels on opposite sides of a transversal, are known as *alternate angles.*

(e) *The Construction of Parallel Lines*

If we require in a drawing a series of parallel lines which happen to be horizontal or vertical (*i.e.*, parallel to the edges of the drawing-board) we do best to use a T-square. But otherwise

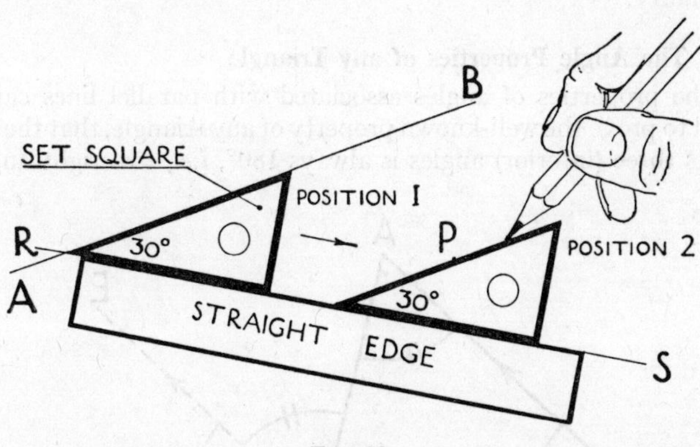

FIG. 63.

we may apply the property of " corresponding angles " to construct parallels with a set-square and a straight-edge.

Fig. 63 illustrates the construction involved. We wish to draw through a point P a line parallel to AB. The longest edge (the hypotenuse) of the set-square is placed against the line AB —Position 1 in our diagram—the straight-edge is placed firmly against a second edge of the set-square, and the latter is made to slide along the straight-edge until its longest edge is just below P. Using the set-square as a guide (as in Fig. 63) the required parallel through P can now be drawn. A 30°/60° set square has been used in the present diagram, but a 45° one would do equally well.

This construction is probably familiar to you already. Now why must it produce an exactly parallel line? Consider the straight-edge *RS* as acting the part of a " transversal " between the lines we hope are parallel, while the 30° angle of the set-square gives us a pair of *corresponding* angles in its two positions. As these corresponding angles are beyond all doubt equal (both 30°, in fact) the lines we have drawn must of necessity be true parallels.

In the next section we shall see that both *alternate* and *corresponding* angles are used to prove important propositions in geometry.

7.4. The Angle Properties of any Triangle

The properties of angles associated with parallel lines can be used to prove the well-known property of any triangle, that the sum of its three (interior) angles is always 180°, *i.e.*, two right angles.

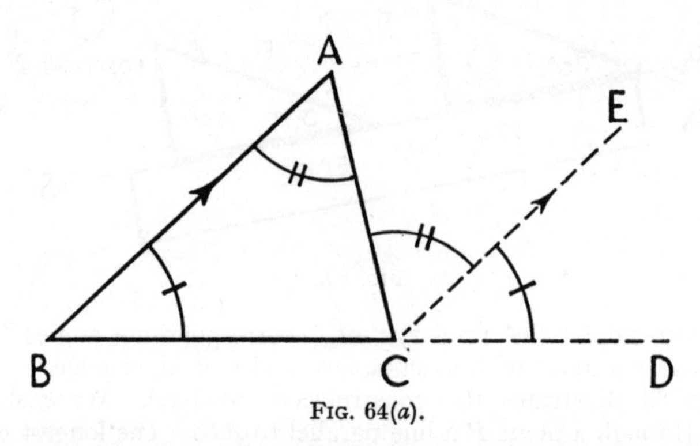

FIG. 64(*a*).

In Fig. 64(*a*) *ABC* represents any triangle. We have produced the line *BC* to any point *D*, and then through *C* have drawn a fresh line *CE* parallel to *BA*. The construction lines, extra to the triangle itself, are shown dotted; while the parallel lines are indicated by arrow-heads. Both these conventions are useful in showing how a geometrical figure is " built up ".

We now have a pair of parallels CE and BA and two transversals BC and AC; using the angle properties of parallel lines

(1) $\angle ABC = \angle ECD$ [corres. angles : $CE \parallel BA$]
(2) $\angle BAC = \angle ACE$ [alt. angles : $CE \parallel BA$]

Adding these results together,

$$\angle ABC + \angle BAC = \angle ACD. \quad . \quad . \quad . \quad (a)$$

Thus the interior angles at A and B add up to the single angle ACD at C outside the triangle. This angle ACD we call the *exterior* angle at C.

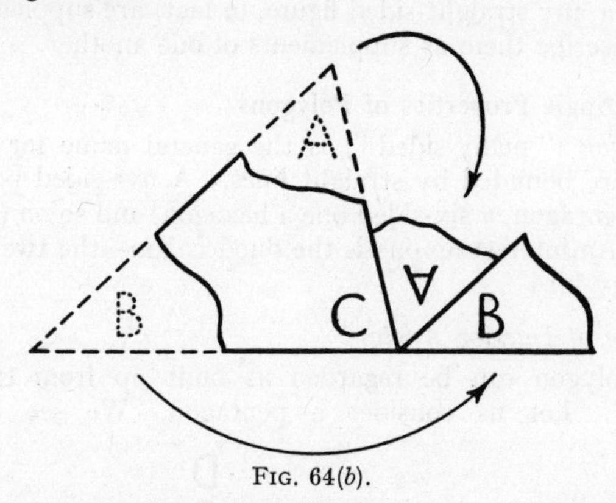

FIG. 64(*b*).

Further, if we add together *all three* interior angles of the triangle,

$$\begin{aligned}\angle ABC + \angle BAC + \angle ACB &= \angle ACD + \angle ACB \quad \text{[using (a)]} \\ &= \angle BCD \\ &= 180° \quad . \quad . \quad . \quad . \quad . \quad (b)\end{aligned}$$

since BCD is a straight line, and $\angle BCD$ must represent half a complete turn, *i.e.*, 180°.

So we have proved two important " propositions " about a typical triangle:

(a) *An exterior angle of a triangle is equal to the sum of the other two interior angles.*

(b) *The sum of the three interior angles of a triangle is* 180° (2 *right angles*).

In Fig. 64(*b*) we suggest a simple experiment to verify the above results. Cut out a paper triangle ABC, tear off the " corners " A and B; on a contrasting paper backing stick the rest of the triangle and the two " corners " A and B adjacent to the " corner " C, as shown in the diagram. Check that the three angles at C do make 180°, *i.e.*, a " straight angle ".

Two angles which add up to 180° are called *supplementary*. For instance, the interior and exterior angles at any vertex of a triangle (or any straight-sided figure, in fact) are supplementary: we can describe them as supplements of one another.

7.5. The Angle Properties of Polygons

A *polygon* (" many-sided ") is the general name for a closed plane figure bounded by straight lines. A five-sided polygon is called a pentagon, a six-sided one a hexagon, and so on (*see* table below). An interesting one is the duodecagon—the twelve-sided threepenny bit !

(a) *Sum of Interior Angles*

Any polygon can be regarded as built up from triangular " bricks ". Let us consider a pentagon. We see how the

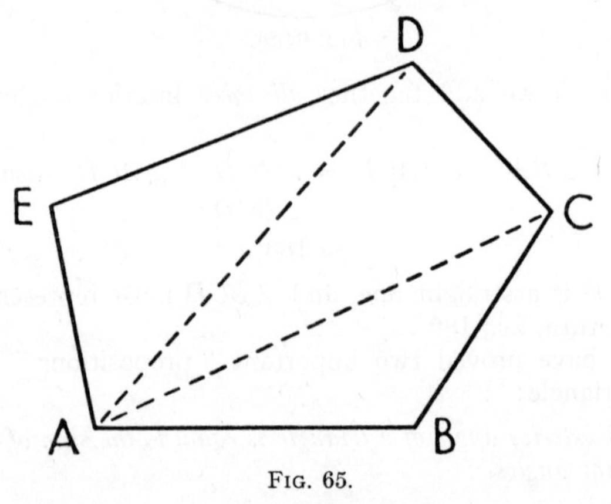

FIG. 65.

pentagon $ABCDE$ in Fig. 65 divides easily into *three* triangles. If we consider the sum total of all the interior angles of these three triangles (*i.e.*, $3 \times 180° = 540°$), we note that they exactly contain the five interior angles of the pentagon. In other words, the five interior angles of a pentagon must add up to 540° (or 6 right angles).

In the table below are the results of repeating this process for other polygons:

Name of polygon.	Sides.	Triangles.	Sum of interior angles.
Quadrilateral .	4	2	$180° \times 2 = 360° = 4$ rt. $\angle$s
Pentagon . .	5	3	$180° \times 3 = 540° = 6$ rt. $\angle$s
Hexagon . .	6	4	$180° \times 4 = 720° = 8$ rt. $\angle$s
Octagon . .	8	6	$180° \times 6 = 1080° = 12$ rt. $\angle$s
Decagon . .	10	8	$180° \times 8 = 1440° = 16$ rt. $\angle$s
Duodecagon .	12	10	$180° \times 10 = 1800° = 20$ rt. $\angle$s
General Polygon	n	$(n - 2)$	$180\,(n - 2)° = 2\,(n - 2)$ rt. $\angle$s

Notice that there are always two less triangles than the number of sides. So that if we generalise (to obtain a formula), an n-sided polygon will divide into $(n - 2)$ triangles; thus its interior angle-sum will be twice that number of right angles, *i.e.*, $2n - 4$ right angles.

(b) *Sum of Exterior Angles of a Polygon*

An even neater result to remember for the polygon is the sum of its *exterior* angles. We could deduce this from the interior angle-sum, but a more direct and intuitive way is as follows:

A man lives at No. 43, Long Lane, and takes his evening stroll " round the block " (indicated in Fig. 66(*a*) below). As he enters Myrtle Way he turns through 30° approximately, then through about 100° as he turns into Meadow Road, and so on at each corner until he is back home again. Observe that he turns round completely ONCE in his stroll.

In Fig. 66(*b*) we show how he turns at each of the five corners of the " block ", which are the vertices A, B, C, D, and E of a pentagon. Notice that the angle turned through at the vertex A is the external angle of the pentagon at A. So from a bird's eye

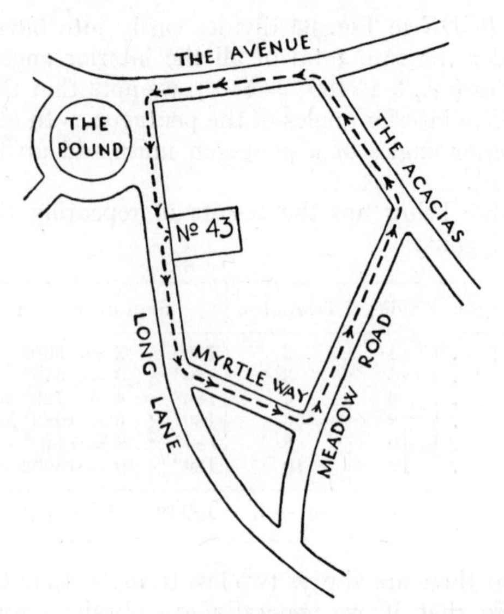

FIG. 66(*a*).

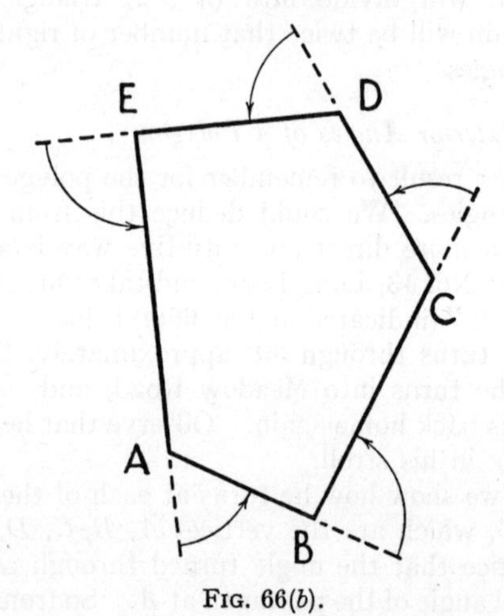

FIG. 66(*b*).

view he turns anti-clockwise at each corner, the total angle turned through—360° or 4 right angles—giving us the sum of the external angles of the pentagon.

The same argument would apply, no matter how many roads formed the " block ", *i.e.*, for a hexagon, octagon, and so on. Thus in general

the sum of the exterior angles of any polygon is 360°.

(This is very much easier to remember than the formula for the interior angle-sum !)

Example 7(a). *Calculate the size of each interior angle of a regular duodecagon.*

If the figure is *regular*, it means that its sides are all equal (*e.g.*, a square is a regular quadrilateral).

Now, for a regular duodecagon,

the sum of the twelve exterior angles = 360°
∴ each *exterior* angle = 30°

But each interior angle will be the supplement of its exterior angle.

So the size of each interior angle = 180° − 30°
= 150°

Now check this result by using the " $2n - 4$ right angles " formula ! (*i.e.*, using $n = 12$.)

EXERCISE 7(a)

A. *Angles and Triangles.*

1. What is the angle between the hands of a watch at (*a*) 1 o'clock, (*b*) 5 o'clock, (*c*) x o'clock, x being a whole number.

2. Find the angle between the hands of a clock at: (*a*) 12.20 a.m., (*b*) 3.40 p.m. Is any of this information irrelevant ?

3. The modern way of giving compass bearings is to state the number of degrees *Eastwards* from the North direction. Thus S.E. becomes a bearing of 135°.

Convert to similar bearings the following directions: N.E., S.W., due West, S.S.E., and S. 15° W.

4. On the Ordnance Survey of the Midlands the following information is to be found: " Magnetic Variation 15° 27′ West in 1921: annual decrease 8′ approximately."

Find: (*a*) the approximate magnetic variation in 1951. (*b*) the true bearing (nearest degree) of a landmark, whose (magnetic) compass bearing is 128°. (Use a rough sketch to help you.) (*c*) When will the magnetic and geographical Norths in this district be most nearly the same? (Assuming constant decrease each year.)

5. It takes 8 complete turns of a windlass handle to wind up a bucket from the bottom of a well 24 ft. deep. Through what angle must the handle be turned to raise the bucket: (*a*) 5 ft., (*b*) 1 in.?

6. In a right-angled triangle, calculate the third angle if one angle is: (*a*) 72°, (*b*) 33° 37′, (*c*) y degrees.

7. In the $\triangle LMN$, $\angle L = 104°$, $\angle M = 27°$; find $\angle N$.

8. In the $\triangle ABC$, $\angle C = 32°$, the exterior angle at $B = 132°$; calculate $\angle A$.

9. In a $\triangle XYZ$, $\angle X = 124°$ and angles Y and Z are equal. Calculate them. If $\angle X$ were $x°$ express $\angle Y$ in terms of x.

10. In a $\triangle ABC$ the bisector BP of $\angle ABC$ meets the side AC at P. Calculate angles A and C if $\angle ABC = 82°$ and $\angle BPA = 110°$.

The following examples (Nos. 11 to 14) are to be solved by scale drawing. It is well worth making a quick freehand sketch before you embark on accurate drawing.

11. Alderley Edge is 20 miles due North of Newcastle under Lyme, and Leek is 12 miles from Alderley Edge on a bearing of 148°. Find the distance and bearing of Newcastle under Lyme from Leek.

12. A yacht race is run on a triangular course—the legs being 1500 yd. N. 70° E., 2000 yd. N. 80° W., and then back to the starting point.

Draw a plan of the course to scale, marking and measuring the angles turned through by the yachts at each main turn, and specify by distance and bearing the third leg of the course.

What is the sum of the three angles turned through? (The *exterior* angles of the triangle concerned.)

13. An aeroplane flies 56 miles N.W., then 72 miles on a bearing of 212°, and then straight home. What is the direction and distance of the last lap?

14. Two planes leave an airfield at the same time, one flying due South at 240 m.p.h. and the other on a bearing of 150°. Half an hour later they are 70 miles apart. At what speeds could the second plane be flying?

15. When it is 12 noon (G.M.T.) at Greenwich (longitude 0°):

(*a*) What is the true time by the sun in New Orleans (longitude 90° W.) Durban (longitude 30° E.), Tokyo (longitude 140° E.), and Los Angeles (148° W.)? And on longitude $L°$ East?

(*b*) On what line of longitude would it then be 6 p.m., 7.30 a.m., t p.m.?

(*c*) What importance has this discussion for the student of telecommunications?

B. *Angles of Polygons.*

16. Reason out the interior angle-sum of: (*a*) a nonagon (9 sides), (*b*) a heptagon (7 sides).

17. Calculate the size of an exterior and an interior angle of the regular pentagon, octagon, and decagon.

18. If the sides of a regular pentagon are each produced in both directions, we obtain a " pentagram ", the five-pointed star which was used by certain Greek philosophers as a mystic symbol or badge. Calculate the angle at each point of this star.

19. Sketch the shape of a groundsheet made to fit exactly the floor space within a bell-tent with 16 panels (including the door). At what angle should the material be cut at each corner to make this groundsheet?

20. Each interior angle of a regular polygon is 140°. How many sides has it? Repeat the process for an interior angle of x degrees.

21. The sum of the interior angles of an n-sided polygon is 6 times the sum of its exterior angles. What is n?

(Form a simple equation and solve it, working in right angles.)

22. Is it possible to have regular polygons whose exterior angles are: (*a*) 18°, (*b*) 11°, (*c*) $4\frac{1}{2}°$, (*d*) $2\frac{1}{2}°$?

23. One angle of a pentagon is a right angle; the other four angles are equal to one another. What size is each of these?

24. If the angles of a quadrilateral taken in order are in the ratio 1 : 4 : 6 : 9 show that two sides of this figure are parallel (*i.e.*, it is a *trapezium*). It helps to consider the smallest angle as being x degrees, and work out an equation from that.

25. An overhead telephone route under construction has to alter direction by 56°. If the change of direction at any angle pole is not to exceed 15°, what is the least number of angle poles required, and the change of direction at each of them? Make a sketch diagram of your proposed siting of the angle poles.

26. With the restriction given in the previous question, what would be the least number of poles needed to build a complete " circular " overhead route? What geometrical figure would the poles mark out?

C. *Problems mainly about Loci.*

27. What is the locus of: (*a*) the centre of a half-crown rolled along a straight groove in a horizontal table-top; (*b*) the centre of a halfpenny which is rolled around a florin (both flat and always in contact) on a flat surface; (*c*) the tip of the pendulum of a grandfather clock when working?

28. A halfpenny (diameter 1 in.) rolls once completely round the outside of a rectangular box 3 in. by 4 in. Sketch the locus of, and calculate the distance travelled by, the centre of the coin. (Both are flat on a table and in contact all the time.)

29. An unmarked underground cable joint is said to be 10 ft. from the edge of a certain track, and 25 ft. from a pine-tree, which is 12 ft. from the same track on the same side as the joint.

Show on a drawing the possible positions of the joint. What geometrical loci are involved?

30. In the previous example, if you are not told on which side of the track the joint lies, how many possible positions are there?

31. A police short-range " walkie-talkie " control set is sited 3 miles

from a straight road on which a police car is patrolling. If the extreme range of intercommunication is 5 miles, find the greatest length of road on which the patrol car can be in radio contact with the control set. Can you verify your result by simple calculation? (See Problem B in Section 7.1 above.)

32. Two listening posts A and B are trying to locate the source of radio interference from a " pirate " station. B is 6·2 miles from A in a direction N. 22° E.

Using direction-finding equipment, B reports to A that the interfering station is on bearing S. 79° E. from B, while A observes that its bearing is N. 63° E.

By drawing to scale find the distance of the " pirate " from A.

7.6. More about Triangles

(a) *The Stable Unit of Construction*

The triangle might be regarded as the basic " brick " of geometrical structure, just as in girder bridges and other engineering forms it is the stable unit of framework construction.

In Fig. 67 we have suggested in very simple structures (your garden gate, an inn-sign, or rural power-lines) that only if these structures involve triangular units will they be stable and keep their shape in service—and not depend on glue or " rust and paint "!

So in " Trigeometry " the basis of our thinking is the *triangle.* We saw in the previous section how the angle properties of any polygon can be deduced from our knowledge of the angles of a triangle.

We must be clear then about the properties of triangles, both generally and in special cases. Our first concern is to be able to *specify* a triangle with certainty, so clearly that, for example, if we were to describe a triangular building site over the transatlantic radio-telephone, colleagues in New York would be able to visualise an exact replica of our triangle here in London.

(b) *Specifications—Data Sufficient to Determine a Triangle*

(i) *Three Sides Given.* If we hark back for one moment to days of Meccano models, a little reflection will bring home to us this fundamental idea—that if we bolt together three metal strips of specified lengths they form exactly the same triangle every time, which cannot be deformed without buckling the strips.

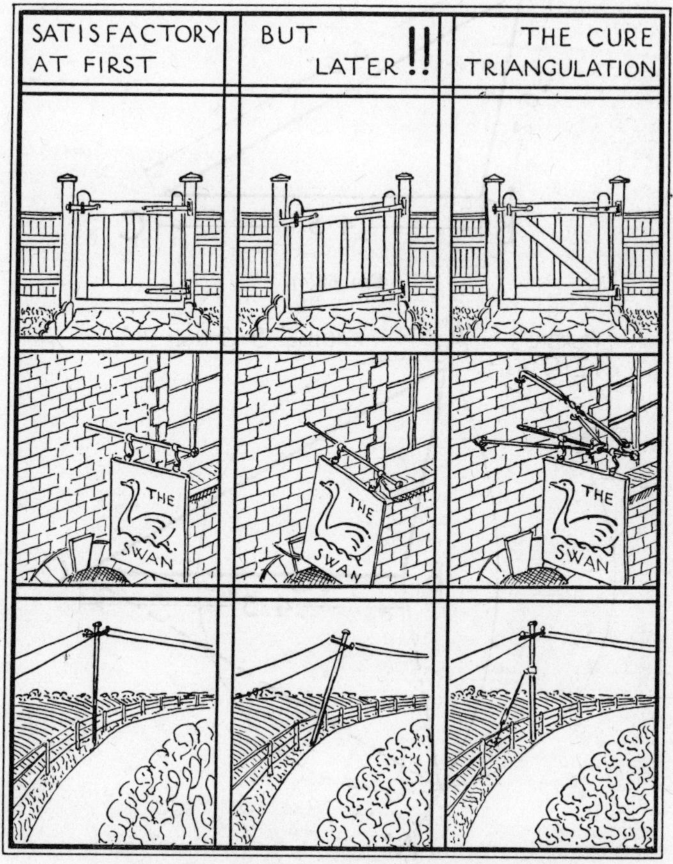

FIG. 67.—THE TRIANGLE AS THE STABLE FRAMEWORK.

Similarly, if we are given the lengths of the three sides of a triangle ABC, *e.g.*, $AB = 2{\cdot}9$ in., $BC = 4{\cdot}5$ in., and $CA = 5{\cdot}5$ in., we can construct it uniquely, as shown in Fig. 68(a) below.

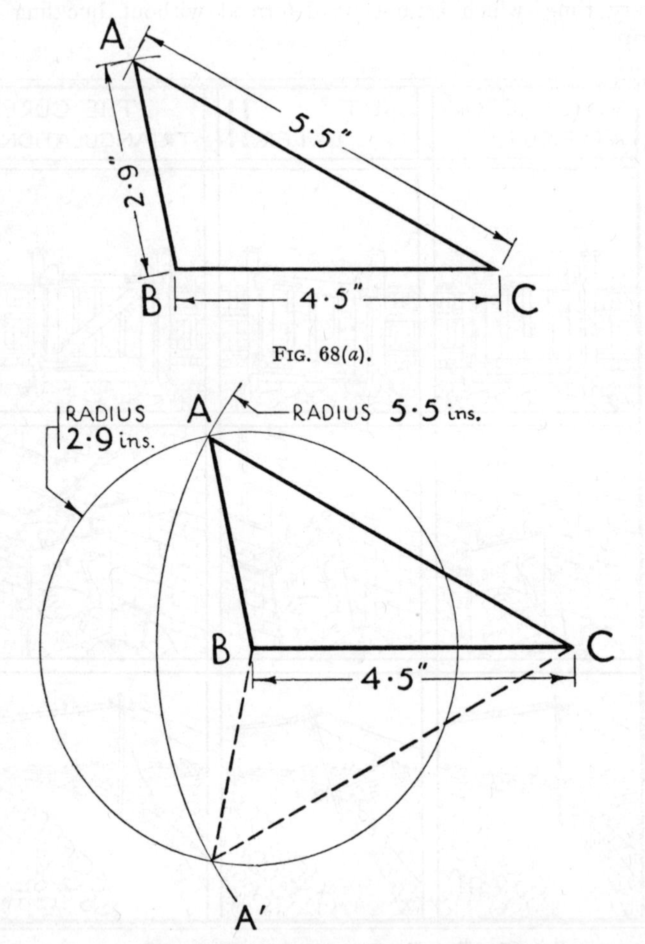

FIG. 68(*a*).

FIG. 68(*b*).—SPECIFICATION—THREE SIDES.

Taking any side as base, say *BC*, draw *BC* 4·5 in. long; with centres *B* and *C* describe arcs of radius 2·9 and 5·5 in. respectively, intersecting at *A*. Then *ABC* is our required triangle.

If we had drawn the complete circles in the above construction, we should have obtained a second triangle $A'BC$ (Fig. 68(*b*)), which is the " mirror image " of ABC, the line BC being the " mirror ". It is clear that the two triangles ABC and $A'BC$ are identical.

This " three sides " specification has an advantage over those that follow in that it requires only ruler and compasses to construct the triangle.

(ii) *Two Sides and the Included Angle Given.*—In Fig. 69(*a*) OAB represents a valuable corner site, available for sale in a busy town. A surveyor wishes to measure the site accurately for

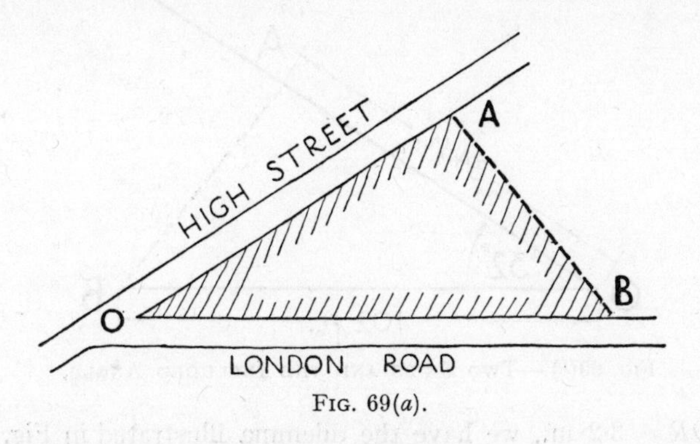

FIG. 69(*a*).

his client. Access is impossible between A and B, so the only measurements he can make are the distances $OA = 84$ ft., $OB = 102$ ft. (measured by a surveyor's chain) and the angle at the corner of High Street and London Rd., *i.e.*, $\angle AOB = 32°$, which he measures with a theodolite.

Test for yourself that you can construct an accurate scale drawing of ΔAOB with just these measurements. (A scale of 1 in. $\equiv$ 20 ft., or 1 cm. $\equiv$ 10 ft., does admirably.)

On a convenient straight base-line mark out with compasses or dividers a length $OB \equiv 102$ ft. (*i.e.*, 5·1 in. on the first scale). With a protractor construct $\angle POB = 32°$, making the line OP longer than is needed for OA. Lastly, with compasses or dividers mark off along OP a length $OA \equiv 84$ ft. (*i.e.*, 4·2 in.);

AB may then be joined and its length determined by measurement (Fig. 69(*b*)).

Now draw and measure the perpendicular from A to OB, to find the height of the triangle, and thence calculate the area of the site AOB; it should be approximately 252 sq. yd.

Note in the above example that the angle measured was that formed or " included " by the two measured sides of the triangle. This is important, for if we were given an angle not so included by the given sides, *e.g.*, a triangle PQR with $\angle P = 41°$, $PQ = 4$ in.

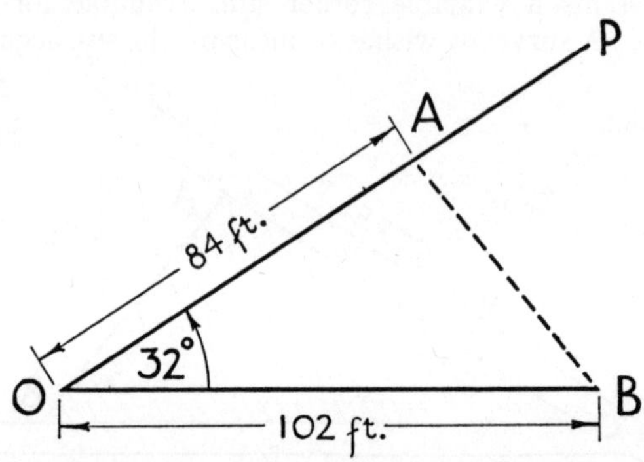

FIG. 69(*b*).—TWO SIDES AND THE INCLUDED ANGLE.

and $QR = 3{\cdot}2$ in., we have the dilemma illustrated in Fig. 70: there are *two* possible triangles which equally well fit these data, quite different in shape and size! We have labelled the alternative positions of R as R_1 and R_2.

Because of this dilemma, we call this non-included angle set of data the " ambiguous " case, and carefully avoid such a specification in practice.

There is a further possibility with non-included angles. For example, if we are told a ΔLMN has $LM = 5$ in., $LN = 3$ in., and $\angle LMN = 58°$ we get the irritating result shown in Fig. 71. Try it yourself!

See how N lies on *two loci*—a straight line MX and a circle centre L, 3 in. in radius—*which do not intersect*! So the point N and consequently the ΔLMN also cannot exist.

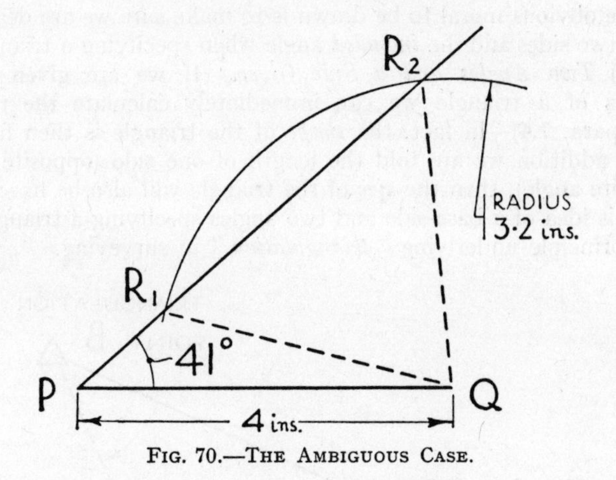

FIG. 70.—THE AMBIGUOUS CASE.

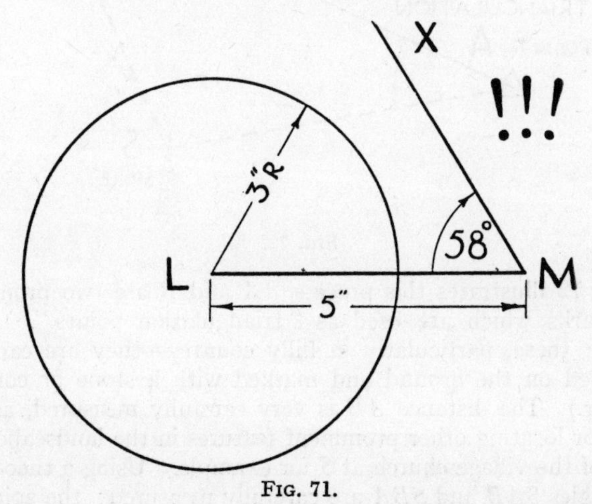

FIG. 71.

(A unique case is on the rare occasion when the circle-locus exactly touches the straight-line-locus, and then only one triangle can be drawn. The angle at the third vertex is then a right angle.)

The obvious moral to be drawn is to make sure we are dealing with two sides and the *included* angle when specifying a triangle !

(iii) *Two Angles and a Side Given.* If we are given two angles of a triangle we can immediately calculate the third (see para. 7.4)—in fact, the *shape* of the triangle is then fixed. If in addition we are told the length of one side (opposite one definite angle), then the *size* of the triangle will also be fixed.

This idea of a base-side and two angles specifying a triangle is the principle underlying "*triangulation*" in surveying.

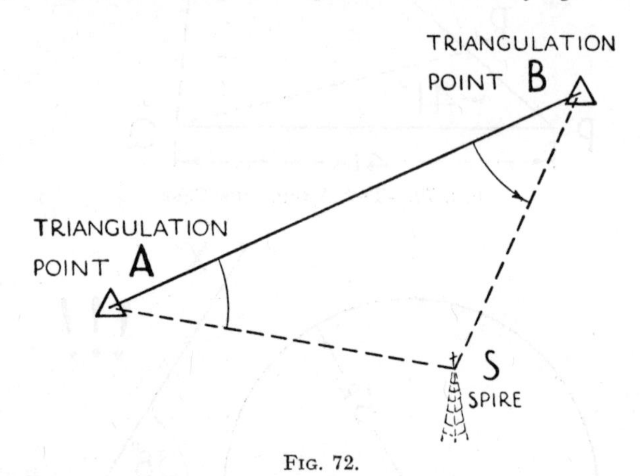

FIG. 72.

Fig. 72 illustrates this process. A and B are two prominent landmarks which are used as "triangulation points". (Look out for these, particularly in hilly country—they are carefully surveyed on the ground and marked with a stone or concrete marker.) The distance AB is very carefully measured, as it is used for locating other prominent features in the landscape—the spire of the village church at S for example. Using a theodolite, the angles SAB and SBA are carefully measured; the spire can then be accurately pin-pointed on an outline map.

If AB is 1540 yd., $\angle SAB = 36°$, $\angle SBA = 41°$, and the bearing of B from A is 81°, draw to scale the $\triangle ASB$ and find the distances and bearings of A and B from S. Show the North direction and your scale on the drawing, and verify the bearings by calculation.

In practical map-making the process above is repeated for many other prominent features—the gable of a farm-house, a conspicuous road-junction, a bridge, and so on.

We see now that if the length of one side is specified and two of its angles known both in size and in position relative to that side, then a triangle is uniquely determined.

(iv) *A Right Angle, the Hypotenuse, and One Side Given.* A

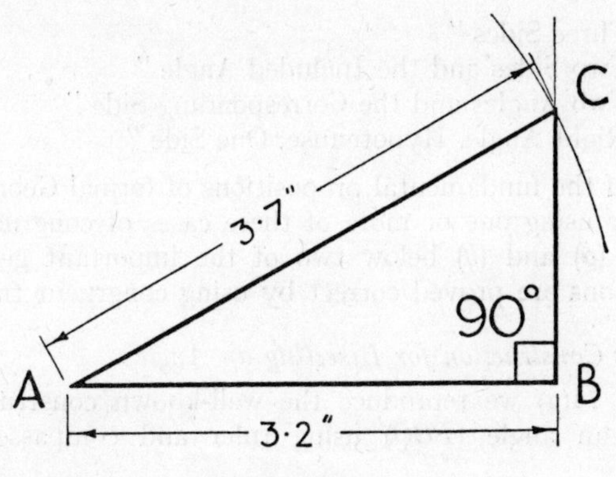

FIG. 73.

triangle ABC is specified as having $\angle ABC$ a right angle, $AB = 3{\cdot}2$ in., and the hypotenuse $AC = 3{\cdot}7$ in. If we start (Fig. 73) with AB, then draw our right angle at B, finally describing an arc, centre A and radius 3·7 in., we obtain a unique point C. The specification is thus a sound one.

This is very much a special case; but it is useful in proving results in formal Geometry (see " Congruence " in para. 7.7).

7.7. The Technique of " Congruent Triangles "—and Two or More Loci

In the previous section (7.6) we discussed four ways of " specifying " a triangle. As a consequence, if two or more triangles bear the same specification they must be exact replicas of one another. Such identical triangles are said to be CONGRUENT. In some Geometry books the identity symbol ≡ is used as a shorthand symbol for " congruent with " or " is congruent with " :

e.g., $$\triangle ABC \equiv \triangle EFG$$

is translated as " the triangle ABC is congruent with the triangle EFG ".

There are four recognised " Cases of Congruence ", which correspond to the four valid specifications for a triangle discussed above. They are:

" Three Sides "
" Two Sides and the Included Angle "
" Two Angles and the Corresponding Side "
" Right Angle, Hypotenuse, One Side "

Many of the fundamental propositions of formal Geometry are proved by using one or more of these cases of congruence. In examples (*a*) and (*b*) below two of the important geometrical constructions are proved correct by using congruent triangles.

(a) *The Construction for Bisecting an Angle*

In Fig. 74(*a*) we reproduce the well-known construction for bisecting an angle (POQ) using ruler and compasses. With

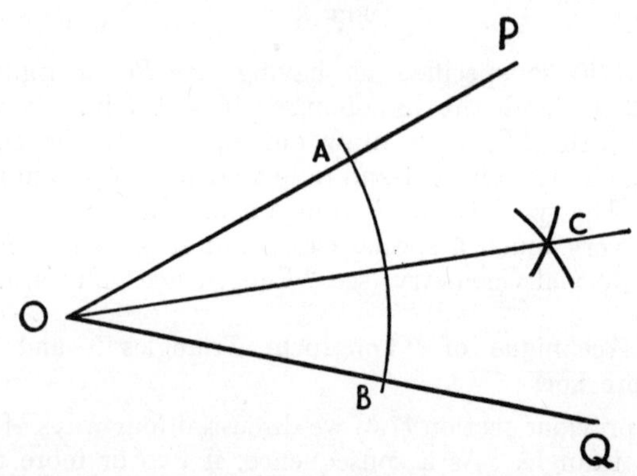

FIG. 74(*a*).—BISECTING AN ANGLE—CONSTRUCTION.

centre O and any radius an arc AB is described, cutting the arms OP, OQ of our angle POQ at A and B. With centres A and B and any equal radii, describe two arcs to intersect at C. Then OC should be the required bisector. We will now endeavour to prove this fact.

Proof: Join CA and CB (Fig. 74(*b*)).

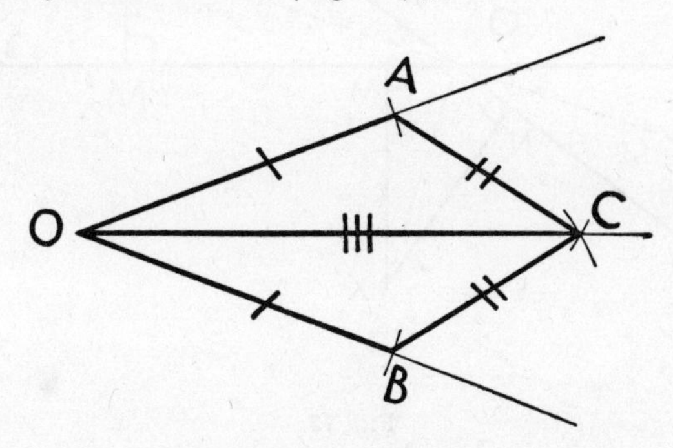

FIG. 74(*b*).—BISECTING AN ANGLE—PROOF

Then in the triangles ACO, BCO

$$\begin{cases} OA = OB & \text{(same radius)} \\ AC = BC & \text{(equal radii)} \\ OC = OC & \text{(a “ common ” side)} \end{cases}$$

$$\therefore \quad \triangle ACO \equiv \triangle BCO \text{ (Case: 3 sides)}$$

Hence the *angles* of the two triangles must be equal in pairs. In particular, $\angle AOC = \angle BOC$ (angles opposite equal sides AC, BC), *i.e.*, OC is the bisector of the original angle AOB, which justifies the construction.

The Bisector as a Locus. *Two straight lines* PQ, RS *cross at O.* X *is a point which is equidistant from the two straight lines. Prove that* X *must lie on the bisector of one of the angles formed by the given straight lines.*

In Fig. 75 two possible positions X and X' are shown: their perpendicular distances from the lines PQ and RS must be equal; *i.e.*, $XM = XN$ and $X'M' = X'N'$. It appears highly probable that X lies on the bisector of $\angle QOS$ and X' on the bisector of $\angle SOP$.

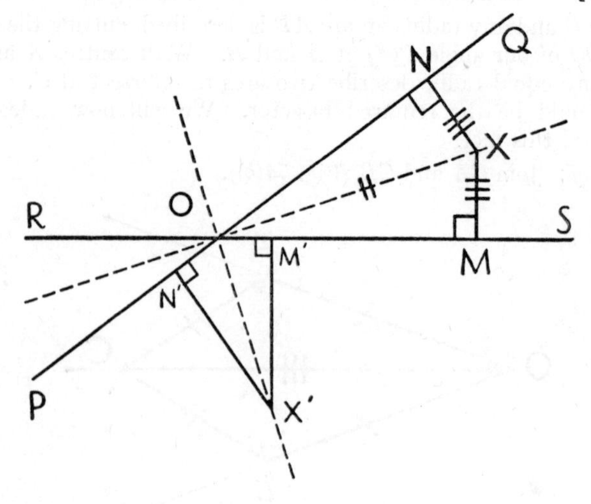

FIG. 75.

To prove : X lies on the bisector of $\angle QOS$.
Construction : Join OX.
Proof : In the $\triangle$s OXM, OXN

$$\begin{cases} OMX \text{ and } ONX \text{ are right angles } (XM \text{ and } XN \text{ are perp. distances}) \\ OX = OX \text{ (a ``common'' hypotenuse)} \\ XM = XN \text{ (given equal distances)} \end{cases}$$

$\therefore \quad \Delta OXM \equiv \Delta OXN$ (Case : right angle, hypotenuse, 1 side)

Hence the remaining sides and angles of these triangles are equal in pairs.

In particular, $\angle XOM = \angle XON$

i.e., X *does* lie on the bisector of $\angle QOS$.

Similarly, X' lies on the bisector of $\angle SOP$, and in general, points equidistant from the original two lines will lie on the bisector of one of the four angles made by these lines. (These bisectors together form the *locus* of X.)

(b) *A Point Equidistant from two Points—and the Bisector of a Line*

In Fig. 76, P is a point which moves in a plane so as to be equidistant from two fixed points A and B in that plane. On what path does P move? (*i.e.*, what is the locus of P?).

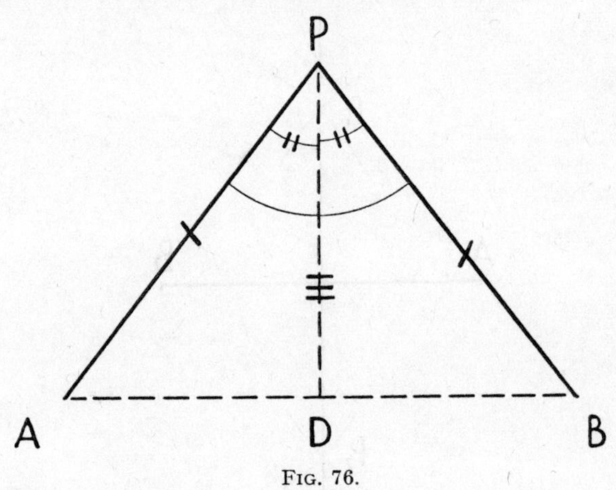

FIG. 76.

Consider one position of the point P.

Join AB and draw the bisector of $\angle APB$, to intersect AB at D.

Then in the triangles APD, BPD

$$\left\{\begin{array}{lll} AP = BP & \text{(given } P \text{ equidistant from } A \text{ and } B) \\ \angle APD = \angle BPD & (PD \text{ bisects } \angle APB) \\ PD = PD & \text{(common to both triangles)} \end{array}\right.$$

$\therefore\ \triangle APD \equiv \triangle BPD$ (Case: 2 sides and the incl. angle)

Hence $AD = BD$, *i.e.*, D is the mid-pt. of AB.

Also $\angle ADP = \angle BDP$, and as ADB is a straight line each of these must be a right angle.

So PD is perpendicular to AB and bisects AB at D. This special line is called the *perpendicular bisector* of AB and is the *locus* of our point P.

Note : The above discussion is the basis of the well-known construction for bisecting a line (or drawing its perpendicular bisector) as follows :

With centres A and B equal arcs are drawn, intersecting at P_1 and P_2 (Fig. 77). Then P_1P_2 is the perpendicular bisector of AB.

For P_1 and P_2 are simply alternative positions of our point P in the previous argument. They each must lie on the perpendicular bisector of AB, therefore P_1P_2 is the required perpendicular bisector.

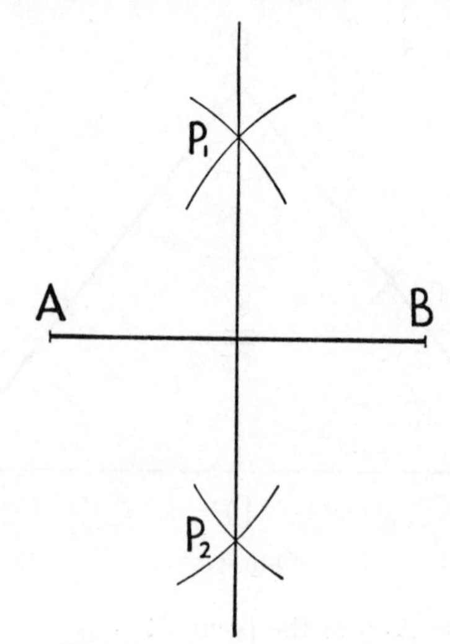

FIG. 77.—BISECTING A LINE.

7.8. Special Triangles

(a) *The Isosceles Triangle* is one having two of its sides equal; thus in $\triangle PAB$, $PA = PB$ (Fig. 78). The word *iso-sceles* means " the same scale "; compare the use of the prefix " iso " in such technical terms as the meteorologist's *isobar* (line joining points with the same barometric pressure) and the chemist's *isotopes* (substances with the same chemical behaviour but with slight differences in molecular structure).

In para. 7.7 we proved that P must lie on the perpendicular bisector PD of the line AB, as a result of *triangles* APD *and* BPD *being congruent.*

It thus follows that $\angle A = \angle B$ (angles opposite the common side PD in these congruent triangles), i.e., *the " base angles " of any isosceles triangle must be equal.*

Notice that the whole triangle PAB balances (is " symmetrical ") about the line PD, which is called the " axis of symmetry " of the isosceles triangle. A good analogy is the use of " shear-legs " in field engineering—the geometrical

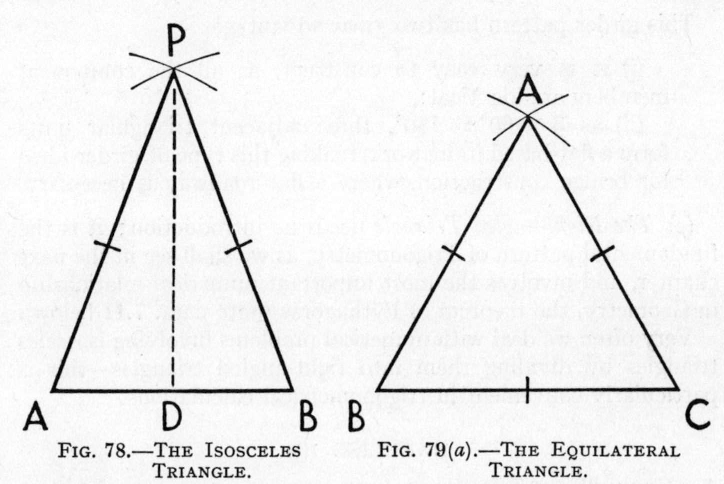

FIG. 78.—THE ISOSCELES TRIANGLE.

FIG. 79(*a*).—THE EQUILATERAL TRIANGLE.

symmetry of equal shear-legs is accompanied by equal distribution of the mechanical strains involved.

(b) *The Equilateral Triangle,* as its name suggests, has all three sides equal (Fig. 79(*a*)).

Whichever " base " we choose with such a triangle, the figure is isosceles and the " base angles " are equal : it follows that *all three angles of an equilateral triangle are equal.* Each must then be 60°. In fact, we could regard Fig. 79(*a*) as a ruler-and-compasses construction for an angle of 60°—it is indeed the most accurate way of constructing this particular angle.

In girder structures extensive use is made of equilateral triangles, *e.g.*, in the " Warren " girder construction suggested in Fig. 79(*b*).

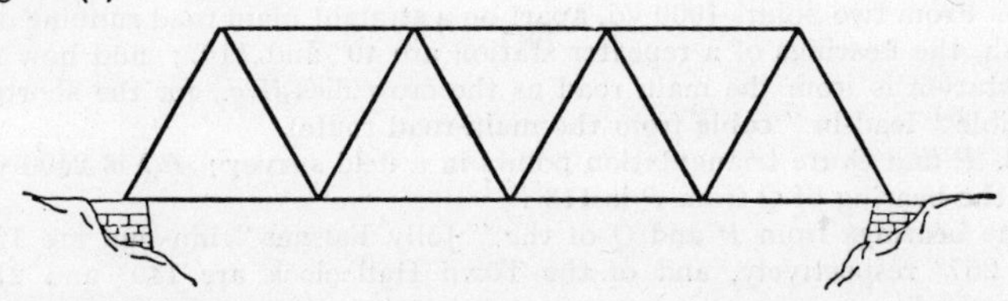

FIG. 79(*b*).—A WARREN GIRDER FRAMEWORK.

This girder pattern has two great advantages:

(i) it is very easy to construct, as all the component members are identical;

(ii) as $3 \times 60° = 180°$, three adjacent triangular units form a flat-based framework, making this type of girder ideal for bridge construction, where a flat roadway is necessary.

(c) *The Right-angled Triangle* needs no introduction; it is the fundamental pattern of Trigonometry, as we shall see in the next chapter, and involves the most important numerical relationship in Geometry, the theorem of Pythagoras (note para. 7.11 below).

Very often we deal with numerical problems involving isosceles triangles by dividing them into right-angled triangles—this is particularly convenient in trigonometrical calculations.

EXERCISE 7(*b*)

A. *Construction of Triangles*

Which of the following specifications for a $\triangle ABC$ are valid? Construct to scale, and measure the remaining angle(s) and/or side(s). Check the angle-sum in each case.

1. $AB = 4{\cdot}2$ in., $BC = 3{\cdot}7$ in., $CA = 5{\cdot}1$ in.
2. $BC = 3{\cdot}5$ in., $\angle BCA = 62°$, $AB = 2{\cdot}4$ in.
3. $AB = 6{\cdot}3$ cm., $BC = 3{\cdot}2$ cm., $CA = 2{\cdot}8$ cm.
4. $AB = 4{\cdot}1$ cm., $\angle ABC = 32°$, $BC = 5{\cdot}3$ cm.
5. $\angle ABC = 32°$, $\angle BAC = 48°$, $BC = 4{\cdot}2$ in.
6. $\angle BCA = 43°$, $BC = 4{\cdot}4$ in., $AB = 3{\cdot}9$ in.
7. $\angle ABC = 41°$, $\angle BCA = 49°$, $\angle CAB = 69°$.
8. $AB = 6{\cdot}4$ cm., $\angle BCA = 90°$, $BC = 4{\cdot}9$ cm.
9. $BC = 6{\cdot}9$ cm., $\angle ABC = 63°$, $\angle BCA = 117°$.
10. $\angle ABC = 41°$, $\angle BCA = 58°$, $\angle CAB = 67°$.

The following examples are to be solved by scale drawing and measurement. (A preliminary rough sketch will help.)

11. From two points 1000 yd. apart on a straight main road running due North, the bearings of a repeater station are 40° and 110°; find how far the station is from the main road as the crow flies (*e.g.*, for the shortest possible " lead-in " cable from the main-road route).

12. P and Q are triangulation points in a field survey; PQ is 2400 yd. and the bearing of Q from P is 113°.

The bearings from P and Q of the " Jolly Farmer " inn-sign are 191° and 257° respectively, and of the Town Hall clock are 149° and 211° respectively.

Find the distance and direction of the " Jolly Farmer " from the Town Hall.

13. Radar Station A picks up an unidentified surface vessel on a bearing of 282° and at a distance of 12,600 yd. A second Radar Station B is asked to confirm the vessel's position. If B is 19,400 yd. from A on a bearing of 223° what readings would you expect B to report?

14. Dorchester is 55 miles due South of Bristol, while Exeter is 42 miles from Dorchester and 64 miles from Bristol.

Find the distance and direction of Dorchester from Barnstaple, given that Exeter is 33 miles South-west of Barnstaple. (Exeter lies West of the line Bristol–Dorchester.)

15. An engineer has two alternatives in installing a new cable:

(a) to suspend aerial cable on an existing poled route, from A to B (4200 yd. S.W.), and then on a second existing route from B to C (5800 yd. in a direction N. 18° W.);

(b) to bury cable direct from A to C.

He estimates it will cost approximately twice as much per yard to instal buried cable. Find from a scale diagram which method will be cheaper and by how much per cent.

16. In level country the elevation of the top of a radio transmitting mast is 20° from a certain spot; 120 yd. nearer, it is 30°. Find the height of the mast.

17. A boat is steered due North at 8 knots, but in addition a strong current carries her North-east at 3 knots. Find how far she has travelled (in nautical miles) and in what direction at the end of 20 min. sailing. What is her actual speed in knots?

18. One end of a 15-ft. cord is fastened to a nail, and a weight attached to its free end; this weight is set swinging to and fro through 15° on each side of the vertical. How high does the weight rise vertically? What is the horizontal distance between its extreme positions?

B. *Use of Congruent Triangles*

19. D is the mid-point of the side BC of $\triangle ABC$. Prove that B and C are equidistant from the line AD (produced if necessary).

20. In the quadrilateral $ABCD$, AB is equal and parallel to DC; prove that AD must be equal and parallel to BC.

(*Hint:* Joint AC.)

21. The parallelogram is defined as " a quadrilateral with both pairs of opposite sides *parallel* ". Using this definition, prove that the *opposite sides and angles of a parallelogram must be equal.* (Show also that a diagonal bisects the area of the parallelogram.)

22. Both pairs of opposite sides of a quadrilateral are equal. Prove it to be a parallelogram (*i.e.*, opposite sides parallel).

23. $ABCD$ is a square; P is the mid-point of AB. A circle with centre P cuts AD, BC at Q, R respectively. Prove that $\angle AQP = \angle BRP$.

24. A kite-shaped figure $ABCD$ has $AB = AD$ and $CB = CD$. Prove

that AC bisects the interior angles at A and C, and also bisects the other diagonal BD at right angles.

What is the " axis of symmetry " of this figure?

25. In any $\triangle ABC$, the perpendicular AD from A to the bisector BD of $\angle ABC$ meets the base BC (produced if necessary) at E. Prove that $AD = DE$.

26. If the diagonal of a quadrilateral bisects the angles which it joins, prove that it must be a " kite ". (See Question 24 above, of which this is the " converse ".)

27. D is any point inside an equilateral $\triangle ABC$. On DC as base a second equilateral $\triangle DEC$ is drawn, with A and E on the same side of BC. Show that $\angle ACE = \angle BCD$, and prove that $BD = AE$.

28. Assuming the properties of a parallelogram proved in Question 21 above, prove that the *diagonals of a parallelogram bisect each other.*

29. $ABCD$ is a square. A point P is taken on BC and a point Q on CD such that $AP = BQ$. Prove that $\angle PAB = \angle QBC$, and deduce that AP is perpendicular to BQ.

30. The diagonals of a parallelogram $ABCD$ intersect at P. A line through P meets AB at X and CD at Y. Prove $PX = PY$. (Assume results in Question 28 above.)

C. *Special Triangles—Isosceles, Equilateral, Right-angled*

31. Find from a good dictionary the meaning and derivation of the following terms: isoclines; an isometric drawing; isomorphism; isothermal.

Look out for other scientific terms beginning " iso- ".

In the next set of examples (Nos. 32–39), calculate the unspecified interior angles of each triangle, and check angle-sums.

32. $\triangle ABC$: $AB = AC$, $\angle ABC = 62°$.

33. $\triangle PQR$: $PQ = PR$, $\angle QPR = 108°$.

34. $\triangle LMN$: $LM = LN$, $\angle LMN = x°$.

35. $\triangle DEF$: $DF = EF$, exterior angle at $F = 74°$.

36. $\triangle XYZ$: $XY = YZ$, $\angle XYZ = y°$.

37. $\triangle UVW$: $VW = UW$, exterior angle at $V = y°$.

38. $\triangle ABC$: $\angle ABC = 90°$, $\angle A = 4\angle C$.

39. $\triangle ABC$: $\angle BAC = 90°$, $\angle B = x°$.

40. Using the methods suggested in paras. 7.7(*a*) and 7.8(*b*) above, construct, with a straight edge and compasses only, angles of 30°, 120°, 135°, 75°. Check by protractor.

41. If $x°$ and $y°$ are the exterior angles at the vertices A and B in a $\triangle ABC$ having $AB = AC$, express x in terms of y as a " formula ".

42. $ABCDE$ is a regular pentagon; prove that the bisector of $\angle BAC$ is perpendicular to AE.

43. The sides of a regular hexagon are produced to form a six-pointed star. Find the angle at each " point " of this star.

44. In the $\triangle ABC$, $AB = AC$ and $\angle BAC = 32°$. BC is produced to

D, and the bisectors of angles ABC and ACB meet at I. Show that $\angle BIC = \angle ACD$.

45. Prove that the result in Question 44 is true with *any* angle at A ($x°$ say).

46. ABC is an equilateral triangle; BC is produced to D making $BC = CD$. Show that BAD is a right-angled triangle.

47. Test the following construction for a regular hexagon of side **3** in.: " With centre O describe a circle radius 3 in. Taking any point A on this circle as centre mark off an equal arc (*i.e.*, radius 3 in.) cutting the circle at B. Repeat with centre B to obtain a point C on the circle, and so on until six points A, B, C, D, E, and F have been marked."

Prove that $ABCDEF$ must be a regular hexagon.

48. Two equal equilateral triangles ABC, CDE have a common vertex C, and overlap so that CD cuts AB and $\angle BCD = 48°$. Calculate angles DBA and DEB and prove that $AD = BE$.

49. Through the vertex C of any ΔABC, CD and CE are drawn parallel to the bisectors of the angles BAC and ABC, meeting BA produced at D and E respectively. Show that DE measures the *perimeter* of the triangle.

50. The accompanying Fig. 80 shows a roof truss framework, in which BD and DC are equal and horizontal; AD is vertical. The remaining

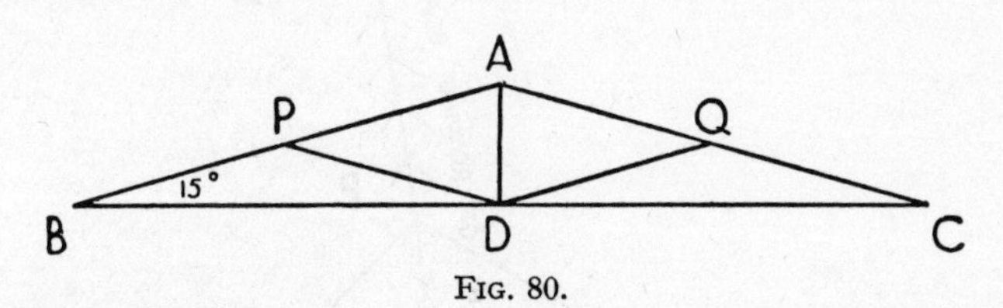

FIG. 80.

members are all inclined at 15° to the horizontal. Prove that ΔAPD is isosceles, and P is the mid-point of AB. How many equal members are there in this framework?

D. *More Loci Problems*

51. Using the diagram of Fig. 75 above, show that the complete locus of a point equidistant from two intersecting straight lines is a *pair of perpendicular straight lines*.

52. A terminal pole on a cross-country overhead telephone route is an equal distance from two straight paths which cross at A at an angle of 30°, and is also known to be 60 yd. from a third straight path which crosses the other two at B and C. If AB = 450 yd. and AC = 200 yd., find by scale drawing how many possible positions the pole may have, and their distances from A.

53. Construct a ΔPQR with $PQ = 5{\cdot}2$ in., $QR = 4{\cdot}8$ in., and $RP = 4{\cdot}3$ in.

Think out how to find the centre of the circle through P, Q, R (the "circum-circle" of the triangle). (*Clue:* The centre of a circle through P and Q is equidistant from P and Q: it must lie on . . .?)

Construct this circle and measure its radius.

54. On what path lie the centres of all circles which touch a given pair of intersecting straight lines?

Construct the circle which touches all three sides (not produced) of the triangle specified in Question 53. Measure the radius of this "inscribed circle".

55. Draw the ΔABC with $AB = 3{\cdot}4$ in., $BC = 4{\cdot}1$ in., $CA = 4{\cdot}7$ in., and construct the circle which touches AB and AC *both produced* and also the side BC.

This is called an "escribed circle". How many escribed circles are there for any triangle?

56. A and B (Fig. 81) are two posts on the straight bank of a river,

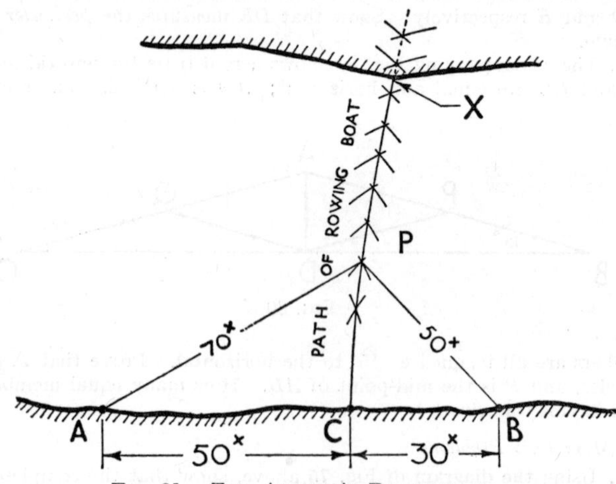

FIG. 81.—THE ANGLER'S BRAINWAVE.

80 yd. apart. A party of anglers wishes to cross the stream in darkness so as to land at a point X on the opposite bank, which is 110 yd. from A and 90 yd. from B. One of the party has a brain-wave: he suggests that as the *difference* in these two distances is 20 yd. they have only to start at the point C (where $AC = 50$ yd. and $BC = 30$ yd., *i.e.*, differing by 20 yd.), tie the ends of two lines to the posts A and B, and starting with these lines taut pay out the two together as they row across, steering such a course that

they both remain taut. Thus at any subsequent position P of the boat $PA - PB = 20$ yd.

Choosing simple values, *e.g.*, $PA = 70$ yd., $PB = 50$ yd., construct a series of positions of the boat on a scale drawing, and draw the " locus " of P. How wide is the river at X?

Mathematically, the locus of P with $PA - PB$ constant ($= 20$) is known as a " hyperbola ".

(This highly improbable story illustrates a practical method of guiding an aircraft equipped with radar on to its target, using two ground radar stations A and B (corresponding to the river posts above) and maintaining a constant difference in the aircraft's distances from the ground stations.)

57. Draw a line PQ 6 in. long, with mid-point O. Draw OX perpendicular to PQ and mark off along it $OA = 2$ in.

By constructing individual points on it, draw the locus of a point which is equidistant from the *line* PQ and the *point* A. This is called a " parabola "; it is, for example, the principal section of a " parabolic " reflecting mirror, used for searchlights.

58. Fix two pins A and B 4 in. apart in your drawing-paper, and round them stretch a loop of string (more than 8 in. long). Keeping the string taut, move a pencil round inside the loop, tracing out a smooth curve. Now vary the length of the string and draw other curves of the same family, keeping A and B fixed.

What shape is each curve of our family? What name do we give to the fixed points A and B?

What geometrical relationship holds between the lengths PA and PB for any point P on one of these curves ? (Cf. Question 56.)

7.9. The Properties of Circles

(a) *Symmetry*

A circle is defined as the locus of a point in a plane which is a fixed distance—the radius—from a fixed point—the centre. Thus in using a pair of compasses, we open out the legs to a specified radius, and use the " sharp end " as the fixed centre while the " pencil end " describes the circumference of the required circle.

The most obvious property of a circle is its symmetry. Through the ages the perfection of the circle has been associated with folk-lore—for instance, in the wedding-ring, symbolising perfect unity and balance. Thus an isosceles triangle has only one axis of symmetry, but any diameter of a circle is an axis of symmetry. A further axis of symmetry (in three dimensions) is the perpendicular to the plane of the circle through its centre—

" wheel-and-axle " symmetry. Similarly, with solid bodies the sphere is perfectly symmetrical about any diameter, and is thus the symbol of perfection in three dimensions. (As in " music from the spheres "?)

In Fig. 82 A and B are any two points on the circumference of a circle, centre O.

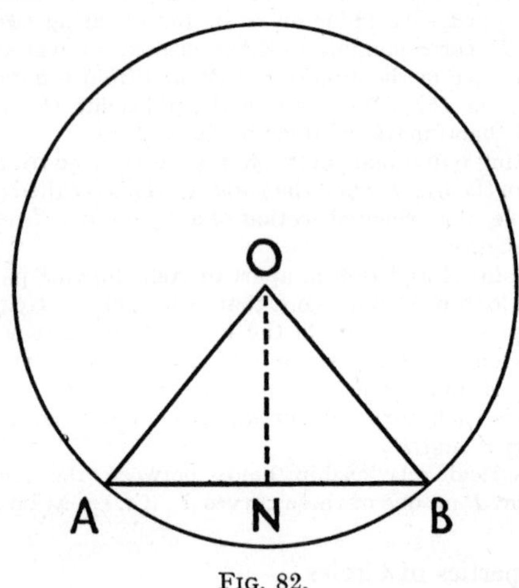

FIG. 82.

Then $OA = OB$ (radii).

So $\triangle OAB$ is isosceles.

$\therefore$ the bisector ON of angle AOB will also bisect the chord AB and be perpendicular to it.

This symmetrical property of the circle is often useful in calculation, for $\triangle AON$ is right-angled, and if we know the length of the chord and the radius of the circle we can readily calculate (using Pythagoras' Theorem) the distance ON of the chord from the centre.

Example 7(*b*). *Fig.* 83(a) *represents the principal section of a planoconcave lens,* 4 *cm. in diameter,* 3 *mm. thick at the periphery, and* 1·8 *mm. thick in the centre. We wish to find the radius of curvature* (r *mm.*) *of the concave face of the lens (assumed part of a sphere).*

Fig. 83(*b*) shows the geometrical essentials of the problem. Arc AB is the section of the concave face of the lens; the chord AB is then 40 mm., O is the centre of curvature of the face, and ONM the axis of symmetry of

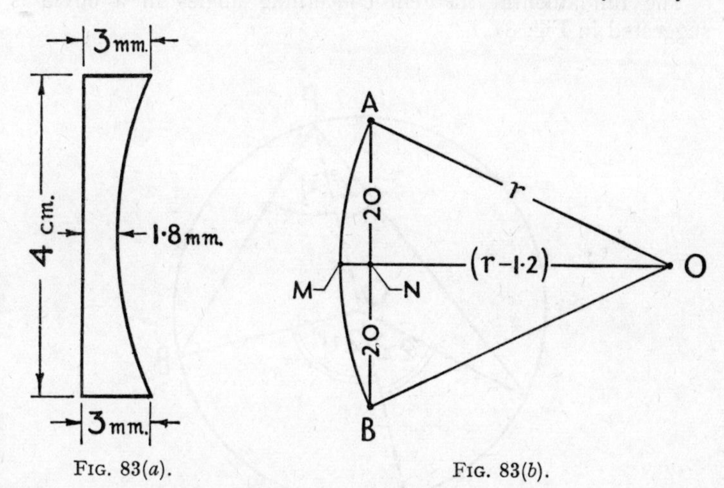

FIG. 83(*a*). FIG. 83(*b*).

the lens, meeting the arc and chord at M and N respectively. ON bisects the chord AB at right angles. $MN = 3 - 1{\cdot}8 = 1{\cdot}2$ mm.

Then in the right-angled $\triangle ONA$, by *Pythagoras' Theorem*

$$OA^2 = ON^2 + AN^2$$

Then

$$r^2 = (r - 1{\cdot}2)^2 + 20^2$$

$$r^2 = r^2 - 2{\cdot}4r + 1{\cdot}44 + 400$$

$$\therefore \quad 2{\cdot}4r = 401{\cdot}44$$

$$r = \frac{401{\cdot}44}{2{\cdot}4}$$

$$= 167 \text{ (mm.) to 3 significant figures.}$$

So the radius of curvature of the concave face of the lens is 16·7 *cm.*

Further symmetrical properties of circles can be deduced from the one enunciated above, providing simple exercises in proofs by congruent triangles:

(i) Equal chords in a circle are equidistant from the centre (and conversely).

(ii) Equal chords in equal circles are equidistant from their centres (and conversely).

(b) *Angle Properties of Circles*

The fundamental theorem concerning angles in a circle is suggested in Fig. 84.

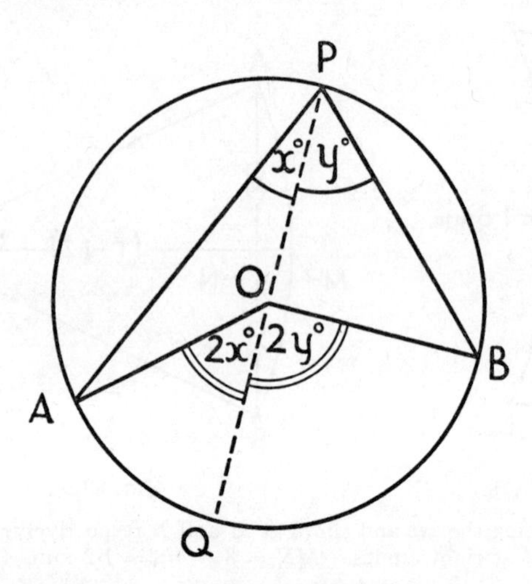

FIG. 84.

Given : AOB is any angle at the centre O of a circle, and APB is any angle at the circumference " subtended " by the same arc AB.

To Prove : $\angle AOB = 2\angle APB$.

Construction : Join PO and produce to Q.

Proof : In the $\triangle AOP$, $OA = OP$ (radii).

$\therefore$ the triangle is isosceles, and the " base " angles OAP, OPA must be equal. (Let each be $x°$.)

Then exterior $\angle QOA = x° + x°$

$$= \underline{2x°}$$

Similarly, $\triangle BOP$ is isosceles, and if $\angle OPB = y°$ the exterior $\angle QOB = 2y°$.

Adding, $\angle QOA + \angle QOB = 2(x + y)°$

i.e., $\underline{\angle AOB = 2\,\angle APB}$

Thus the angle at the centre of a circle is twice any angle at the circumference subtended by the same arc.

Other results follow readily from the above theorem:

(i) *Angles subtended at the circumference by the same arc are equal* (*i.e.,* they must each be half the same angle at the centre).

(ii) *The angle in a semicircle is a right angle* (for it must be half the angle at the centre subtended by half the circumference, *i.e.,* half of 180°).

(iii) *Opposite angles of a cyclic quadrilateral* * *are supplementary* (add up to 180°).

For if such a pair of opposite angles are $p°$ and $q°$, each is half the angle at the centre subtended by the arc concerned (see Fig.

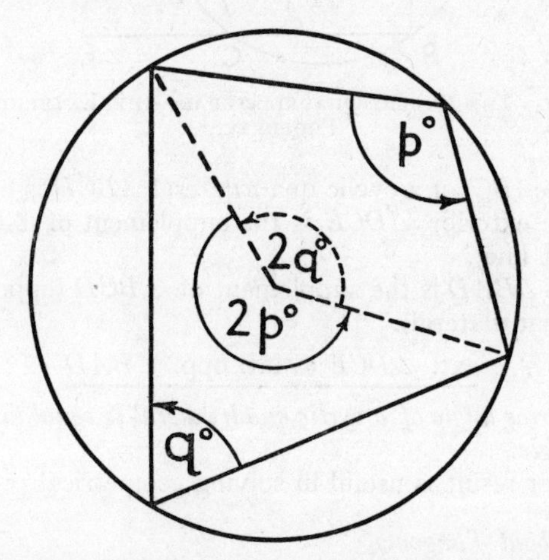

FIG. 85(*a*).—THE OPPOSITE ANGLES OF A CYCLIC QUADRILATERAL.

85(*a*)). But these angles at the centre, $2p°$ and $2q°$ respectively, must add up to one complete revolution, *i.e.,* 360°.

* A quadrilateral whose four vertices lie on a circle.

$$\therefore \quad 2p + 2q = 360$$

i.e.,

$$p + q = 180$$

Hence this pair of opposite angles of the cyclic quadrilateral are supplementary (similarly for the other pair of opposite angles).

A variant of this last proposition is shown in Fig. 85(*b*).

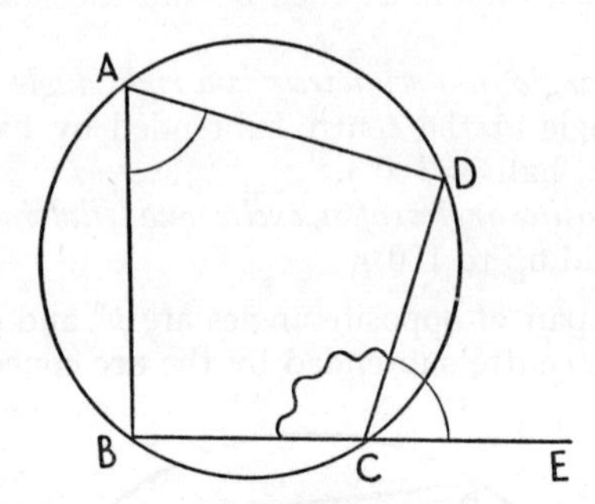

FIG. 85(*b*).—THE CYCLIC QUADRILATERAL—THE EXTERIOR ANGLE PROPERTY.

If one side (BC) of a cyclic quadrilateral $ABCD$ is produced to E, then the exterior $\angle DCE$ is the supplement of $\angle BCD$ (BCE is a straight line).

But also $\angle BAD$ is the supplement of $\angle BCD$ (opposite angles of cyclic quadrilateral).

$$\therefore \quad \underline{\text{ext. } \angle DCE = \text{int. opp. } \angle BAD}$$

i.e., *an exterior angle of a cyclic quadrilateral is equal to the interior opposite angle.*

This latter result is useful in solving geometrical problems.

(c) *Facts about Tangents*

(i) *A tangent must be perpendicular to the radius through its point of contact.*

In Fig. 86(*a*) the cutting line or " secant " XY meets the circle centre O at P and Q. If we join O to the mid-point M of the

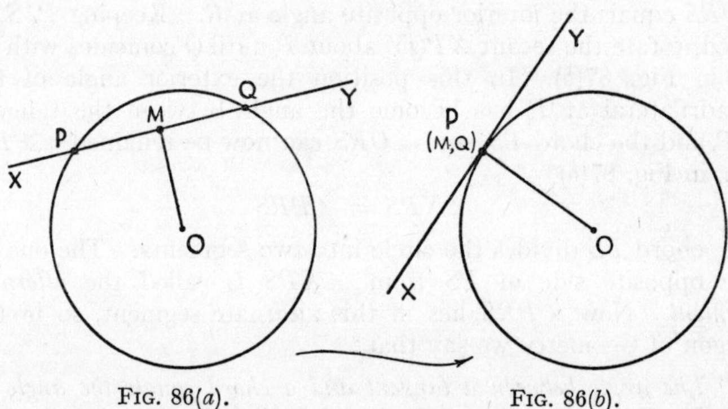

FIG. 86(*a*). FIG. 86(*b*).

chord PQ, we know that OM *must be perpendicular to the chord* PQ, *i.e.*, to the secant XY.

This is true however near Q is to P. If Q were to *coincide* with P, the *secant* $XPQY$ would then become the TANGENT XPY, touching the circle at P (Fig. 86(*b*)). As M is mid-way between P and Q, M will also coincide with P in Fig. 86(*b*), and as OM is perpendicular to XY, this means that OP is perpendicular to the tangent XY in this second diagram. So the tangent at P is perpendicular to the radius OP.

(ii) *The angle between tangent and chord.*

In Fig. 87(*a*) the exterior angle at P of the cyclic quadrilateral

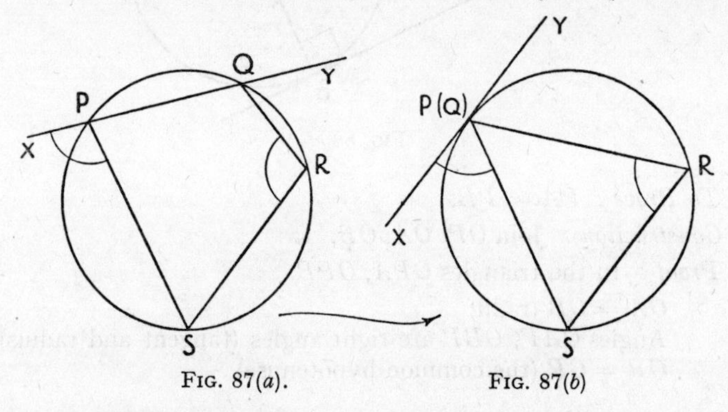

FIG. 87(*a*). FIG. 87(*b*)

$PQRS$ equals the interior opposite angle at R. Keeping P, S, R fixed, rotate the secant $XPQY$ about P until Q coincides with P, as in Fig. 87(*b*). In this position the exterior angle of the quadrilateral at P has become the angle between the tangent XP and the chord PS, and $\angle QRS$ can now be renamed $\angle PRS$, *i.e.*, in Fig. 87(*b*),

$$\angle XPS = \angle PRS$$

The chord PS divides the circle into two segments. The one on the opposite side of PS from $\angle XPS$ is called the *alternate segment.* Now $\angle PRS$ lies in this alternate segment, so in the jargon of Geometry we say that:

" *The angle between a tangent and a chord equals the angle in the alternate segment.*" (See para. 7.10(*b*) below.)

(iii) *Tangents from an external point to a circle are equal.*

In Fig. 88, A and B are the points of contact of the tangents from an external point P to a circle, centre O.

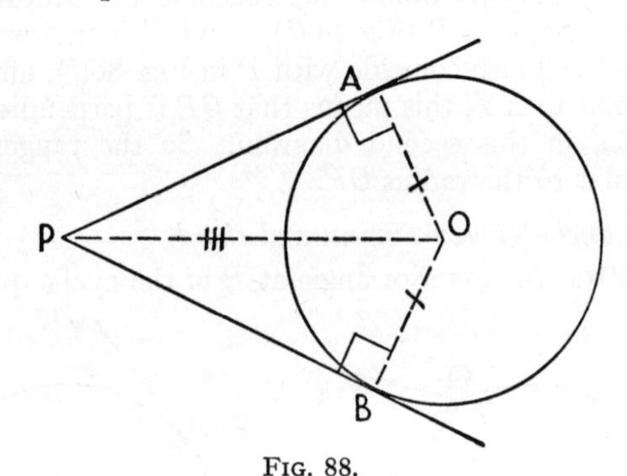

FIG. 88.

To Prove : $PA = PB$.

Construction : Join OP, OA, OB.

Proof : In the triangles OPA, OPB

$OA = OB$ (radii)

Angles OAP, OBP are right angles (tangent and radius)

$OP = OP$ (the common hypotenuse)

∴ the triangles are congruent (right angle, hypotenuse, one side.)

So $PA = PB$

(*Sometimes useful :* The angles at P and O are also equal; *i.e.*, OP bisects the angles at P and O.)

A Construction. As angles OAP, OBP are right angles, A and B both lie on the circle OP as diameter. So to construct accurately the points of contact of the tangents from a given point P to a given circle centre O, it is only necessary to bisect OP and draw the circle on OP as diameter, cutting the original circle at the required points of contact. (Try it !)

(iv) *Common Tangents to two circles.*

We sometimes need to construct the common tangents to two unequal circles (*e.g.*, of radii 2 in. and 5 in., and centres A and B, where $AB = 8$ in.).

Thinking it out. Let us imagine the job done (Fig. 89(*a*)); and

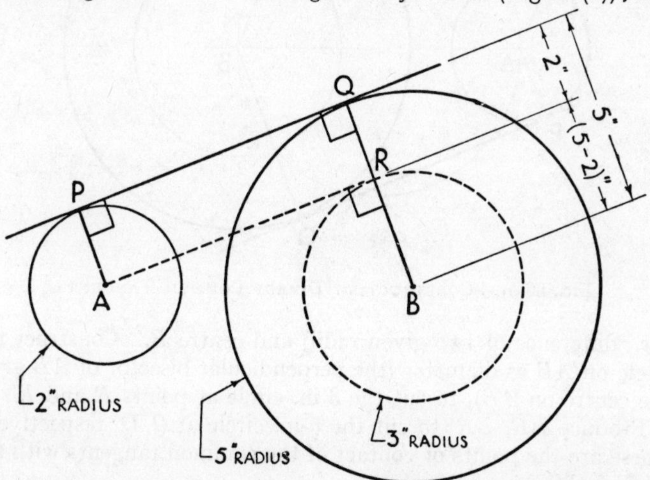

FIG. 89(*a*).—COMMON TANGENTS—THINKING OUT THE PROBLEM.

PQ to be one of the desired common tangents. Then angles APQ, BQP must both be right angles.

If we draw AR parallel to PQ it will cut QB (also at right angles) and complete a rectangle $PQRA$.

So $$QR = PA = 2 \text{ in.}$$

$$\therefore \quad BR = 5 - 2 = 3 \text{ in.}$$

Imagine a circle (dotted in Fig. 89 (*a*)) drawn with centre B and radius 3 in. (*i.e.*, BR), then AR is the tangent at R to the dotted circle.

Doing the Job. After this preliminary detective work, we can now start (Fig. 89(*b*)) by drawing the third circle, radius 3 in.

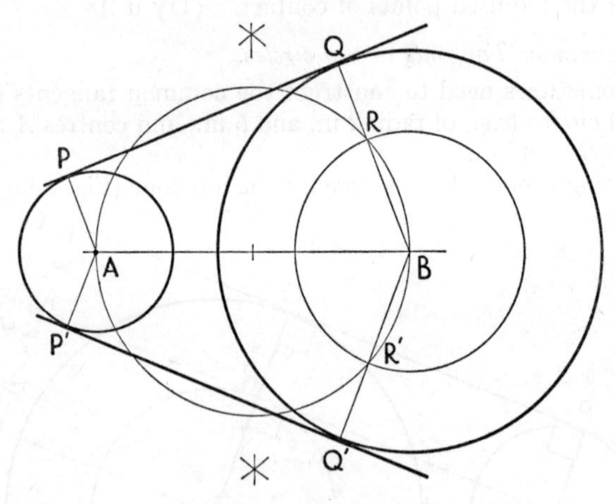

FIG. 89(*b*).—CONSTRUCTING DIRECT COMMON TANGENTS.

(*i.e.*, difference of two given radii) and centre B. Construct the circle on AB as diameter (the perpendicular bisector of AB gives the centre on AB), to cut the 3 in. circle at points R and R'.

Produce BR, BR' to cut the 5-in. circle at Q, Q' respectively. These are the points of contact of the common tangents with the 5-in. circle.

Draw AP, AP' parallel to BQ, BQ' to cut the 2-in. circle at P, P' respectively.

Then PQ, P'Q' *are our required common tangents.*

The above discussion concerns direct common tangents. If transverse common tangents are needed, as in a crossed-belting diagram, a similar argument may be used. The radius of our first construction circle would then be the *sum* of the two given radii. (Try this yourself !)

7.10. Similar Figures

(a) *Scale Models and Scale Drawings*

The keen model engineer, the city architect, or the expert photographer are all familiar with the processes of making models, drawings, or enlargements to scale. Let us remind ourselves of some of the features of scale proportions in both two and three dimensions.

Example 7(*c*). *A model of a new telephone trunk exchange building is built on a scale of* 1 *in.* = 1 *ft. If the model were made of the same materials as the full-size exchange, it would weigh* 4 *cwt. What would be the total weight of materials for the actual building?*

This is an important consideration. Clearly 1 cu. in. of the model is equivalent to 1 cu. ft. of the building. As 1 cu. ft. = 12^3 = 1728 cu. in. the volume, and therefore the weight, of materials for the real building is 1728 times that of the model.

So the estimated weight of materials needed

$$= 1728 \times 4 \text{ cwt.}$$
$$= 345{\cdot}6 \text{ tons}$$

or about 115 3-ton lorry loads.

We see here that the " volume scale " is the cube of the " linear scale ".

Example 7(*d*). *A copper conductor weighs* 200 *lb. per mile, and has a resistance of* 8·8 *ohms per mile loop, and is* 0·29 *cm. in diameter. What would be the weight* (W *lb.*) *per mile, and resistance* (R *ohms*) *per loop mile of a similar conductor* 0·12 *cm. in diameter?*

The linear proportions of the two specimens are in the ratio 12 : 29.

The weight per mile will be proportional to the cross-sectional area.

So $$\frac{W}{200} = \frac{12^2}{29^2} = \frac{144}{841} = 0{\cdot}171$$

$$\therefore \ W = 34{\cdot}2 \text{ (lb. per mile)}$$

On the other hand, the greater the cross-sectional area, the smaller the resistance per loop mile, *i.e.*, it is " inversely " proportional to the cross-sectional areas concerned :

i.e.,
$$\frac{R}{8{\cdot}8} = \frac{841}{144}$$
$$\therefore \quad R = \frac{8{\cdot}8 \times 841}{144} = 51{\cdot}7 \text{ (ohms per mile)}$$

Example 7(e). *Preliminary tests show that a certain coil is capable of dissipating (with the permissible rise in temperature) only one half of the watts loading which it has to carry. In what proportion must its linear dimensions be increased if it is to remain of the same design and proportion? How is its weight affected?*

Since the power dissipated for a given temperature is proportional to the surface area, the surface area must be doubled. To double the surface area the linear dimensions must be increased in the proportion $\sqrt{2} : 1$. Then every unit of surface in the original design is increased to $\sqrt{2} \times \sqrt{2} = 2$ units.

And every unit of volume is increased to $\sqrt{2} \times \sqrt{2} \times \sqrt{2} = 2\sqrt{2}$ or 2·83. So the weight will be 2·83 times the original weight.

(b) *Similar Figures in Two Dimensions*

The above discussion concerned solid, practical things. But similar (scale) figures in two dimensions are also of great importance in geometrical argument.

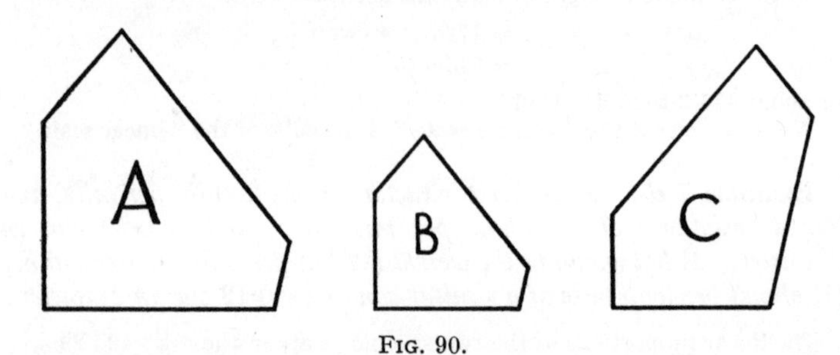

FIG. 90.

Consider the three pentagons in Fig. 90. All three have the same angles at their vertices. But clearly A and B only are the same " shape "—the sides of the pentagon C are " out of proportion ".

Thus for plane figures to be " similar " (scale reproductions of one another) two requirements must be met:

(i) corresponding angles must be equal;

(ii) corresponding sides must be in the same proportion.

We saw in para. 7.6 above that to specify a triangle by its angles (two being sufficient, the third then being fixed) did not fix the *size* of the triangle, but *did* fix its *shape*. So for triangles to be similar it is only necessary for them to be equiangular—the sides will then of necessity be proportional. Several interesting propositions in Geometry can be proved by using this fact.

Example 7(*f*). TP *is a tangent to a circle* PQR, *and* TQR *is a secant, meeting the circle at* Q *and* R. *Prove that* $TP^2 = TQ \cdot TR$*

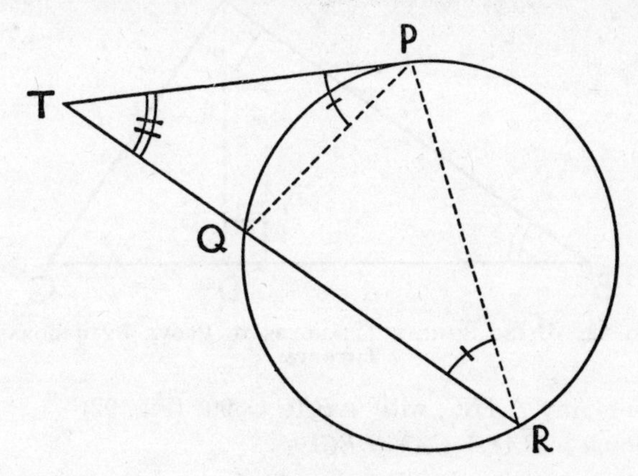

FIG. 91.

Construction: In Fig. 91, join *PQ* and *PR*.

Proof: In the triangles *TPQ*, *TRP*

(i) $\angle TPQ = \angle TRP$ (" tangent and chord ")

(ii) $\angle PTQ = \angle RTP$ (the same angle !)

So the triangles are equiangular, and therefore similar.
So their sides must be in the same proportion.

i.e.,
$$\frac{TP}{TR} = \frac{TQ}{TP}$$
$$\therefore \quad TP^2 = TQ \cdot TR$$

* *Note:* $TQ \cdot TR$ means " *TQ* multiplied by *TR* ".

7.11. A Very Important Theorem of Pythagoras

At the beginning of this chapter we mentioned a very useful property of the sides of a right-angled triangle : that the square of the length of the hypotenuse (the longest side) is equal to the sum of the squares of the other two sides. This, in one particular case—the $3:4:5$ triangle—was known to the Egyptians. Their surveyors were the " rope-stretchers ", who used knotted ropes, marked off in equal divisions, to set out a true right angle by making a $3:4:5$ triangle.

Here is a neat way of proving this result generally.

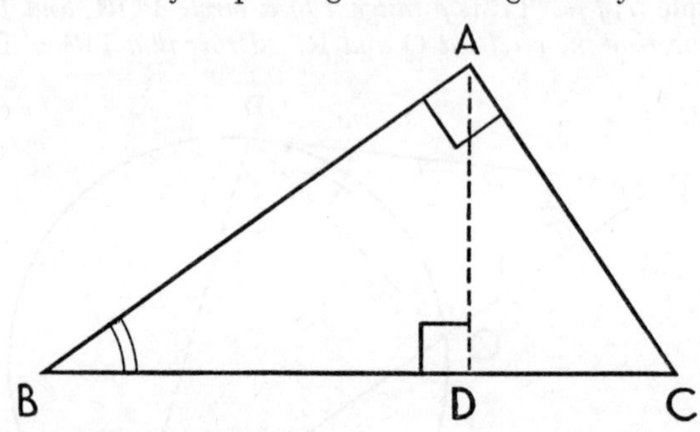

FIG. 92.—USING SIMILAR TRIANGLES TO PROVE PYTHAGORAS' THEOREM.

Given : Any $\triangle ABC$, with $\angle BAC = 90°$ (Fig. 92)

To Prove : $BA^2 + CA^2 = BC^2$

Construction : Draw AD perpendicular to BC, meeting BC at D.

Proof : In the triangles ABC, DBA

$$\angle BAC = \angle BDA \text{ (both right angles)}$$
$$\angle ABC = \angle DBA \text{ (the same angle)}$$

The triangles are equiangular, and hence similar.

The corresponding sides are in the same proportion.

In particular

$$\frac{BA}{BD} = \frac{BC}{BA}$$

whence (multiplying b.s. by $BD \times BA$)

$$\underline{BA^2 = BC \,.\, BD} \quad . \quad . \quad . \quad . \quad . \quad \text{(i)}$$

Similarly (using triangles ACB, DCA),

$$\underline{CA^2 = CB \,.\, CD} \quad . \quad . \quad . \quad . \quad . \quad \text{(ii)}$$

Adding the two results

$$\begin{aligned} \underline{BA^2 + CA^2} &= BC \,.\, BD + CB \,.\, CD \\ &= BC \,.\, BD + BC \,.\, DC \\ &= BC(BD + DC) \text{ (" taking out the common factor ")} \\ &= BC \,.\, BC \\ &\underline{= BC^2} \end{aligned}$$

which proves Pythagoras' Theorem.

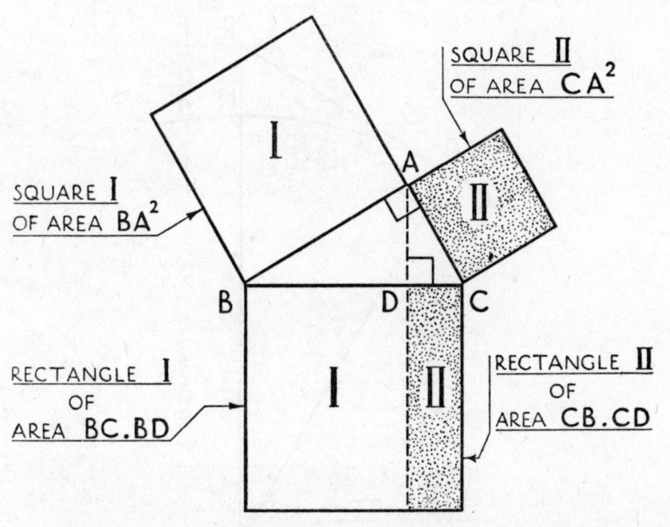

FIG. 93.—A PICTORIAL VIEW OF PYTHAGORAS' THEOREM.

Fig. 93 suggests how the main steps of the above proof may be represented pictorially. BA^2 and CA^2 represent the area of the squares I and II. While the products $BC . BD$ and $CB . CD$ represent the areas of the two rectangles I and II into which AD produced divides the square on the hypotenuse.

7.12. A Practical Illustration of the Use of Pythagoras' Theorem

Example 7(g). *What would you expect to be the extreme range seawards of a radar set installed at the top of a cliff* 540 *ft. above the sea ?*

We will assume the radius of the earth to be 3960 miles. Then in Fig. 94 RN represents the height of the radar installation above sea-level. RP represents the distance x miles from R of the visible horizon (presumably the extreme range for radio-location). Then RP is the tangent from R to the sections of the earth's surface—the arc PN of a circle radius 3960 miles.

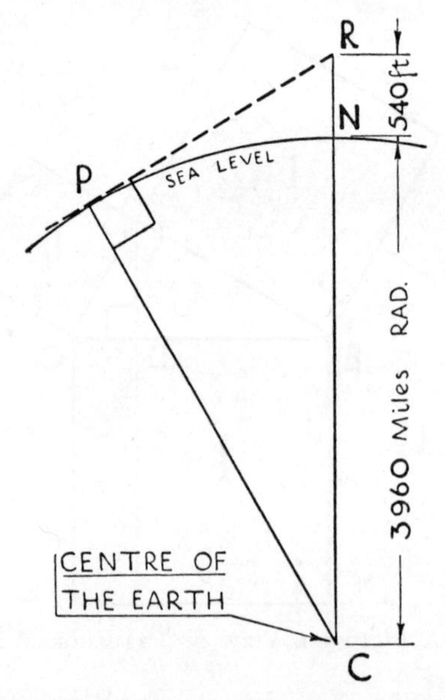

FIG. 94.—THE RANGE OF A RADAR STATION.
(Diagram NOT to scale.)

Now $RN = 540$ ft.
$= 180$ yd.
$= \frac{9}{88}$ miles

Using the Theorem of Pythagoras in the right-angled $\triangle CPR$

$$PR^2 + CP^2 = CR^2$$
$$\therefore \quad x^2 + (3960)^2 = (3960\tfrac{9}{88})^2$$
$$\therefore \quad x^2 = (3960\tfrac{9}{88})^2 - (3960)^2$$
$$= 7920\tfrac{9}{88} \times \tfrac{9}{88} \qquad \text{(difference of squares)}$$
$$\simeq 7920 \times 0{\cdot}102$$

Whence $x = 28{\cdot}4$ (3 significant figures)

i.e., *the extreme range of operations is approximately* 28 *miles from the transmitter.*

(See Exercise 7(c), Nos. 36, 37.)

7.13. A Practical Use of the Similar Triangles Technique in Optics

Example 7(*h*). *A lens has a focal length of* 6 *in.* (i.e., *parallel light is brought to a focus* 6 *in. from the lens*). *If a bright object is placed* 8 *in. from the lens, find the correct position of a screen to give a clear image, and the magnification produced.*

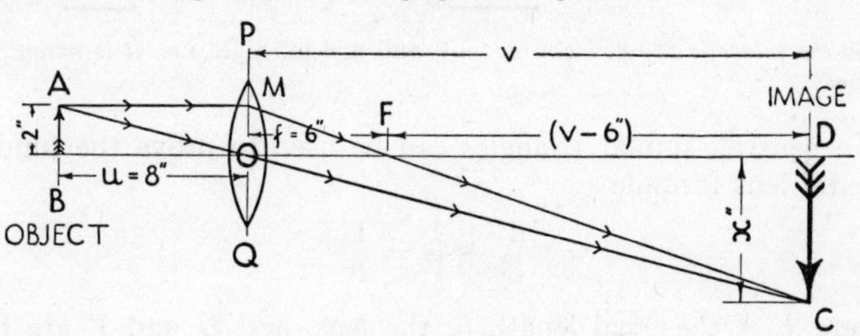

FIG. 95.—SIMILAR TRIANGLES APPLIED TO A PROBLEM IN OPTICS.

Fig. 95 shows a side-view of the required set-up. AB (2 in. high) represents the object, and PQ the cross-section of the lens.

A ray of light (AM) at right angles to the plane of the lens will pass through the focus F (6 in. from the lens). A second ray of light, passing through the centre O of the lens, will not alter course, and will meet the ray MF produced at a point C. CD is then the required image, and the "in-focus" position of the screen.

We could determine the distance of the screen (V in.) from the lens by scale drawing, and also measure CD (x in.) and find its magnification.

But we could also calculate these data from the geometry of the diagram:

(*a*) Triangles ODC, OBA are similar

$$\therefore \quad \frac{DC}{BA} = \frac{OD}{OB}$$

i.e.,

$$\frac{x}{2} = \frac{V}{8}$$

or

$$\underline{x = \frac{V}{4}} \quad \ldots \ldots \ldots \quad \text{(i)}$$

(*b*) Triangles CDF, MOF are also similar

$$\frac{CD}{MO} = \frac{DF}{OF}$$

i.e.,

$$\frac{x}{2} = \frac{V-6}{6}$$

$$\therefore \quad \underline{x = \frac{V-6}{3}} \quad \ldots \ldots \ldots \quad \text{(ii)}$$

Equating the two values of x, from (i) and (ii),

$$\frac{V-6}{3} = \frac{V}{4}$$

$$\therefore \quad 4V - 24 = 3V$$

$\therefore$ $\underline{V = 24}$ and substituting in (i), $\underline{x = 6}$

So *the image is* 24 *in. from the lens,* and is 6 in. high. *i.e., it is magnified* 3 *times.*

In general, similar triangles can be used to prove the fundamental lens formula

$$\frac{1}{U} + \frac{1}{V} = \frac{1}{F}$$

where F is the focal length of the lens, and U and V are the object and image distances from the lens. (See if you can prove this !)

7.14. Further Areas

In Chapter 1 we discussed the areas of the commoner geometrical figures: rectangles, triangles, and circles. All practical problems involving polygons or circles need no further knowledge. There are, however, two important types of area which deserve further thought.

(a) *The Area of a Trapezium*

A trapezium is a quadrilateral with one pair of opposite sides parallel, such as the end view of a lean-to shed or the longitudinal cross-section of a swimming-pool.

$ABCD$ in Fig. 96 is a typical trapezium with AB (a units) parallel to DC (b units). It is clearly divisible into two triangles, ABC and ACD with bases a and b units and the same perpendicular height h.

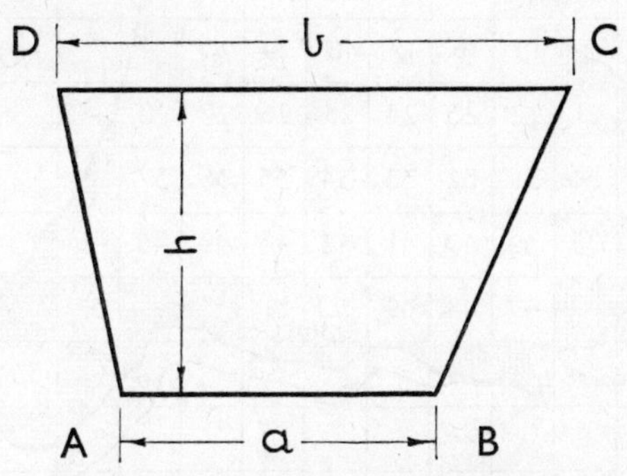

FIG. 96.—THE AREA OF A TRAPEZIUM.

$$\begin{aligned}\text{Then the Area } ABCD &= \text{Area } ABC + \text{Area } ACD\\ &= \tfrac{1}{2} \times ah + \tfrac{1}{2} \times bh\\ &= \frac{h}{2}(a + b) \text{ or } \frac{a + b}{2} \times h\end{aligned}$$

i.e., the area is " *half the sum of the parallel sides times the distance between them* ".

Alternatively, it is " the average width times the height ".

(b) *The Area of an Irregular Figure*

There are three practical methods suggested at this stage:

(i) *Trace the area on to squared paper and " count squares "*. For example, to find the area of an English county (*e.g.*, East Riding of Yorkshire) trace the county boundary from a $\frac{1}{4}$-in. map on to $\frac{1}{10}$-in. squared paper (Fig. 97) and count squares,

The small squares are each 0·01 of a square inch: all portions or small squares less than one-half are ignored in the counting.

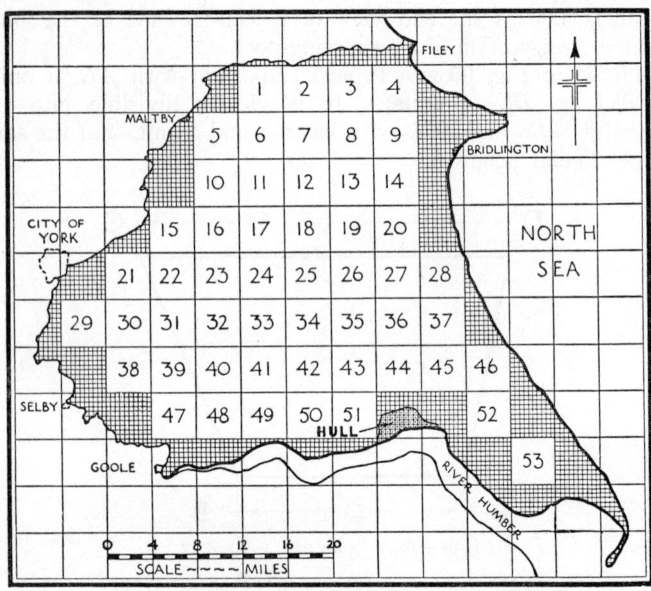

FIG. 97.—THE EAST RIDING OF YORKSHIRE.

Counting the Squares

	No.	Area on Map.
Inch-squares . .	53	53 sq. in.
Small squares . .	1952	19·52 sq. in.
Total . .		72·52 sq. in.

But 1 sq. in. on the map represents 4×4 sq. miles on the ground.

* Area of East Riding $= 16 \times 72{\cdot}52$ sq. miles

$= \underline{1160 \text{ sq. miles}}$ (to 3 sig. figs.)

* The correct area is 741,172 acres, or 1158 sq. miles, to the nearest sq. mile.

(ii) *By Weighing.* Cut out the area concerned in cardboard or tinplate, and weigh. Compare with the weight of a known rectangular area (*e.g.*, $2 \times 5 = 10$ sq. in.) of cardboard; hence find by proportion the unknown area. This is quite a quick and accurate method.

(iii) *The Trapezoidal Rule.* This method is particularly useful when we have an area to find under a graph, bounded by the graph axes or definite ordinates. Figs. 98(*a*) and 98(*b*) demonstrate the method of working, and Fig. 98(*c*) gives a practical example of its use.

In Fig. 98(*a*) we have divided the area *OAPN* we wish to find into a convenient number of strips (seven in this case) of equal width *K* units; *i.e.*, $ON = 7K$. The parallel sides of these strips are $y_0, y_1, y_2, \ldots$ etc., up to y_7.

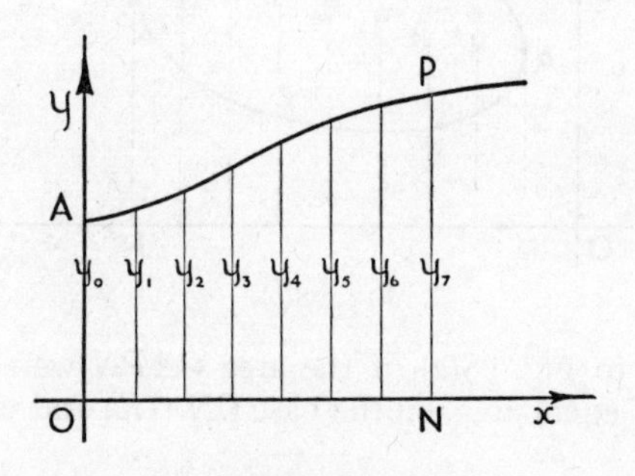

FIG. 98(*a*).

Each strip is very nearly a *trapezium*. So as a close approximation the first strip has an area $\frac{K(y_0 + y_1)}{2}$, by the formula for the area of a trapezium given on p. 229.

Adding up the areas of all seven strips,

$$\text{Area } OAPN \simeq \frac{K(y_0 + y_1)}{2} + \frac{K(y_1 + y_2)}{2} + \ldots + \frac{K(y_6 + y_7)}{2}$$

$$= K\left[\frac{y_0 + y_7}{2} + y_1 + y_2 + y_3 + \ldots + y_6\right]$$

Let us express this result in words—it looks rather formidable in algebra.

$$\begin{matrix}\textbf{The} \\ \textbf{required} \\ \textbf{area}\end{matrix} = \left\{\begin{matrix}\textbf{Width of} \\ \textbf{each} \\ \textbf{strip}\end{matrix}\right\} \times \left\{\begin{matrix}\textbf{Average of} \\ \textbf{first and last} \\ \textbf{ordinates}\end{matrix} + \begin{matrix}\textbf{Sum of the} \\ \textbf{remaining} \\ \textbf{ordinates}\end{matrix}\right\} \quad (1)$$

With the closed area of Fig. 98(*b*) the same argument applies, but in this case $y_0 = y_8 = 0$, so the approximation to this closed area reduces to

$$K[y_1 + y_2 + y_3 + y_4 + y_5 + y_6 + y_7]$$

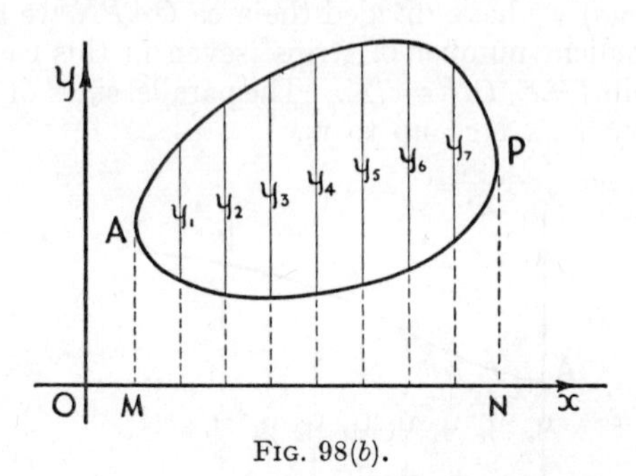

FIG. 98(*b*).

Returning to Fig. 98(*a*), if the area *OAPN* were replaced by a rectangle of equal area, on the base *ON* (7*K*) and with height *h*, then

$$7Kh = K\left\{\frac{y_0 + y_7}{2} + y_1 + y_2 + y_3 + y_4 + y_5 + y_6\right\}$$

So this " average height " *h* is given by

$$h = \frac{\left\{\frac{y_0 + y_7}{2} + y_1 + y_2 + y_3 + y_4 + y_5 + y_6\right\}}{7}$$

In words:

$$\begin{matrix}\textbf{The average} \\ \textbf{height of} \\ \textbf{the curve}\end{matrix} = \frac{\begin{matrix}\textbf{Average of first} \\ \textbf{and last ordinates}\end{matrix} + \begin{matrix}\textbf{Sum of the} \\ \textbf{remaining ordinates}\end{matrix}}{\textbf{Number of strips}} \quad . \quad (2)$$

In practice, it is often a little neater to calculate the average height of the curve above the base-line first, and then multiply by the length of the base-line to obtain the area under the curve. This we shall do in the example which follows.

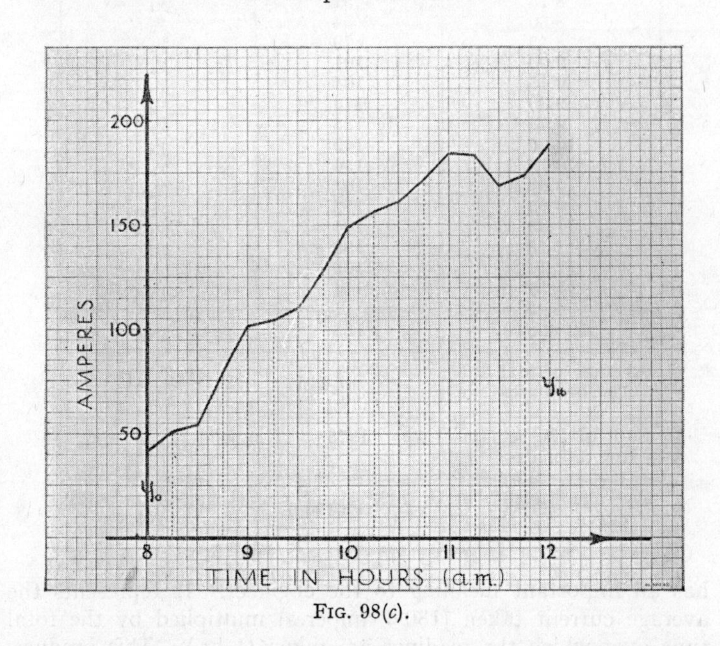

FIG. 98(c).

Fig. 98(c) was obtained by plotting the ammeter reading for an exchange battery at quarter-hour intervals during a particular morning. The ordinates $y_0, y_1, \ldots y_{16}$ are thus the current in amperes at times 8 a.m., 8.15, etc., etc., up to 12 noon, as shown in the table below.

As the ordinate at any given time represents the current drain from the battery at that instant, it means that the average height of the curve represents the average current, 131 amperes (3 significant figures), taken from the battery during this 4-hr. period.

The area under the curve, between 8 a.m. and 12 noon, also

Time (a.m.)	Ammeter reading	Ordinates
8.00	42	y_0
8·15	51	y_1
8·30	54	y_2
8.45	79	y_3
9.00	101	y_4
9.15	104	y_5
9.30	110	y_6
9.45	128	y_7
10.00	149	y_8
10.15	156	y_9
10.30	161	y_{10}
10.45	172	y_{11}
11·00	184	y_{12}
11·15	183	y_{13}
11.30	169	y_{14}
11·45	174	y_{15}
12 noon	188	y_{16}
Average of first and last ordinates	$\frac{230}{2} = 115$	$\frac{y_0 + y_{16}}{2}$
Sum of the remaining ordinates	1975	$y_1 + \cdot\cdot\cdot + y_{15}$
Average height of curve	$\frac{2090}{16} \simeq 130{\cdot}6$	(Using formula (2) above.)

has an important meaning to the engineer. It represents the average current taken (130·6 amperes) multiplied by the total time over which the readings are taken (4 hr.). This product, 522 ampere-hr. (3 significant figures) is a measure of the total drain from the exchange battery during this 4-hr. period.

Now use the formula (1) above to check that the area under the curve is approximately 522 ampere-hr. (The width of each strip is $\frac{1}{4}$ hr.)

EXERCISE 7(*c*)

1. A chord of length 12 cm. is at a distance of 8 cm. from the centre of a circle; calculate its radius.

2. A circle is drawn circumscribing a $\triangle ABC$ in which $AB = AC = 13$ cm., $BC = 10$ cm. Calculate its radius.

3. Two circles of radii 3 in. and 5 in. are described with the same centre O. A line $PQRS$ cuts one at P, S and the other at Q, R. If $QR = 2$ in., calculate PQ. How far is this line from O?

4. A straight stick 4 in. long rests inside a rough hemispherical bowl, 5 in. in diameter. How far from the centre of the bowl is the mid-point of the stick? On what surface does this mid-point move as the stick is moved in contact with the bowl?

5. How many axes of symmetry have the following: (*a*) an equilateral triangle; (*b*) a square; (*c*) a cone; (*d*) a rectangular block; (*e*) a cylinder; (*f*) a cube?

6. A, B, C are three points on a circle, centre O. If $\triangle AOB = 100°$, $BOC = 130°$, prove that $\triangle ABC$ is isosceles.

7. By constructing suitable angles at the centre, inscribe in a circle radius 4·2 cm. a triangle having angles 32° and 106°.

8. In Fig. 84 above, supposing the line AP were to cut the radius OB (between O and B), how would the proof of this fundamental theorem be altered?

9. $PQRS$ is a quadrilateral inscribed in a circle; PQ is a diameter; $\angle PSR = 127°$; calculate $\angle QPR$.

10. Two chords DE, FG when produced meet at X; $\angle XDG = 31°$, $\angle DXF = 42°$; find $\angle XEF$.

11. $APQRB$ is a pentagon inscribed in a circle, with AB as a diameter. Prove $\angle APQ + \angle QRB = 270°$.

12. AP and AQ are diameters of two circles which intersect at A and B. Prove that P, B, and Q lie on a straight line.

13. AB is any chord of a circle, centre O. A second circle is drawn with AO as diameter cutting AB at P. Show that P must be the mid-point of AB.

14. Fig. 99 represents a "spacer" used in building up a contact bank for an automatic switch. The inner curve is supposed to be a semicircle

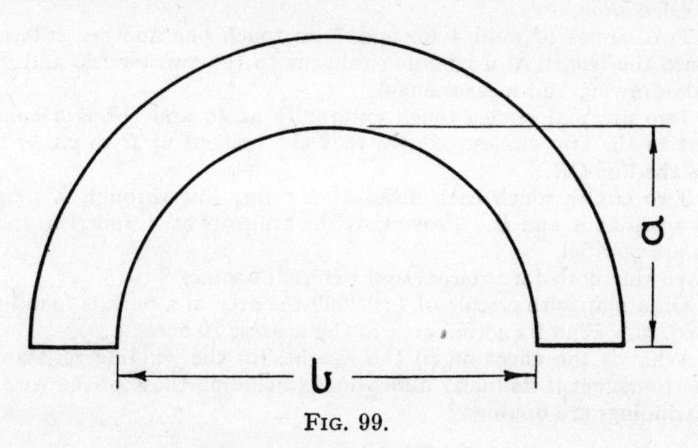

FIG. 99.

2·5 in. in radius. The dimensions a and b have been checked and found to be correct, *i.e.*, 2·500 and 5·000 in. respectively.

Suggest a method of checking whether the inner curve is truly circular, without making a special gauge for the purpose.

15. Two unequal circles intersect at A and B, and parallel lines PAQ, RBS are drawn cutting the circles at P, R and Q, S. Using the exterior-angle property of a cyclic quadrilateral (or otherwise) prove that PR must be parallel to QS (*i.e.*, $PQSR$ is a parallelogram).

16. PQR is a minor arc of a circle, centre O; the tangents at P and R to this circle meet at S. If $\angle PQR = 117°$, calculate $\angle PSR$.

17. TP is a tangent to a circle PQR, and TQR is a straight line (a "secant") cutting the circle at Q and R. If $\angle TPR = 114°$, calculate $\angle PQR$.

18. From a point P outside a circle a tangent PT and a secant PQR are drawn. Prove that $\angle PQT = \angle PTR$.

19. A, B, C are points on a circle. Through B a line is drawn parallel to CA to cut the tangent at A in the point P. Prove that $\angle APB = \angle ABC$.

20. The sides BC, CA, AB of a triangle touch a circle at L, M, N. If $\angle ABC = 48°$, $\angle BCA = 76°$, calculate the angles of $\triangle LMN$.

21. Construct the two tangents from a point P to a circle, centre C and radius 4·2 cm., if $PC = 7·9$ cm. Measure their lengths, and check these by calculation.

22. Two pulleys are 8 in. and 6 in. in diameter, and they are mounted with their centres 11 in. apart. By drawing to scale and measurement find the total length of belting required (*a*) with direct belt drive, (*b*) with crossed belting.

(Measure the angle subtended at the centre of each pulley by the curved portion of belting at the pulley concerned; then this curved length of belting is a proportion of the total circumference of this pulley.)

23. Check the lengths of the *straight* portions of the belting in Question 22 by *calculation.*

24. Two circles of radii 4 in. and 7 in. touch one another externally. Calculate the length of a common tangent to the two circles, and verify by scale drawing and measurement.

25. Two unequal circles touch externally at P, and QR is a common tangent to the two circles. Prove that the tangent at P to either circle bisects the line QR.

26. Two circles touch each other at X; any line through X cuts the circles again at A and B. Prove that the tangents at A and B to the two circles are parallel.

(Prove this both for internal and external contact !)

27. On a map with a scale of 1 : 25,000 the area of a park is found to be 10·24 sq. in. Find its actual area to the nearest 10 acres.

28. What is the effect on (i) the weight, (ii) the winding resistance of an electromagnet if its linear dimensions (including those of the wire used in its winding) are doubled?

29. The dimensions of a pre-war dining-room table were: length 6 ft., width 4 ft., height 3 ft. 6 in. For a utility range of furniture a table of similar proportions is to be made 4 ft. in length. What are to be its width and height? In what ratio is the area of the table-top reduced? If the same hardwood is to be used, calculate the probable weight of the utility table, if the pre-war edition weighed 108 lb.

30. The linear dimensions of a radio valve are increased by 50%. Assuming its proportions remain unaltered, by what percentage will the maximum emission of the cathode be increased? (This emission is proportional to the *area* of the emitting surface of the cathode.) By how much is the weight of the valve increased?

31. ABC is an isosceles triangle inscribed in a circle centre O; $AB = AC$. The diameter AON is drawn, cutting BC at M, and BN, CN are joined.

Prove that $\triangle$s BMN, ABN are similar; hence deduce that BN would be a tangent to the circle through A, B, M.

How many more triangles in your diagram are similar to $\triangle BMN$?

32. (See para. 7.13.) If a lens has a focal length of 8 cm., find by scale drawing the size and position of the sharp image of an object 3 cm. high placed (*a*) 12 cm., (*b*) 18 cm. from the lens. What is the magnification produced in each case?

Verify by calculation, using similar triangles.

33. With the lens of Question 32, where will the image be when the object is placed 6 cm. from the lens? What physical interpretation can you give to this result? What use of the lens does this illustrate?

34. Check the results of the two previous examples using the formula

$$\frac{1}{U} + \frac{1}{V} = \frac{1}{F}.$$

What algebraic result does Question 33 give?

35. Prove that the diagonals of a rhombus (a parallelogram with adjacent sides equal) are perpendicular. If its diagonals are 16 and 12 in. long, calculate the length of a side of the rhombus. What is its area?

36. How high above the water must a ship's "look-out" man be if he is to see 12 miles to the horizon? (Assume data of para. 7.12.)

37. Show that if this look-out be h ft. above the water, this horizon is d miles away, where to a very close approximation $d = \sqrt{\frac{3h}{2}}$. Verify your answer to Question 36 by substituting in this formula.

38. A brick arch with a total span of 12 ft. is to be an arc of a circle, and have its mid-point 2 ft. above the extremities. Calculate the radius of the required arc (let it be r ft.).

39. If the foot of a 12-ft. ladder is 2 ft. from a wall, how high will it reach up the wall? If the foot of the ladder is drawn 18 in. farther away from the wall, how much lower will the top of the ladder be?

40. Fig. 100 illustrates the operation of a simple switchboard key. The springs A, B and C, D are brought into contact when the key lever turns the cam K, whose cross-section is a circle with 2 equal and opposite segments removed. Each of these segments subtends 90° at the centre of the circle.

If the "throw" or travel of the inner springs is to be 2 mm., what radius must be chosen for the circular parts of the cam? What must be the normal separation between the inner springs B and C? What is the least angle the cam must turn through in order to achieve the full "throw"?

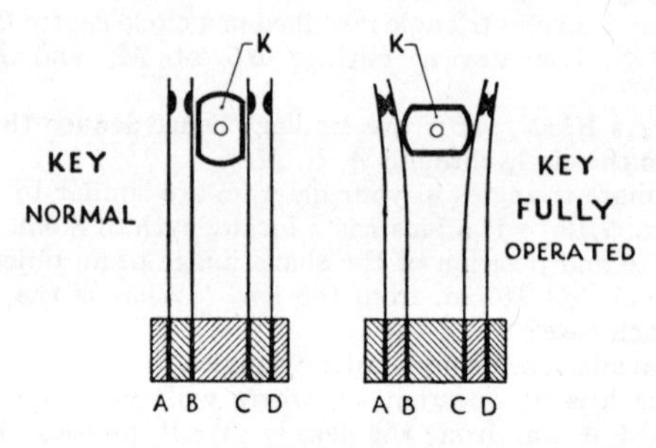

FIG. 100.

41. If the parallel sides of a trapezium are 8 cm. and 10 cm., and its area is 72 sq. cm., how far apart are these parallel sides?

42. The vertical cross-section of a 150-ft. length of sea-wall tapers from 12 ft. wide at the bottom to 4 ft. wide at the top. Its vertical height is 16 ft. Calculate the volume of material used in constructing this length of wall.

43. A swimming-pool is 25 yd. wide by 40 yd. long, and its depth varies uniformly from 9 ft. at the deep end to 2 ft. 6 in. at the shallow end. How many gallons of water are required to fill the pool? (A cu. ft. of water is approximately $6\frac{1}{4}$ gall.)

44. $ABCDE$ is a field of irregular shape bounded by straight hedges. AB, BC, CD, DE, EA, AD, BD are 221, 188, 74, 192, 134, 231, 227 yd. respectively.

Draw the field to scale on graph paper, and hence find its area in acres and sq. yd.

45. Using two pins, string, and a pencil, construct an ellipse (see Exercise 7(*b*), Question 58) on squared paper. By counting squares verify that its area is πab, where $2a$ and $2b$ are the lengths of the major and minor axes of the ellipse.

46. Repeat Question 45, using the trapezoidal rule to calculate the area.

47. The speed of an electric train is recorded every minute in its run between two suburban stations.

t min. . .	0	1	2	3	4	5	6	7	8
v m.p.h. . .	0	6·8	16·5	24·2	28·1	33·7	23·8	7·2	0

Plot a graph of t against v. Use the trapezoidal rule to find the area under this graph between $t = 0$ and $t = 8$. Hence estimate the train's average speed between the two stations.

From your result, calculate the distance in miles between the stations, and show this is approximately the area under the curve, if the time-base be measured in *hours*.

48. Check your results in Question 47 by counting squares on the graph paper. (Note that the area under a v, t curve is *always* equal to the distance travelled: it represents *average speed* $\times$ *time*.)

49. The volume v cu. in. of an amount of gas is recorded for different values of its pressure measured as p in. of mercury.

v . .	160	140	120	100	80	60
p . .	18·7	21·4	25	30	37·5	50

Plot p against v. To compress the gas from 160 to 60 cu. in. requires an expenditure of energy proportional to the area under the graph between $v = 60$ and $v = 160$. Use the trapezoidal rule to evaluate this area.

50. The depth of an old cable conduit below the ground is found to vary as follows:

Horizontal distance x ft. from test point .	0	50	100	150	200	250	300	350
Depth of conduit, y ft.	2·0	2·5	2·8	3·1	3·7	3·2	2·5	1·6

Use the trapezoidal rule to calculate the average depth of the conduit below the present ground surface, and hence find the number of cu. yd. of earth to be excavated to lay bare this 350-ft. length of conduit, assuming a trench width of 14 in.

Note on the Mid-Ordinate Rule

An alternative to using the Trapezoidal Rule is to measure (or calculate) the middle ordinate of each strip. The average height of the curve is then taken to be the average of these "mid-ordinates".

The reader is invited to check the solutions of Nos. 47, 49, 50 above by this method. In practice the trapezoidal rule is preferable.

CHAPTER 8

AN INTRODUCTION TO TRIGONOMETRY

8.1. The Sine Ratio

We are all familiar with gradients of steep hills (*e.g.*, Countisbury Hill is 1 in 3½), and of the more gradual railway inclines

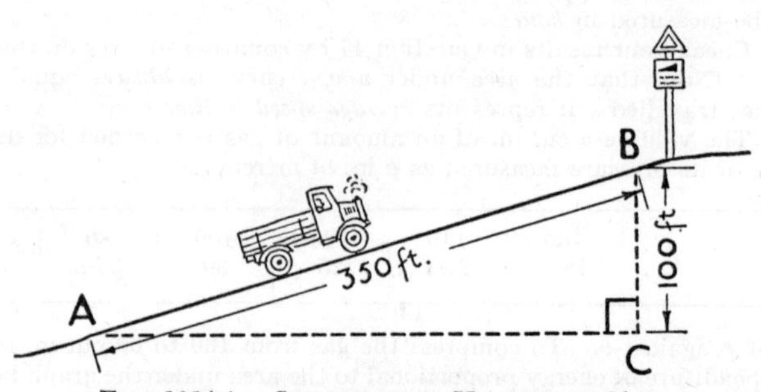

FIG. 101(*a*).—COUNTISBURY HILL—1 IN 3½.

(*e.g.*, 1 in 60), being specified by a *ratio* rather than by an angle of elevation. (Figs. 101(*a*) and (*b*) refer to this.)

The use of this " gradient ratio " is adopted because it is so

FIG. 101(*b*).—A RAILWAY GRADIENT OF 1 IN 60.

simple to measure. Thus a railway engineer measures 120 ft. (AB) with a chain along the railway track and then uses sighting rods to measure the vertical rise BC (exactly 2 ft.). So the track rises a foot for every 60 ft. of track—or just " 1 in 60 ". A

French engineer would observe it rose a metre for every 60 metres of track—note that the " 1 in 60 " ratio is quite independent of the units of measurement.

In studying Trigonometry we call this gradient-ratio the SINE of the angle of slope of the hill or railway.

Thus in Fig. 101(*a*) (Countisbury Hill) this ratio

$$= \frac{100}{350} \text{ (or 1 in } 3\tfrac{1}{2})$$
$$= \frac{2}{7}$$
$$= 0{\cdot}2857 \text{ (4 significant figures)}$$

From a Table of Sines (see the end of this volume) we find that the angle of slope of Countisbury Hill is 16° 36′.

In mathematical shorthand,

$$\underline{\sin BAC = 0{\cdot}2857}$$

Example 8(*a*). *A mountain path with average gradient* 1 *in* 4 *is* 1320 *yd. long. What height does one rise in walking up it? What is the angle of elevation of the top as viewed from the bottom of the hill?*

In this case, for each yard vertical rise one walks 4 yd. up the hill.
So 4 yd. up the hill means 1 yd. rise.
So by proportion 1320 yd. up the hill means $\frac{1320}{4}$ yd., *i.e.*, 330 yd. rise.
So the rise is 990 ft.
If the angle of elevation is α,* then we may write

$$\sin \alpha = \tfrac{1}{4} = 0{\cdot}2500$$

and from the Table of Sines, α is found to be 14° 29′ (nearest minute). (See para. 8.6 below on the use of these tables.)

8.2. How to Make a Home-made Sine Table

From the above paragraph we see that the SINE of any angle is a ratio in a right-angled triangle. Thus in Fig. 102, if $\triangle ABC$ is right-angled at C, and we denote $\angle BAC$ by α, then

$$\sin \alpha = \frac{BC}{AB} = \frac{\text{opp. side}}{\text{hypotenuse}}$$

Note: BC is the side opposite to the marked angle α.

---

* It is customary to use the Greek letters α, β, θ, ϕ, etc., to denote angles.

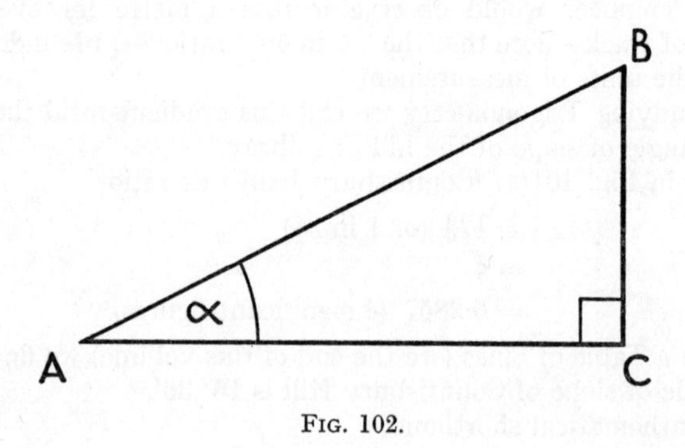

FIG. 102.

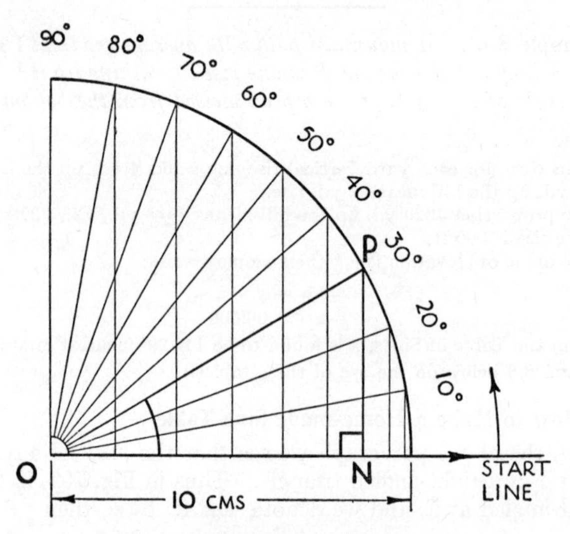

FIG. 103.—MAKING A TABLE OF SINES.

In Fig. 103 we have chosen a convenient moving hypotenuse OP, 10 cm. in length, to use in different positions and thus find the SINE of several angles—*i.e.*, for different angles the opposite side PN may be measured with dividers. Thus when the angle is 30°, PN is found to be exactly 5 cm., so

$$\sin 30° = \frac{PN}{OP} = \frac{5}{10} = \frac{1}{2} \text{ exactly}$$

Whereas for 70°, PN is 9·40 cm. (3 significant figures).

$$\therefore \sin 70° = \frac{9{\cdot}40}{10} = 0{\cdot}940$$

We obtain the following table for intervals of 10°:

Angle	0	10°	20°	30°	40°	50°	60°	70°	80°	90°
Sine	0	0·174	0·342	0·5	0·643	0·766	0·866	0·940	0·985	1

(Compare these values with the table printed at the end of the book.) Two interesting values should be noted:

(i) When $\alpha = 0$, $PN = 0$, so $\sin 0° = 0$
(ii) ,, $\alpha = 90°$, $PN = 10$, so $\sin 90° = 1$

From such a diagram we can obtain directly the sine of any angle between 0° and 90°.

A graph may be drawn of our values (see Fig. 116), from which we can read off the sines of intermediate values of α.

In practice, we rely on printed tables to find the sines of given angles, or conversely to translate a gradient-ratio into an angle.

Find by construction the sines of 45°, 78°, 32°, and check your results from Sine Tables.

8.3. The Tool-maker's Sine-bar

The sine-ratio and the printed Tables of Sines are not to be regarded as just a mathematician's " toy ". They are valuable in precision work in the tool-room in setting out or checking angles with high accuracy. A typical example is the use of the " sine-bar " and its associated Johannsen gauges (or " Joeys ").

Several patterns of sine-bar are in common use; they are all designed on the principle of Fig. 104(*a*). Their characteristics are:

(*a*) the top and bottom edges must be true and parallel with the centre line of the two plugs A and B;

(*b*) the centres of the plugs must be exactly 10 in. (in smaller models, 5 in.) apart;

(*c*) the diameters of the plugs must be exactly equal.

Fig. 104(*b*) illustrates the setting out of a particular angle (18° 42′) on a piece of work. From the tables sin 18° 42′ = 0·3206.

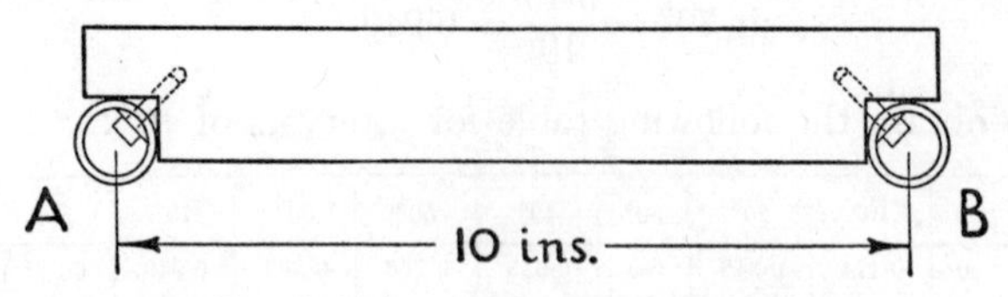

FIG. 104(*a*).—A 10-INCH SINE-BAR.

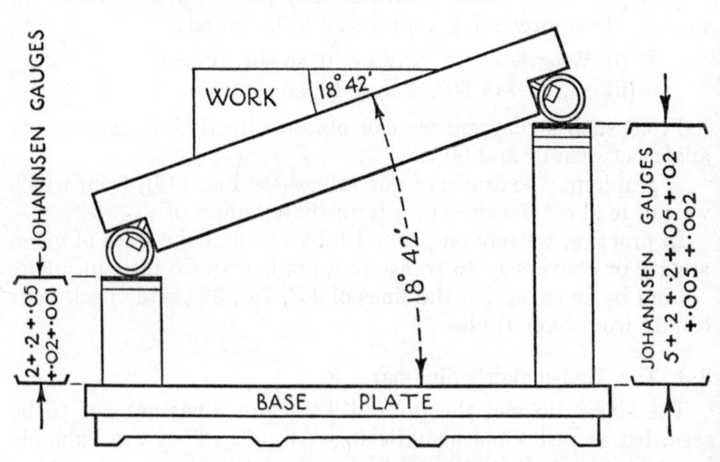

FIG. 104(*b*).—THE SINE-BAR IN ACTION.

The top edge of the work is parallel to the machined face of the base-plate (see Fig. 104(*b*)). If we lift plug *B* exactly 3·206 in. above *A* (Fig. 104(*c*)) the ratio $\dfrac{BC}{AB} = \dfrac{3{\cdot}206}{10}$ or 0·3206, so the slope α of the line AB (and therefore of the top face of the sine-bar) is then given by $\sin \alpha = 0{\cdot}3206$. Thus α is exactly 18° 42′.

The Johannsen gauges are " blocks " and " slips ", of thickness 5, 2, 1, 0·5, 0·2, 0·1 in.; 0·05, 0·02, 0·01 in., and so on down to a thousandth of an inch. (The same arrangement as the weights

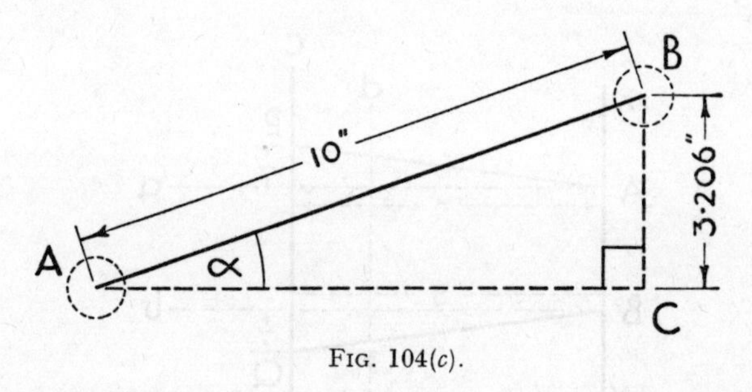

FIG. 104(c).

of a chemical balance.) In Fig. 104(b), stud A needs to be raised by the gauges a height 2·271 in. above the base-plate; then stud B must be raised a further 3·206 in., *i.e.*, 5·477 in. in all. The top edge of the sine-bar will then be inclined at exactly 18° 42′ to the top edge of the work.

8.4. The Mathematical Meaning of " Slope " or " Gradient " : the Tangent Ratio

(a) *The Cabinet-maker Sets out his Dove-tail Joints*

In making dove-tail joints it is essential that the tongues and grooves should be cut at precisely the same angle with the surface of the wood, to ensure perfect fit and a true right angle at the corner under construction.

Fig. 105 suggests how the cabinet-maker sets out his work. XY represents in section the surface of the wood to be joined. Pencil lines a and b are drawn perpendicular to the edge a measured distance apart (according to the size of the job). A third line c is drawn parallel to XY and at a distance 8 units (*e.g.*, $8 \times \frac{1}{4}$ in.) from XY. Along this third line points P and Q are marked at a distance 1 unit (*e.g.*, $\frac{1}{4}$ in.) above and below lines a and b. The slope of the lines AP and BQ are thus accurately fixed as " 1 in 8 ", and $APQB$ will mark the groove to be cut.

A similar " 1 in 8 " procedure is used to mark the tongue in the other piece of work.

The above " 1 in 8 " slope is usually employed for hardwood joints; for softwood work " 1 in 6 " is the rule.

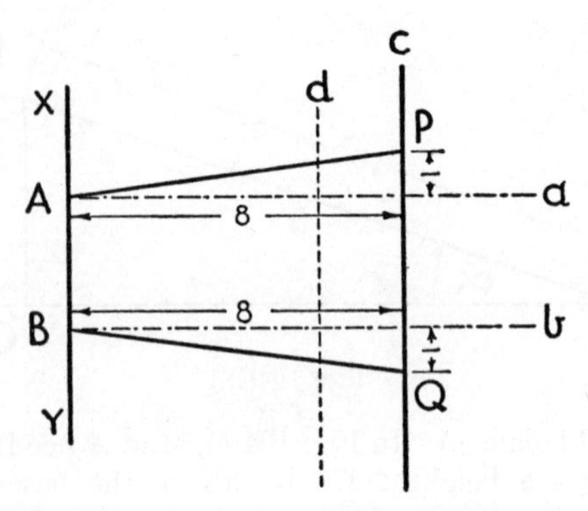

FIG. 105.—HARDWOOD DOVETAIL—1 IN 8.

If a smaller groove is required a parallel line d can be drawn (or marked with a gauge)—but the slope " 1 in 8 " is unaltered.

When many dove-tails have to be made a mask or template may be cut in sheet metal and used stencil-fashion to mark out each joint.

(b) *The " Slope " of a Straight Line*

The mathematician uses the same idea of " slope " as the cabinet-maker in the previous discussion. In Chapter 6 (para. 6.1) we saw that the relation $y = ax + b$ could be represented by a straight-line graph, which cut the y-axis at a height b, and whose steepness of slope was indicated by $a = \frac{QN}{PN}$ (see Fig. 106).

Thus if a straight line cuts the y-axis at $y = 3$, and slopes at " 1 in 6 ", $a = \frac{1}{6}$, and $b = 3$, so the " equation " of such a line is

$$y = \frac{1}{6} \times x + 3 \quad \text{or} \quad y = \frac{x}{6} + 3$$

The ratio $\frac{QN}{PN}$ gives us then a measure of the angle of slope α. So important is this ratio that in trigonometry we call it the

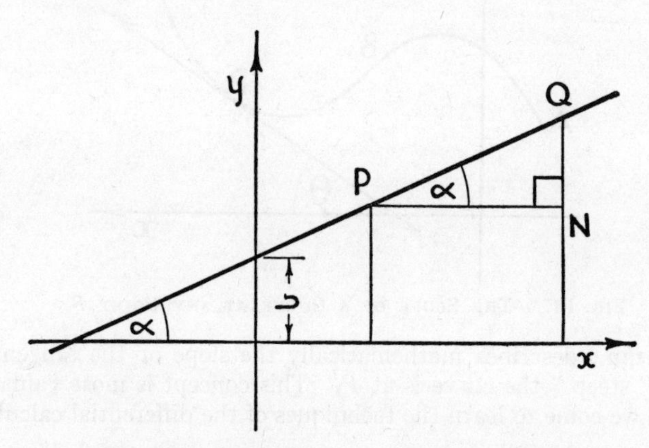

FIG. 106.—THE SLOPE OF A STRAIGHT LINE.

" tangent-ratio " or just the " tangent " of the angle of slope. In our shorthand

$$\tan \alpha = \frac{QN}{PN} = \frac{1}{6} \text{ say.}$$

A " downhill " slope of 1 in 8 would be represented by $\tan \alpha = -\frac{1}{8}$.

Notice also that the line makes the same angle α with the x-axis (" corresponding " angles !).

(c) *The Steepness of a Curve at any point*

Fig. 107 explains why we call this mathematical slope-ratio a " tangent ".

The slope at P, a point on the curve ABC, is the " *steepness* " of the curve at P. The curve at this point is " pointing " in the direction of the arrow—*i.e.*, of the *tangent* to the curve at P.

If this tangent-line meets the x-axis at an angle θ, then the

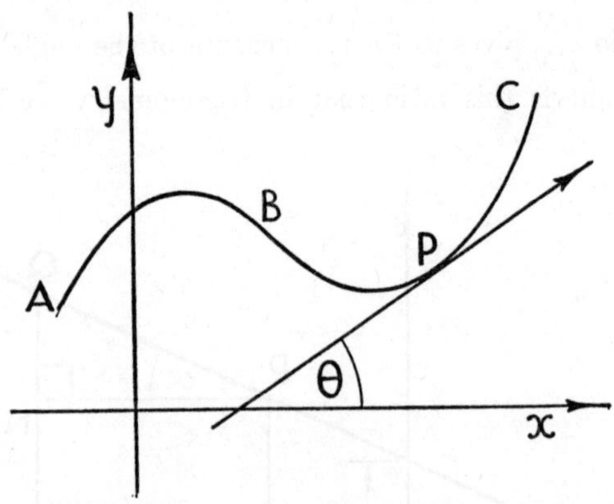

FIG. 107.—THE SLOPE OF A CURVE AT ANY POINT P.

ratio $\tan \theta$ describes mathematically the slope of the tangent—how " steep " the curve is at P. This concept is most valuable when we come to learn the techniques of the differential calculus.

(d) *A Further Practical Illustration*

Example 8(*b*). *On the* 1-*in. Ordnance Map of North-west Kent the spot-height at Bunker's Farm, on the ridge of the North Downs, is* 612 *ft. The main road to the East from this farm crosses the* 250-*ft. contour line on the map* 1·1 *in. from the farm. What is the average slope of this stretch of road?*

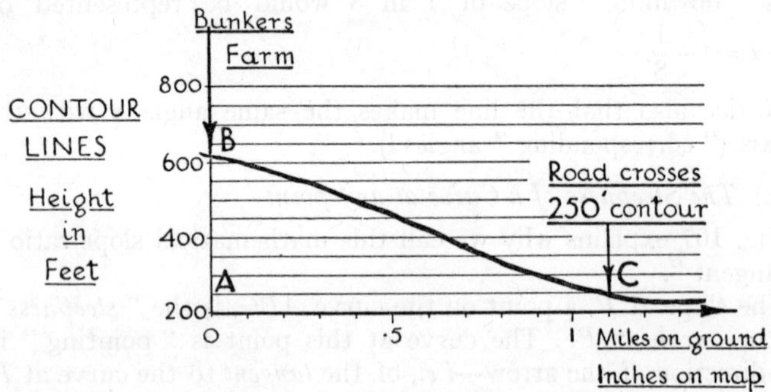

FIG. 108(*a*).—VERTICAL SECTION THROUGH BUNKER'S FARM.

Fig. 108(a) shows a vertical section through the road from Bunker's Farm (at B) to where it cuts the 250-ft. contour (at C). (*Note :* The distances on the map represent the *horizontal* distances on the ground.) In Fig. 108(b) we have abstracted the trigonometrical facts of the problem, in the $\triangle ABC$. Describing the average slope of the road from B to C by the angle α,

$$\tan \alpha = \frac{362}{5808}$$

$$= 0{\cdot}0622 \text{ (slide-rule)}$$

From the Tangent Tables

$$\alpha = 3^\circ\ 33'$$

So the road BC *has an average slope of about* $3\frac{1}{2}^\circ$.

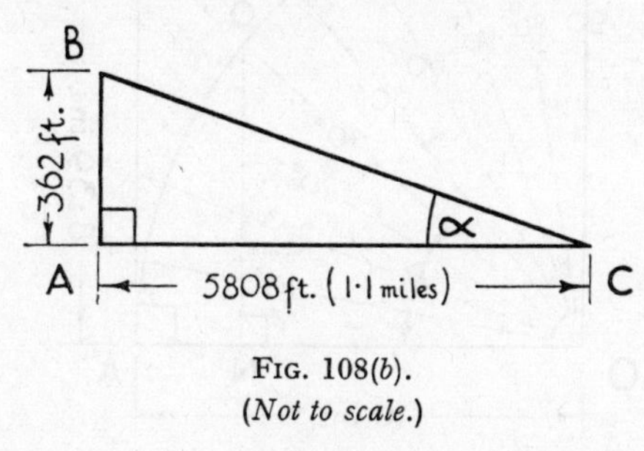

FIG. 108(b).
(*Not to scale.*)

From the Sine Tables $\sin 3^\circ\ 33' = 0{\cdot}0619 = \frac{1}{16{\cdot}2}$, which is almost indistinguishable from the tangent-ratio of the same angle $\left(0{\cdot}0622 = \frac{1}{16{\cdot}1}\right)$. In the usual road-sign language the whole hill has an *average gradient of* 1 *in* 16.

Note that for fairly small angles there is very little to choose between the tangent and sine ratios as a measure of slope.

8.5. Making a Table of Tangents

Fig. 109 indicates a simple method of finding the tangent-ratio of various angles by scale drawing.

AB is a quadrant of the circle centre O and radius 10 in.; AT is the tangent at A to this circle. Angles 10°, 20°, 30°, up to 80° are constructed at O as shown. We will consider the 40° angle POA as a typical example.

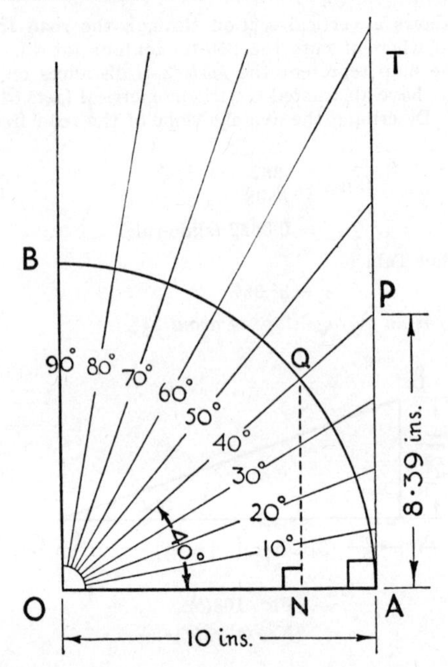

FIG. 109.—CONSTRUCTING A TABLE OF TANGENTS.

By measurement $AP = 8{\cdot}39$ in. Then in ΔOPA, $\angle OAP = 90°$ (tangent and radius)

$$\therefore \quad \tan 40° = \frac{8{\cdot}39}{10}$$

$$= \underline{0{\cdot}839}$$

(Four-figure tables give 0·8391.)

Similarly, we can calculate the tangents of the other angles, and obtain a short table of values as shown below:

Angle .	0°	10°	20°	30°	40°	50°	60°
Tangent .	0	0·176	0·364	0·577	0·839	1·192	1·732

Our paper is too small to read off values for 70°, 80°, 90°. From the full tables we note that tan 70° = 2·7475, tan 80° = 5·6713, and as we approach 90° the ratio gets very large indeed. Thus tan 89° 54′ = 573·0 ! What value can we give to the tangent of 90° ? Look again at Fig. 109. OB and AT are both perpendicular to our start line OA, so they are parallel—they can never meet ! We could consider them to meet at an " infinite " distance away. So the value of tan 90° is infinitely great ; we say

$$\tan 90° = \infty \text{ (" infinity ")}$$

What special value must tan 45° have? Verify this from the Tangent Table.

Before we leave Fig. 109, note that while AP = 10 tan 40°, the shorter dotted line QN = 10 sin 40° (hypotenuse OQ = radius 10 in.). Thus tan 40° is clearly greater than sin 40°, and as we deal with larger and larger angles the difference between the tangent and sine ratios widens. Only for small angles (*e.g.*, less than 10°) are the two ratios nearly equal.

8.6. Notes on the Use of Natural Sine and Tangent Tables

These are no more difficult to use than are logarithm tables. The ratio is given in the main part of the table in steps of 6 minutes (*e.g.*, sin 46° 18′ = 0·7230). For the intermediate angles (*e.g.*, 46° 22′) the " difference column " on the same row of the table is used to modify the four-figure number in the main table.

Thus sin 46° 22′ = 0·7230 (sin 46° 18′) + 0·0008 (difference for 4′)

Conversely, if we know the trigonometrical ratio of an angle, we can search the appropriate table to find this angle to the nearest minute, *e.g.*, if tan α = 1·2142, we note that tan 50° 30′ = 1·2131 (from the body of the tangent table).

There is a difference of 0·0011 to be accounted for. On the same " 50° " row look at the " differences " columns : a difference of " 7 " indicates 1′, a difference of " 14 " means 2′ ; so our " 11 " is nearer to 2′ than 1′

$$\therefore \quad \alpha = 50° \, 32' \text{ (to the nearest minute).}$$

8.7. A Use of the Sine Ratio in Technical Drawing : A Table of Chords

Figs. 110(a) and 110(b) illustrate a useful application of the sine ratio, in inscribing a regular polygon in a given circle. We have used a regular pentagon in our illustration; and a circle of radius 4 in.

Our problem is to obtain five points A, B, C, D, and E equally

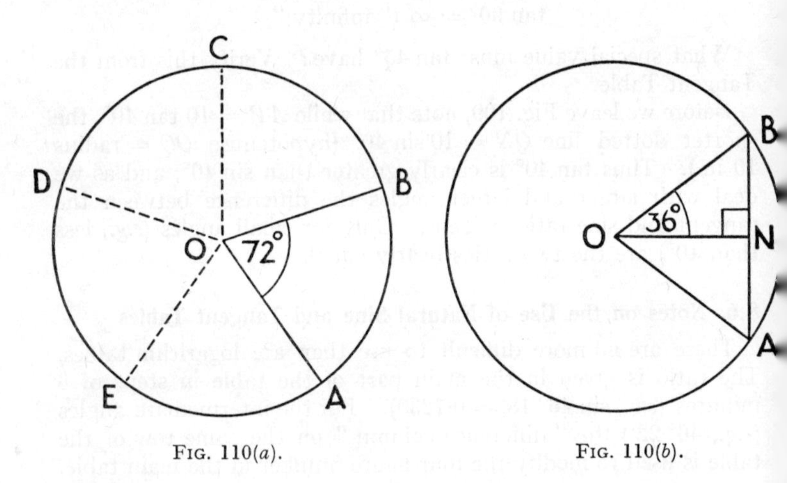

FIG. 110(a). FIG. 110(b).

spaced on the circumference of our circle. The arc AB would clearly subtend an angle $\frac{360°}{5}$, *i.e.*, 72° at the centre O. In theory we could mark off an angle $AOB = 72°$: but a protractor is seldom accurate. If instead we knew the length of the chord AB, we could " step " this out with dividers or compasses, and then obtain points C, D, and E in turn.

We can easily calculate the distance AB, as suggested in Fig. 110(b). The triangle OAB must be isosceles ($OA = OB$, radii), and by symmetry the line ON perpendicular to the chord AB bisects it at N, and $\angle BON = \frac{72°}{2} = 36°$.

Now $\frac{BN}{4} = \sin 36° \text{ in } \triangle OBN$

$\therefore BN = 4 \sin 36°$

$\therefore$ the whole chord $AB = 8 \sin 36°$

$= 8 \times 0{\cdot}5878$

$= 4{\cdot}7024$ in.

Starting with any point A on the circle, successive chords AB, BC, CD, and DE, each 4·70 in. in length would mark out the inscribed pentagon $ABCDE$.

More generally, the chord $AB = 2 \sin 36° \times$ radius

$= 1{\cdot}1756 \times$ radius

Similarly, for any regular polygon, the chord length (one side of the polygon) will be a constant times the radius of the circle. A " table of chords " could be constructed as shown below.

Name.	Pentagon.	Hexagon.	Octagon.	Decagon.	Duodecagon.
No. of sides . .	5	6	8	10	12
a angle at centre .	72°	60°	45°	36°	30°
Sin $\frac{a}{2}$	0·5878	0·5	0·3827	0·3090	0·2588
Chord " constant "	1·1756	1	0·7654	0·6180	0·5176

For example, to construct a regular octagon in a circle of radius 6 in., the side of the octagon would be $0{\cdot}7654 \times 6$ or 4·5924 in. (4·59 in. as a practical measurement).

To sum up, in any regular polygon

Chord AB = Diameter $\times$ Sine (half the angle subtended at the centre by one side).

EXERCISE 8(*a*)

Draw rough sketch-diagrams wherever possible.

1. Find from the Sine and Tangent Tables (*a*) sin 63° 24′, (*b*) tan 42° 16′, (*c*) sin 59° 57′, (*d*) tan 63° 49′.

2. Find the unknown angles, if : (*a*) $\sin \alpha = 0{\cdot}4914$, (*b*) $\tan \beta = 0{\cdot}2031$, (*c*) $\tan A = 1{\cdot}8990$.

3. Calculate x in the following triangles:

(a) $\triangle ABC$; $\angle ABC = 72°$, $\angle ACB = 90°$, $AB = 6$ cm., $BC = x$ cm.

(b) $\triangle PQR$; $\angle PQR = 90°$, $\angle RPQ = 32° 17'$, $PQ = 5{\cdot}3$ in., $QR = x$ in.

(c) $\triangle LMN$; $\angle NLM = 90°$, $\angle LNM = 64° 26'$, $LM = 4{\cdot}8$ cm., $MN = x$ cm.

4. The $\triangle ABC$ is right-angled at B; calculate the other angles of the triangle if: (a) $AB = 6{\cdot}2$ cm., $BC = 3{\cdot}8$ cm., (b) $AC = 4{\cdot}2$ cm., $BC = 2{\cdot}9$ cm.

5. In the triangle PQR, $\angle PQR = 90°$. If $PR = l$ cm. and $\angle QPR = \alpha$, express PQ and RQ in terms of l and α.

6. The base angles of an isosceles triangle are each 63°, and the equal sides are each 4 in. long. Calculate the third (base) side, the height, and the area of the triangle.

7. Two equal poles, 14 ft. long, are lashed together at one end to make " shear-legs ". If they are set up so that the tops of the poles are 11 ft. above the ground, calculate the inclination of each pole to the horizontal.

8. A 16-ft. ladder rests against a factory wall, and its base is 5 ft. from the base of the wall. Calculate the angle the ladder makes with the horizontal ground, and find how far up the wall the ladder reaches.

9. The shadow of a vertical pole 6 ft. high is 8 ft. 8 in. long on level ground. What is the " altitude " (angle of elevation) of the sun at this moment?

10. An electricity pylon stands on level ground and is 60 ft. high. Find the angle of elevation of the top of this pylon from a point 240 yd. from the base of the pylon.

11. A hill is sign-posted as " Steep Hill—1 in $5\frac{1}{2}$ ". Find its approximate slope, as an angle with the horizontal.

Find the error involved if the *tangent* instead of the *sine* ratio were used to find this angle.

12. A coal seam slopes downwards at an angle of 27° with the horizontal in a straight line from A to B. If $AB = 480$ yd., how far is B below A?

13. A hill slopes upwards at an angle of 18° with the horizontal. What height does a man rise in walking a quarter-mile up this hill? What slope would a road sign show for this hill?

14. From the top of a sheer cliff 375 ft. high the angle of depression of a boat is 17°. How far is the boat from the base of the cliff?

15. The steps of a staircase are 9 in. deep and $7\frac{1}{2}$ in. high. What angle does the bannister rail make with the horizontal?

16. A railway tunnel has to be bored between two points P and Q, whose heights above sea-level are 462 and 618 ft. respectively. PQ is 2 miles 218 yd. as the crow flies (*i.e.*, *horizontal* distance). Calculate the angle of elevation of the tunnel workings, if operations begin from the end P.

17. A road 1180 yd. long " zigzags " up a mountain pass from A to B. On a 6 in. to the mile map A is $0{\cdot}86$ in. from B. If A and B have spot heights 672 ft. and 1129 ft., find (a) average slope (α) of the road from A to B. (b) the slope (β) of the cable for a proposed cable railway direct from A to B.

18. A stay for a telegraph pole is inclined at 67° with the horizontal, and is 18 ft. long from its point of attachment to the pole to where the stay-rod enters the ground. If the top of the pole is 27 ft. above ground level, how far down the pole is the stay wire attached?

19. In erecting an overhead line route in rocky terrain a stay-hole is blasted 6 ft. away from a pole-hole. If the stay-wire is to be attached to the pole an estimated 17 ft. above ground level, estimate the length of stay-wire needed, and find the angle the stay will make with the horizontal.

20. A railway cutting is to be 32 ft. wide at the bottom and 50 ft. deep. If it is specified that the sides must not be steeper than 60° with the horizontal, what is the least width of the cutting at the top?

Calculate the cross-sectional area of the cutting, and find how many cubic yards of earth must be excavated for each 100 yd. of the cutting.

21. Using the table of chords in para. 8.7 above, (*a*) construct a regular octagon whose vertices lie on a circle, radius 3·7 in., (*b*) deduce the radius of a circle which *circumscribes* a regular decagon of side 6·8 cm.

22. How long is the side of a regular heptagon (7 sides) to be inscribed in a circle of radius r cm.?

23. Construct a regular pentagon whose sides touch a circle 4 in. in diameter. Measure its side, and confirm by calculation. (*Hint:* First construct the five points of contact.)

24. Calculate a side of a regular nonagon (9 sides) which touches (circumscribes) a circle, radius r cm.

25. Without using a table of chords, inscribe a regular duodecagon in a circle of radius 2·8 in. Measure its side, and confirm by calculation.

26. Obtain a general formula for the side l in. of a regular n-sided polygon (*a*) inscribed in (*b*) circumscribed to a circle of radius r in.

27. A road 400 yd. long is represented on a map with a scale of 1 : 10,000 by a line 1·42 in. long. Find the average slope of this section of the road.

28. A pendulum 42 ft. long swings through 10° either side of the vertical. Through what distance does the bob move horizontally? How high does the bob rise above its lowest position?

29. The centre of a golf ball is 6 ft. from the centre of the hole, which is 3 in. in diameter. Within what angle must the ball be struck if it is to drop into the hole?

30. The wavelength control knob of a radio set turns a spindle which is linked with a pulley of radius 2·8 cm. by means of a taut " belting " of string. As the pulley turns, the moving vanes of an air-spaced variable capacitor move with it. Ignoring the diameter of the spindle, find the length of string required, if the axis of the spindle is 8 cm. from the axis of the pulley.

31. Two pulleys of radii 2 and 3 in., are mounted side by side, with their centres 8 in. apart. If they are linked by straight (direct) belting, calculate: (*a*) the angle between the straight portions of the belting; (*b*) the total length of belting required. (*Hint:* Find the curved length of

belting on each pulley as a simple proportion of the circumference, after calculating angles at the centre.)

32. Repeat the calculations of Question 31 for the case where the belting is *crossed* (transverse).

8.8. The Cosine Ratio

(a) *Meaning of "Cosine"*

In Fig. 102 how are we to describe the ratio $\frac{AC}{AB}$ in terms of the angle α? It represents the ratio of the adjacent side to the hypotenuse. But in terms of the angle at B, the complement of α,

$$\begin{aligned} \frac{AC}{AB} &= \sin B \\ &= \sin (90^\circ - \alpha) \\ &= \textit{sine} \text{ (}\textit{co}\text{mplement of } \alpha\text{)} \end{aligned}$$

We thus call this ratio the *co-sine* of angle α. In practice cosine α is abbreviated to cos α.

Summing up the three important ratios in trigonometry:

$$\sin \alpha = \frac{BC}{AB} \quad \text{or} \quad \frac{\text{opposite side}}{\text{hypotenuse}}$$

$$\cos \alpha = \frac{AC}{AB} \quad \text{or} \quad \frac{\text{adjacent side}}{\text{hypotenuse}}$$

$$\tan \alpha = \frac{BC}{AC} \quad \text{or} \quad \frac{\text{opposite side}}{\text{adjacent side}}$$

(b) *The Snag with Cosine Tables*

Note that if α is a very small angle, AC is very nearly equal to AB, so cos α is very nearly 1. In contrast, as α approaches 90° AC approaches zero. So as α *increases* from 0° to 90°, cos α changes from 1 to 0, *i.e.*, it *decreases*.

Cosine Tables are just as simple to produce as Sine or Tangent

Tables; on closer inspection you will note that it is just the *Sine Table in reverse.* This is so because the cosine of an angle is the sine of its complement.

$$e.g., \cos 15° = \sin 75° = 0{\cdot}9659$$

All is well until we need the cosine of an angle not appearing in the main part of the table.

For example, looking for cos 57° 38′ we note that cos 57° 36′ = 0·5358, while cos 57° 42′ = 0·5344. So cos 57° 38′ must lie between these values. In the " Differences " Column for 2′ we find a difference of 0·0005. For our result to be between the values above we have to SUBTRACT this difference from cos 57° 36′.

i.e.,

$$\begin{aligned} \cos 57° \; 36' &= 0{\cdot}5358 \\ \text{difference for } 2' &= 0{\cdot}0005 \\ \text{SUBTRACTING, } \cos 57° \; 38' &= 0{\cdot}5353 \end{aligned}$$

How can we be sure? We just check in the *Sine Tables.* The complement of 57° 38′ is 32° 22′, so if we are right, then

$$\cos 57° \; 38' = \sin 32° \; 22' = 0{\cdot}5353$$

(Check this for yourself.)

Where space is limited (in diary or pocket-book) this simple connection between Sine and Cosine Tables may be exploited by printing the Sine Table normally (starting at the top left corner and working down) and using the right-hand margin to show " Cosines "—starting at the bottom right corner and working *up* through the same set of figures.

If we have any choice then, it is wiser to use the Sine or Tangent Tables rather than look up Cosines.

8.9. Example involving a Cosine

Example 8(c). *A wire stay of a vertical wireless mast is* 42 *ft. long and is anchored* 33 *ft. from the base of the mast* (*on level ground*).

Find the angle the stay makes with the horizontal and the height above the ground of the point of attachment of its upper end.

The situation is pictured in Fig. 111; the stay AB is inclined at an angle α with the horizontal, and is attached to the mast x ft. above the ground at the point B.

(i) Then in the right-angled $\triangle ABC$

$$\cos \alpha = \tfrac{33}{42} = 0{\cdot}7857$$

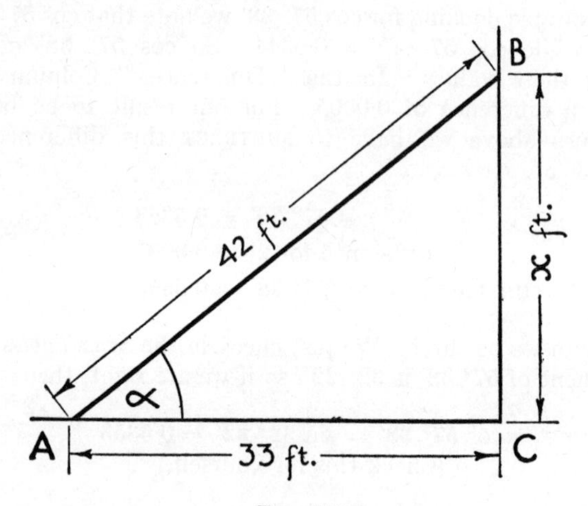

Fig. 111

From the Cosine Tables

$$\begin{array}{rl} 0{\cdot}7848 &= \cos 38^\circ\, 18' \\ 0{\cdot}0009 &= \text{Diff. for } 5' \\ \hline \therefore \quad 0{\cdot}7857 &= \cos 38^\circ\, 13' \quad (\textit{i.e.}, \text{ SUBTRACTING } 5') \\ \therefore \quad \alpha &= 38^\circ\, 13' \end{array}$$

So the stay is inclined at 38° 13′ *to the horizontal*

(ii) In the same $\triangle ABC$,

$$\frac{x}{42} = \sin \alpha = \sin 38^\circ\, 13' = 0{\cdot}6186$$

$$\therefore \quad x = 42 \times 0{\cdot}6186 = 26{\cdot}0 \text{ (slide-rule)}$$

So the stay is attached to the mast at a height of 26 *ft. approx.*

(Check this by Pythagoras' Theorem.)

8.10. Simple Relations between the Trigonometrical Ratios

(*a*) In Fig. 102 we have:

$$\sin \alpha = \frac{BC}{AB} \quad \cos \alpha = \frac{AC}{AB}$$

$$\tan \alpha = \frac{BC}{AC} = \frac{BC}{AB} \times \frac{AB}{AC} = \frac{\sin \alpha}{\cos \alpha}$$

$$\textbf{Thus } \tan \alpha = \frac{\sin \alpha}{\cos \alpha}$$

(*b*) Also by Pythagoras in Fig. 102

$$AC^2 + BC^2 = AB^2$$

Dividing b.s. by AB^2

$$\frac{AC^2}{AB^2} + \frac{BC^2}{AB^2} = 1$$

$$\textbf{Thus } \cos^2 \alpha + \sin^2 \alpha = 1$$

(*c*) The reciprocals of the ratios sine, cosine, tangent are also widely used. They are (referring again to Fig. 102)

$$\text{The cosecant; } \operatorname{cosec} \alpha = \frac{1}{\sin \alpha} = \frac{AB}{BC} \text{ or } \frac{\text{hypotenuse}}{\text{opposite side}}$$

$$\text{The secant; } \sec \alpha = \frac{1}{\cos \alpha} = \frac{AB}{AC} \text{ or } \frac{\text{hypotenuse}}{\text{adjacent side}}$$

$$\text{The cotangent; } \cot \alpha = \frac{1}{\tan \alpha} = \frac{AC}{BC} \text{ or } \frac{\text{adjacent side}}{\text{opposite side}}$$

Be careful to remember that the cosecant is the reciprocal of the sine, not of the cosine.

(*d*) The student should impress upon his mind the cases of the angles 0°, 30°, 45°, 60°, 90°, by studying the triangles shown in

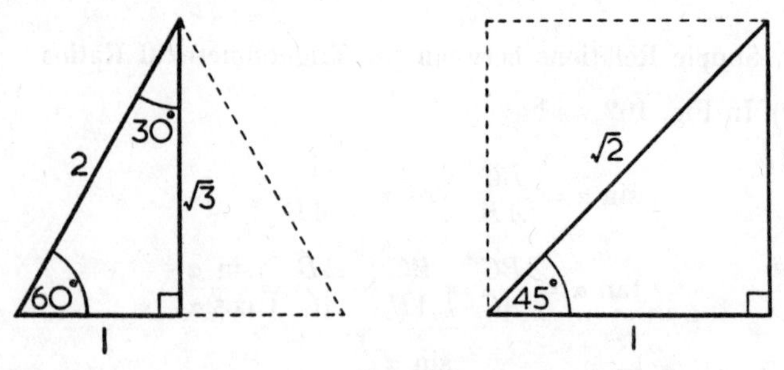

FIG. 112.

Fig. 112, from which (by applying Pythagoras' theorem where necessary) the trigonometrical ratios of these angles are seen to be :

	0°	30°	45°	60°	90°
sin	0	$\frac{1}{2}$	$\frac{1}{\sqrt{2}}$	$\frac{\sqrt{3}}{2}$	1
cos	1	$\frac{\sqrt{3}}{2}$	$\frac{1}{\sqrt{2}}$	$\frac{1}{2}$	0
tan	0	$\frac{1}{\sqrt{3}}$	1	$\sqrt{3}$	∞
cosec	∞	2	$\sqrt{2}$	$\frac{2}{\sqrt{3}}$	1
sec	1	$\frac{2}{\sqrt{3}}$	$\sqrt{2}$	2	∞
cotan	∞	$\sqrt{3}$	1	$\frac{1}{\sqrt{3}}$	0

8.11. Angles Greater than 90°

To give meaning to the sine or cosine of an *obtuse* angle—indeed to *any* angle greater than 90°—we need to widen our definitions of the trigonometrical ratios. For clearly, in a right-angled triangle it is impossible to have an angle greater than 90°.

Fig. **113** suggests how we re-define our three ratios.

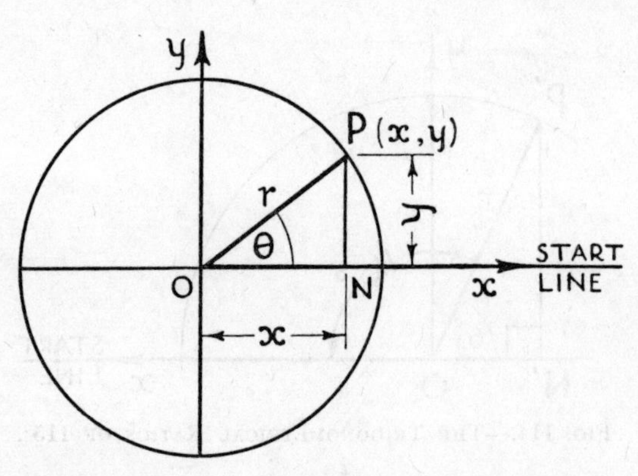

FIG. 113.

The typical angle θ is measured anti-clockwise from the x-axis Ox, which is our START LINE. θ is therefore an *amount of turning*, and so can be greater than 90°—indeed it may exceed 360°. OP is a moving arm, of fixed radius r units. Oy is the usual y-axis. PN is the perpendicular from P on to the x-axis.

ON and PN are the co-ordinates x and y of the point P in *any* position (with the usual sign conventions: x positive *to the right*; y positive *upwards*.)

We can then define the ratios as follows:

$$\sin\theta = \frac{y}{r}$$

$$\cos\theta = \frac{x}{r}$$

$$\tan\theta = \frac{y}{x} = \frac{\sin\theta}{\cos\theta}$$

This is clearly consistent with our previous definitions for angles less than 90°.

Let us see how these new definitions work for an obtuse angle; in Fig. 114 we have chosen 115° as our sample, where OP' is the position of the rotating arm.

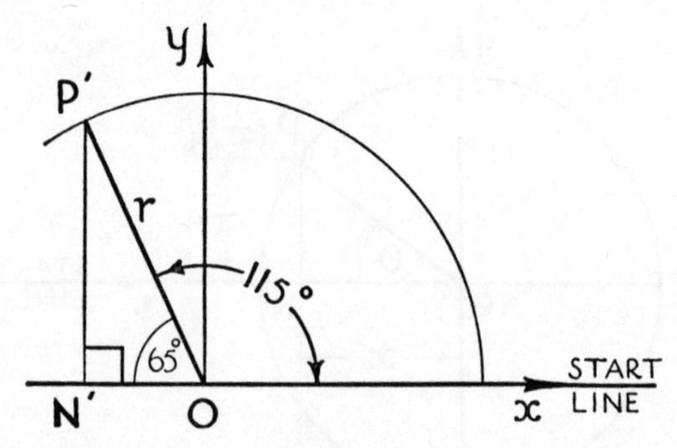

FIG. 114.—THE TRIGONOMETRICAL RATIOS OF 115°.

Now $$\sin 115° = \frac{y}{r} \text{ at } P'$$

$$= \frac{P'N'}{r}$$

But this is the same ratio exactly as $\sin P'ON'$, where

$$\angle P'ON' = 180° - 115° = 65°$$

$$\therefore \quad \underline{\sin 115° = \sin 65° = 0{\cdot}9063}$$

Also $$\cos 115° = \frac{x}{r} \text{ at } P'$$

But in this case x is *negative* (measured to the *left* from O) so we may write:

$$\cos 115° = -\frac{ON'}{r}$$

$$= -\cos P'ON'$$

$$= -\cos 65°$$

i.e., $$\underline{\cos 115° = -0{\cdot}4226}$$

Finally, $\tan 115° = \dfrac{y}{x}$

$$= \frac{P'N'}{-ON'}$$

$$= -\tan P'ON'$$

$$= -\tan 65°$$

i.e., $\underline{\tan 115° = -2{\cdot}1445}$

(Use your slide-rule to check that $\tan 115° = \dfrac{\sin 115°}{\cos 115°}$.)

8.12. The Quadrant Rules for the Trigonometrical Ratios

It is very convenient to divide the complete revolution of 360° into four *quadrants*, as suggested in the table below. Then for *any* angle at all, we have only two questions to ask ourselves:

(a) *In what quadrant does* OP *lie?*
This decides the SIGN (+ or —) of our ratio (see Fig. **115**).
(b) *What* ACUTE angle does OP *make with the* START LINE?
This gives us the angle whose ratio is to be looked up in the trigonometrical tables.

Quadrant.	0° to 90°.	90° to 180°.	180° to 270°.	270° to 360°.
Sign of x Sign of y	+ +	− +	− −	+ −
$\sin\theta = \frac{y}{r}$	+	+	−	−
$\cos\theta = \frac{x}{r}$	+	−	−	+
$\tan\theta = \frac{y}{x}$	+	−	+	−

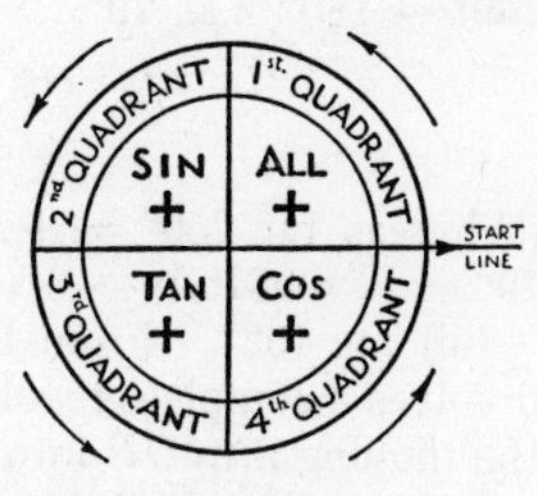

FIG. 115.—THE QUADRANT RULES.

Examples (Fig. **116**)

(i) *sin* 303°. From the diagram (Fig. **116** (i)), for 303° *OP* comes in the 4th quadrant.

Therefore sin 303° is NEGATIVE—("y" is negative, as P is "below the line").

Also the acute angle made by 303° with the Start Line is 360° − 303° = 57°.

$$\therefore \quad \sin 303° = -\sin 57°$$
$$= -0{\cdot}8387$$

(on the other hand, cos 303° = + cos 57° = 0·5446)

(ii) *cos* 259°. From Fig. 116 (ii), 259° arrives in the 3rd quadrant (where only TAN is +).

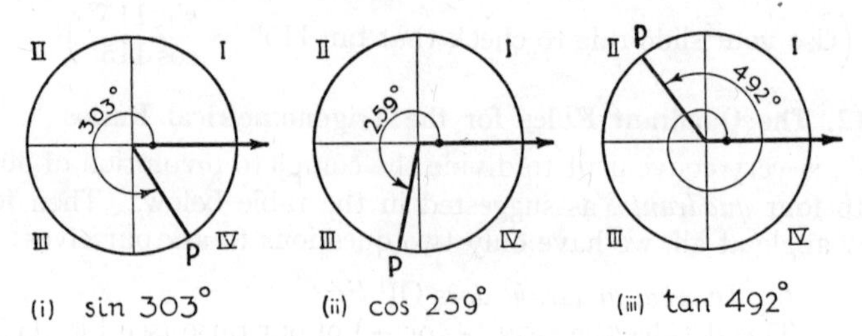

FIG. 116.—DIAGRAMS TO FIND THE TRIGONOMETRICAL RATIOS FOR ANGLES GREATER THAN 90°.

Also from the diagram the acute angle with the START LINE is 259° − 180°, *i.e.*, 79°.

$$\therefore \quad \cos 259° = -\cos 79°$$
$$= -0{\cdot}1908$$

(Whereas tan 259° = + tan 79° = 5·1446 and sin 259° = − sin 79° = − 0·9816.)

(iii) *tan* 492°. From Fig. 116 (iii) we see that a turn of 492° involves a complete revolution (360°) and a further 132°, bringing the moving arm OP into the 2nd quadrant once more; the SINE only is positive in this quadrant. The acute angle with the start line is now 180° − 132° = 48°.

$$\therefore \quad \tan 492° = \tan 132° = -\tan 48° = -1{\cdot}1106$$

8.13. Graphs of the Sine, Cosine, and Tangent Ratios (from 0° to 360° and beyond)

Using the Quadrant Rules and our tables, we can construct a table of the three ratios for selected angles between 0° and 360°, as shown below, and then draw graphs, to show how the three ratios vary during a complete revolution:

Quadrant.													
	I.			II.			III.			IV.			
Angle θ	0°	30°	60°	90°	120°	150°	180°	210°	240°	270°	300°	330°	360°
Sin θ	0	0·5	0·866	1	0·866	0·5	0	−0·5	−0·866	−1	−0·866	−0·5	0
Cos θ	1	0·866	0·5	0	−0·5	−0·866	−1	−0·866	−0·5	0	0·5	0·866	1
Tan θ	0	0·577	1·732	∞	−1·732	−0·577	0	0·577	1·732	∞	−1·732	−0·577	0

For values of θ from 360° to 720°, the same cycle of values of the trigonometrical ratios will be repeated.

Fig. **117** shows the sine, cosine, and tangent ratios plotted between 0° and 720°. It is worth noting that tan 45° $= 1 =$ tan 225°; tan 135° $= -1 =$ tan 315°. (These help in sketching the tangent graph.)

Notice that the sine and cosine graphs are smooth " wave " forms of the same general shape—in fact cos θ is just sin θ moved 90° to the left ! Such graphs are called *sinusoidal* (" sine-like ") and are very important to the communication engineer. Whereas the tangent graph is discontinuous when $\theta = 90°$, $\theta = 270°$, and so on—the value of tan θ is *infinite* for these values of θ.

One further comparison: the sinusoidal graphs follow through a complete cycle of values in 360°; in contrast, the tangent values repeat themselves every 180°, *i.e.*, the tangent graph contains two complete cycles between 0° and 360°.

Find from these graphs the approximate values of: sin 192°, cos 297°, tan 470°, cos 555°, and verify these values, using the Quadrant Rules for sign, and the trigonometrical tables to obtain each numerical value.

8.14. The Graph of 3 sin θ, Obtained by Drawing

Fig. **118** demonstrates a useful method of obtaining a sine wave by projection from a circle. In this case the circle, centre *C*, is 3 units in radius. Using a set-square, angles 30°, 60°, 90°,

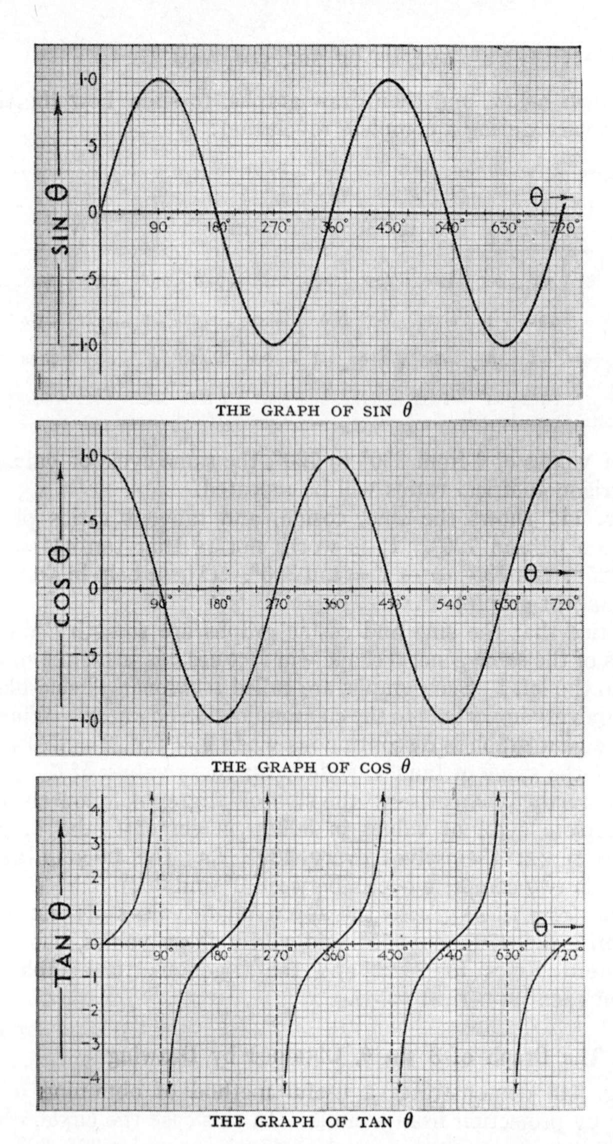

FIG. 117.—THE GRAPHS OF SIN θ, COS θ, TAN θ.

up to 330° are set out with the start line, which is produced to give the base-axis for θ. The *y*-axis *Oy* is drawn at right angles to the start line. Equal dimensions 30°, 60°, 90°, and so on are set off along the θ-axis. Then for $\theta = 30° = \angle PCN$,

$$\sin 30° = \frac{PN}{CP} = \frac{PN}{3}$$

so

$$PN = 3 \sin 30°$$

(Similarly, for all other positions of *P* its height above the START LINE represents 3 sin θ).

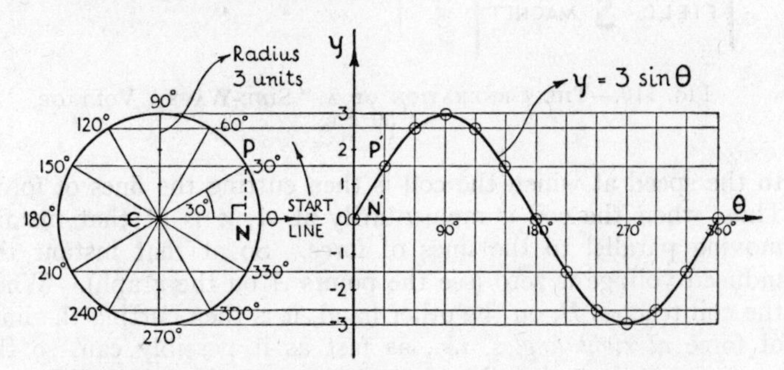

FIG. 118.—CONSTRUCTING THE GRAPH OF $y = 3 \sin \theta$.

Now draw *PP'* parallel to the start line, meeting the line $\theta = 30°$ at *P'*, and similarly for other positions of the moving radius *CP*.

The locus of *P'* is then such that $y = 3 \sin \theta$. It is clearly a sine wave whose greatest height above the start line is 3, *i.e.*, its *amplitude* is 3 units.

8.15. An Electrical Analogy—The Production of an Alternating Voltage

PQ in Fig. 119 represents a section perpendicular to its axis of rotation of a rectangular coil, which is being turned at uniform speed in the magnetic field (also assumed uniform) between the poles N and S of a magnet. The fine lines indicate the " lines of force " in this magnetic field.

The voltage induced in the coil at any instant is proportional

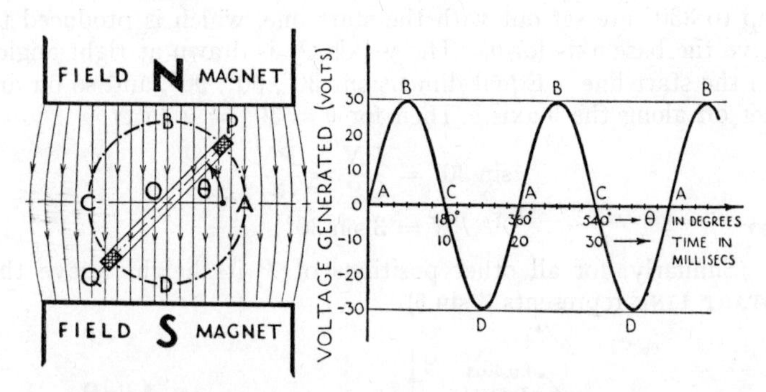

FIG. 119.—THE GENERATION OF A "SINE-WAVE" VOLTAGE AT 50 c/s.

to the speed at which the coil is then cutting the lines of force. Thus, when the coil is momentarily at A, it is at that instant moving parallel to the lines of force. So at that instant the induced voltage is zero (see the points A on the graph). When the coil reaches B, on the other hand, it is then cutting the lines of force *at right angles, i.e.*, as fast as it possibly can, so the maximum voltage is induced at the instant when the coil is at B (as indicated by the points B on the graph. Similarly, for points C and D, the latter representing maximum voltage induced in the OPPOSITE sense.)

We shall prove in the next chapter (para 9.6(b)) that the speed of cutting these lines of force is *proportional to sin* θ, where θ is the angle turned through by the coil from the starting line OA.

Then if the voltage (v volts) at any instant is given by $v = K \sin \theta$, we see that K represents the maximum voltage, *i.e.*, it corresponds (numerically) to the voltage induced when the coil is in position B or D.

In our graph it is clear that $K = 30$, *i.e.*, $v = 30 \sin \theta$.

If the coil were rotated at 50 revolutions per second (*i.e.*, 50 c/s alternating voltage) one complete cycle or change (from A to B and C to D and back to A) would take $\frac{1}{50}$ sec. or 20 millisecs. So we could graduate the base line of our graph "Rotation in Degrees" or "Time in Millisecs", as shown in

Fig. 119. We could express the voltage at any instant t *millisecs* as $v = 30 \sin 18t°$, for each millisec on the scale represents 18°.

In Volume II, we shall see the great value of being able to express alternating currents and voltages in terms of sines and cosines involving the " independent variable " t sec.

8.16. Circular Measure of Angle; the Radian

For scientific purposes and especially in applying the differential calculus to trigonometrical functions (see Volume II) *the circular* measure of angle is indispensable.

Its unit, the *radian*, is easily defined (as in Fig. 120) as the *angle subtended at the centre of a circle by an arc equal to the radius.*

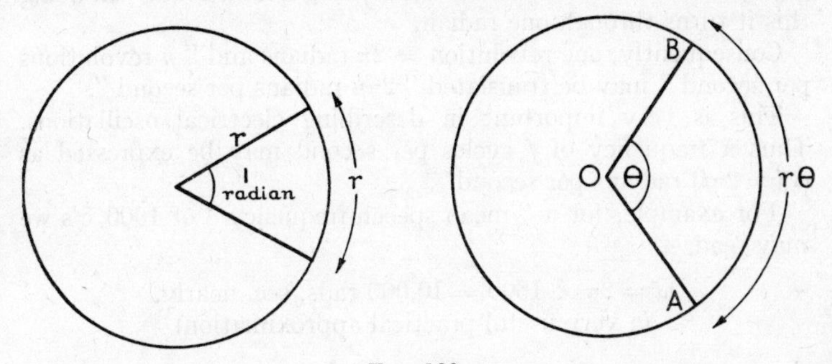

FIG. 120.

It immediately follows, by simple proportion, that any arc AB of a circle of radius r, is of length $r\theta$, where θ radians is subtended at the centre. In particular, the circumference (total arc length) is $2\pi r$, so that a complete 360° is 2π radians.

A little more subtle is the area of the sector OAB in terms of r and θ.

$$\text{By proportion, } \frac{\text{sector } OAB}{\text{whole circle}} = \frac{\theta}{2\pi}$$

But the whole circle is of area πr^2

$$\text{Sector } OAB = \frac{\theta}{2\pi} \times \pi r^2 = \tfrac{1}{2}r^2\theta$$

A fuller discussion of radian measure, with many applications in telecommunication theory, is to be found in Volume II.

At the present stage it is worth noting the following equivalents:

$$\pi \text{ radians} = 180°$$

$$\frac{\pi}{2} \text{ radians} = 90°$$

$$1 \text{ radian} = 57{\cdot}3° \quad \text{(3 significant figures)}$$

Revolutions and Oscillations

A vivid way of thinking of a radian is to imagine a wheel rolling a distance equal to its radius along a level road. In doing this it turns through one radian.

Consequently, one revolution $= 2\pi$ radians and " n revolutions per second " may be translated " $2\pi n$ radians per second ".

This is very important in describing electrical oscillations. Thus a frequency of f cycles per second may be expressed as ω $(= 2\pi f)$ radians per second.

For example, for a " mean speech frequency " of 1600 c/s we may read,

$$\omega = 2\pi \times 1600 = 10{,}000 \text{ rads./sec. nearly.}$$
(a very useful practical approximation)

Conversion of degrees to radians and vice-versa

Since $180° = \pi$ radians, to convert degrees to radians we multiply by $\frac{\pi}{180}$.

Thus $102° \; 16' = 102{\cdot}3°$

$$= \frac{102{\cdot}3 \times \pi}{180} = 0{\cdot}568\pi \text{ or } 1{\cdot}78 \text{ radians.}$$

Similarly, to convert radians to degrees multiply by $\frac{180}{\pi}$.

It is often worth while to leave the angle in radians as a multiple of π. In graphical work it is best to graduate the angle-axis in units of π.

EXERCISE 8(b)

1. Using the Cosine Table, evaluate: (a) cos 52° 18′, (b) cos 47° 51′, (c) cos 82° 37′. Check each one, using the Sine Table.

2. Find the acute angle concerned if: (a) $\cos \alpha = 0{\cdot}3371$, (b) $\cos \beta = 0{\cdot}8850$, (c) $\cos \gamma = 0{\cdot}0777$.

3. Using the quadrant rules, evaluate: (a) sin 137°, (b) tan 163°, (c) sin 313°, (d) cos 204°, (e) tan 293°, (f) cos 111°.

4. If A is the angle of a triangle and $\sin A = 0{\cdot}2996$, what are the possible values for A? If $\cos B = -0{\cdot}6$, what is the size of angle B?

5. If θ has a value between 0° and 360°, give the alternative values of θ when: (a) $\sin \theta = -0{\cdot}4828$, (b) $\cos \theta = 0{\cdot}2222$.

6. Calculate x and y in the triangle ABC, which is right-angled at A. (Fig. 121(a).)

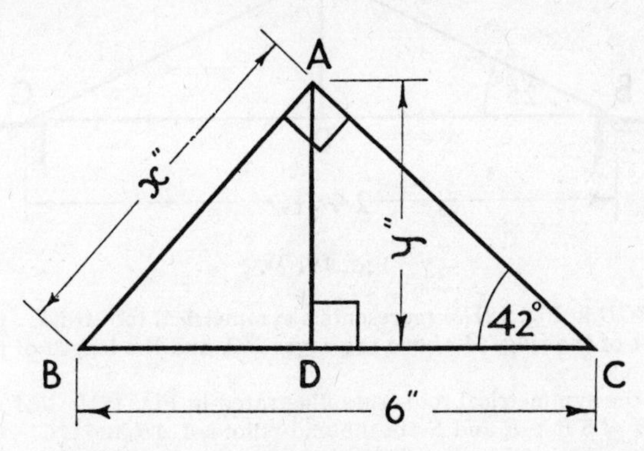

FIG. 121(a).

7. If h in. is the length of the hypotenuse BC of a right-angled triangle ABC, and $\angle ABC$ be denoted by α, express AB and AC in terms of h and α.

8. From a vantage point 60 ft. above ground-level the angle of elevation of the top of a radio mast is 19° and the angle of depression of the base of the mast is 11°.

How far (horizontally) is the observer from the mast? How high is the mast?

Check your answer by drawing to scale.

9. In a coal-mine a gallery leaves the bottom of the pit-shaft (210 ft. below the surface) and slopes downwards at 12° to the horizontal for 240 yd. and then upwards at 26° with the horizontal for a further 108 yd. How far below the surface is the furthest point?

10. Find the height and base radius of a cone which has an angle of 80° at its vertex (in its principal section) and a slant height of 20 in.

11. $ABCD$ is a trapezium in which the parallel sides AB, CD are 4 in. and 5 in. respectively. AD is 3·5 in. and the parallel sides are 3 in. apart. Calculate: (*a*) the angles of the trapezium, (*b*) the length BC.

12. Calculate the radius of the circumcircle of a regular octagon of side 6 cm. and the ratio of the areas of the octagon and the circle.

13. The equal sides of an isosceles triangle are 6·4 cm., the third side is 3·9 cm. Find the angles of the triangle.

14. A telegraph pole is stayed by a wire 30 ft. long. The stay-rod enters the ground 8 ft. from the base of the pole, and its length above the ground is 4 ft. Allowing 5 ft. for "making off", find the inclination of the stay to the horizontal.

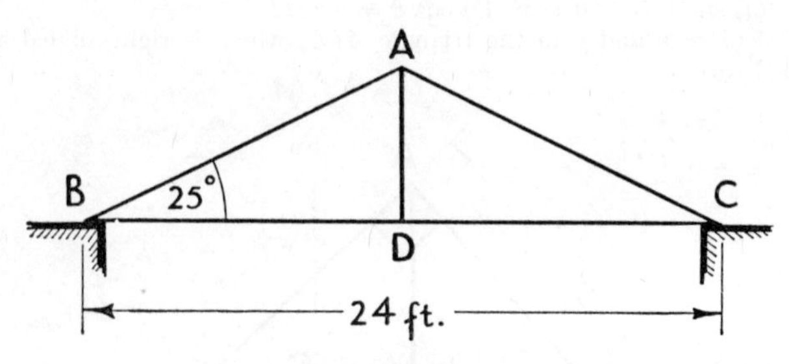

FIG. 121(*b*).

15. $ABCD$ in Fig. 121(*b*) represents a symmetrical roof truss. Calculate the height of the ridge A above the eaves BC, and the length of the beam AB.

16. In the symmetrical roof truss illustrated in Fig. 121(*c*) $PA = PB = QA = QC = 8$ ft.; R and S are the mid-points of AB and AC. Calculate the lengths of the other members of the framework. What is the height of A above BC?

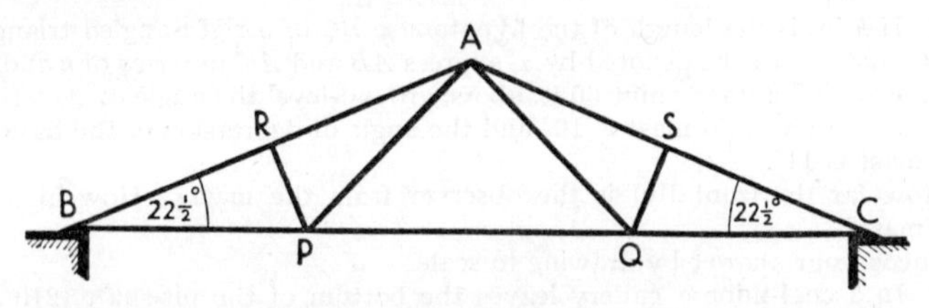

FIG. 121(*c*).

17. AB (8 cm.) is the diameter of a circle. Calculate the length of the chord AP, if $\angle PAB = 38°$.

If AP produced meets the tangent at B at the point Q, calculate BQ.

18. AB, BC, CA are three equal internal struts for a circular rim, of internal diameter 30 in.; how long should each be? (Hint: draw the diameter AD and join BD.)

If they have been cut an inch too long they may be fitted by shortening BC only. By how much must BC be shortened?

19. Two walls of a manual switchroom are at 120° to each other. The switchboard has to be provided with 3 angle sections A, B, C as in Fig. 121(*d*) in order that it may be extended round the corner. What must be the angle (α) of these angle sections? The depth of each switchboard position is 3 ft. 7 in. Calculate the dimensions d and w indicated in the enlargement of section A

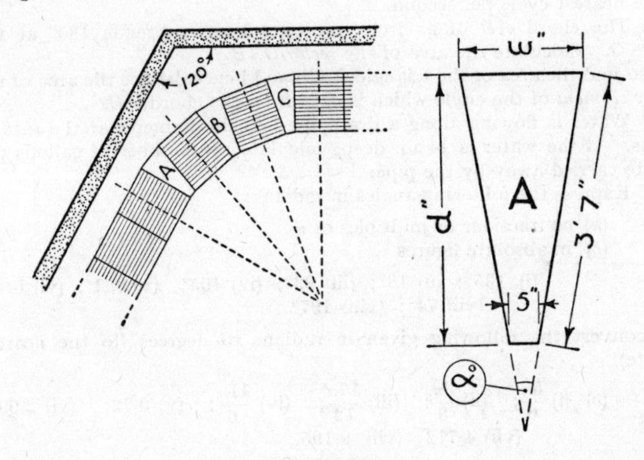

FIG. 121(*d*).

20. The development of the curved surface of a cone is the sector of a circle 8 cm. in radius, which subtends 240° at the centre of the circle. Calculate: (*a*) the base diameter, (*b*) the vertical height, (*c*) the angle at the vertex of this cone.

21. The range on a horizontal plane of a certain gun is approximately 4350 sin 2θ yd. where θ is the angle of elevation of the gun. Calculate this range (R yd.) for values $\theta = 10°, 20°, 30°, 40°, \ldots 80°$, and plot R against θ. What is the greatest range of the gun—at what angle of elevation? What is the likely range if $\theta = 36°$? If a range of 2430 yd. is needed, find from the graph the angle of elevation to the nearest degree. Verify by calculation.

22. Obtaining the necessary data from a diary or almanac, plot the weekly times of sunset and sunrise (Greenwich Mean Time) for a whole year, beginning on or near March 25th.

What do you observe about the curves plotted? What relevance has such data to the radio engineer?

23. Draw carefully the graph of $y = 8 \cos \theta$ for the range $\theta = 0°$ to $\theta = 90°$. Using the Trapezoidal Rule (see para. 7.14) verify that the average value of the ordinate $8 \cos \theta$ for this range is $16/\pi$. (This idea is valuable when discussing the D.C. voltage ideally afforded by full-wave rectification of an A.C. supply.)

24. A voltage is given by the formula $v = 100 \sin 80t°$ at an instant t millisecs. Tabulate its values from $t = 0$ to $t = 10$ at intervals of 1 millisec., and plot v against t.

What is the period of this alternating voltage? State its frequency to the nearest cycle per second.

25. The chord AB of a circle, radius 4·2 in., subtends 130° at the centre O. Calculate the area of the *sector* OAB.

Also find the area of the triangle OAB, and hence deduce the area of the minor *segment* of the circle which is cut off by the chord AB.

26. Water is flowing along a drainpipe 3 ft. in diameter at the rate of 2 f.p.s. If the water is 14 in. deep, calculate the number of gallons per minute carried away by the pipe.

27. Express the following angles in radians:

(*a*) as fractions or multiples of π
(*b*) in absolute figures

(i) 135°; (ii) 18°; (iii) 83°; (iv) 405°; (v) 0° 1′; (vi) 144°; (vii) 74°; (viii) 197°

and convert the following given in radians to degrees (to the nearest minute)

(*c*) (i) $\frac{5\pi}{6}$; (ii) $\frac{3\pi}{8}$; (iii) $\frac{17\pi}{12}$; (iv) $\frac{11\pi}{6}$; (v) 0·72; (vi) 2·094; (vii) 4·712; (viii) 0·195.

28. If an angle is stated correct to the nearest second, to what degree of accuracy must it be given in circular measure?

29. What are the values of the following ratios where the angles are given in radians:

(i) $\sin \frac{2\pi}{5}$; (ii) $\cos \frac{\pi}{3}$; (iii) $\tan \frac{3\pi}{4}$; (iv) $\sin 0{\cdot}278\pi$; (v) $\cos 0{\cdot}723\pi$; (vi) $\sin 1{\cdot}571$; (vii) $\sin 0{\cdot}07$; (viii) $\cos 0{\cdot}07$; (ix) $\tan 0{\cdot}07$; (x) $\tan 3{\cdot}926$.

30. If the coil in Fig. 119 rotates at 50 revolutions per second what is its angular velocity in radians per second? In this case, in how many milliseconds will it have rotated through one radian from its position at A? What is the voltage at this instant?

(NOTE.—Students are NOT recommended at this stage to use general formulæ for the areas of sectors and segments.)

8.17. The Trigonometry of the Scalene Triangle

Many practical situations involve triangles which are *scalene*, *i.e.*, neither isosceles nor right-angled. Two valuable formulæ, the Sine Rule and the Cosine Rule, have been developed to deal with such problems. With them is associated a useful trigonometrical formula for the area of any triangle.

(a) *Area of a Triangle* (Δ)

Denoting any triangle by its vertices A, B, C, a convenient notation is to distinguish the sides opposite these vertices by the small letters a, b, c (Fig. 121(*e*)).

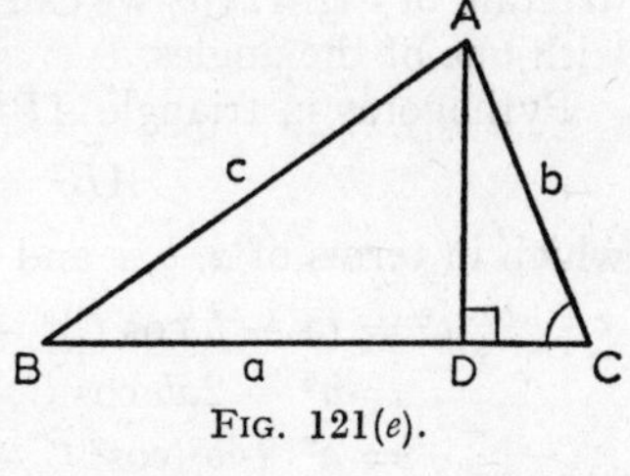

FIG. 121(*e*).

For the purposes of a simple proof we will assume the angles A, B, C to be all acute. Then if AD is the height of this triangle, with BC as its base, we may express the area Δ (delta) of the triangle as

$$\Delta = \tfrac{1}{2}BC \times AD \quad (\text{“}\tfrac{1}{2}\text{ base} \times \text{height”})$$
$$= \tfrac{1}{2}a \times b \sin C$$

using the right-angled triangle ADC to express AD in terms of b and C. Thus we have

$$\boxed{\Delta = \tfrac{1}{2}ab \sin C}$$

as the area of any triangle in terms of two sides and the included angle. Clearly the area can equally well be written as $\tfrac{1}{2}bc \sin A$ or $\tfrac{1}{2}ca \sin B$.

(b) *The Sine Rule*

The three ways of writing the area of a triangle may be equated to give $\Delta = \tfrac{1}{2}bc \sin A = \tfrac{1}{2}ca \sin B = \tfrac{1}{2}ab \sin C$.

Dividing each equal term by abc and multiplying each by 2, we derive

$$\frac{2\Delta}{abc} = \frac{\sin A}{a} = \frac{\sin B}{b} = \frac{\sin C}{c}$$

which shows that the three sides of a triangle are proportional to the sines of the angles opposite to each of them.

The Sine Rule is generally rewritten as

$$\boxed{\frac{a}{\sin A}=\frac{b}{\sin B}=\frac{c}{\sin C}}$$

(c) *The Cosine Rule*

By repeated use of Pythagoras' Theorem in the scalene triangle diagram of Fig. 121(*e*) we can relate the three sides of any triangle with one of the angles.

Pythagoras in triangle ABD gives

$$AB^2 = AD^2 + BD^2$$

which in terms of a, b, c and angle C may be rewritten

$$\begin{aligned} c^2 &= (a - b\cos C)^2 + (b\sin C)^2 \\ &= a^2 - 2ab\cos C + b^2\cos^2 C + b^2\sin^2 C \\ &= a^2 + b^2(\cos^2 C + \sin^2 C) - 2ab\cos C \end{aligned}$$

But for any angle $\cos^2 C + \sin^2 C = 1$ (see 8.10), so we obtain the Cosine Rule

$$\boxed{c^2 = a^2 + b^2 - 2ab\cos C}$$

It can be shown (as in Volume II, page 143) that the Cosine Rule applies whether angle C is acute or obtuse: in the latter case $\cos C$ will, of course, be negative.

The Cosine Rule corresponds to the Extension of Pythagoras in pure geometry, and can be deduced from it.

Example 8(*d*). *Coastguard stations A and B are* 7·2 *miles apart, while a cable ship at C is observed on bearings of* 78° *from A and* 12° *from B. If the bearing of B from A is* 109° (*all bearings East of North*), *find the distance of the cable ship from A.*

Fig. 121(*f*) illustrates this problem. From the given bearings the angles of triangle ABC can be found, by simple subtraction and by using the alternate angles in the diagram. In particular, angles B and C of the triangle are seen to be 83° and 66° respectively.

Then in the triangle the Sine Rule gives

$$\frac{b}{\sin B} = \frac{c}{\sin C}$$

thus

$$b = \frac{c \sin B}{\sin C}$$

$$= \frac{7{\cdot}2 \sin 83°}{\sin 66°}$$

$$= 7{\cdot}82 \qquad \text{[3 sig. fig.]}$$

log
0·8573 $\bar{1}$·9968
0·8541 $\bar{1}$·9607
0·8934

i.e., the ship is distant 7·82 miles from the first coastguard station.

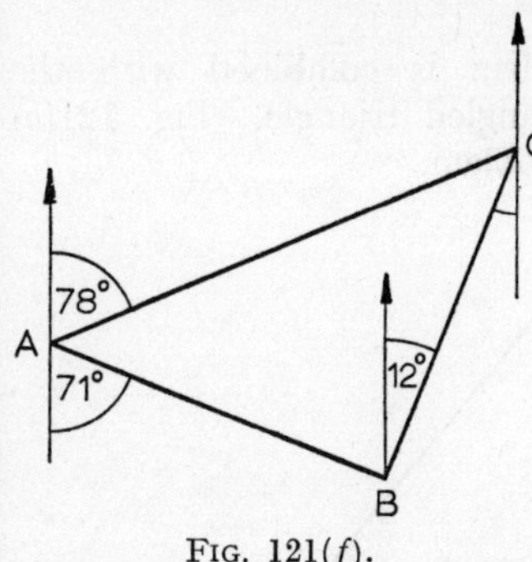

FIG. 121(*f*).

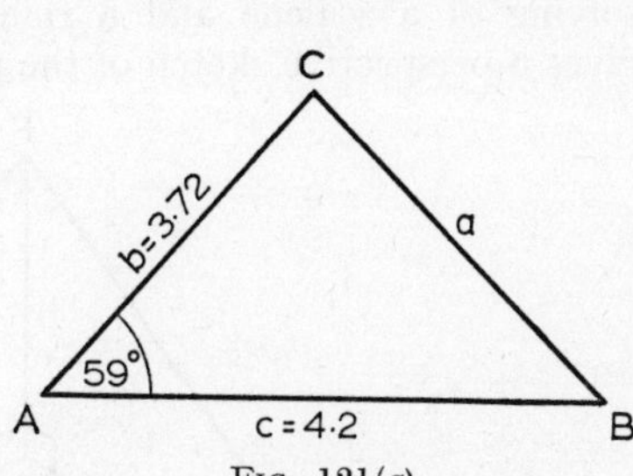

FIG. 121(*g*).

Example 8(*e*). *Two straight roads AB and AC meet at A, so that angle BAC is* 59°, *and B and C are* 4200 *yards and* 3720 *yards respectively from the junction A. Underground cable is to be laid from B to C: find the minimum length needed.*

The geometrical situation here (Fig. 121(*g*) refers) is the " two sides and the included angle " case (cf. page 196) of specifying a triangle.

Using the Cosine Rule, where a is the unknown side

$$a^2 = b^2 + c^2 - 2bc \cos A$$

$$= 13{\cdot}84 + 17{\cdot}64 - 8{\cdot}4 \times 3{\cdot}72 \cos 59°$$

$$= 31{\cdot}48 - 16{\cdot}09$$

$$= 15{\cdot}39$$

$$\therefore\ a = 3{\cdot}92 \text{ (where 1 unit = 1000 yards)}$$

So the shortest length of cable possible is 3920 yards to the nearest 10 yards.

Note the device of taking 1000 yards as a unit in the calculation. Once the third side is determined, angle B could now be found by the Sine Rule, and then the largest angle C from the angle-sum of the triangle (ref. page 184).

Example 8(f). *An aerial mast is observed from two points X and Y at the same horizontal level and distant 600 ft apart. The angle of elevation of the top of the mast from X is 40°. If the foot of the mast Z is at the same level as X and Y with angle $ZXY = 54°$, and the angle $ZYX = 26°$, find the height of the mast and the angle it subtends at Y.*

Here a three-dimensional problem is combined with the solving of a scalene and a right-angled triangle. Fig. 121(h) gives a perspective sketch of the problem.

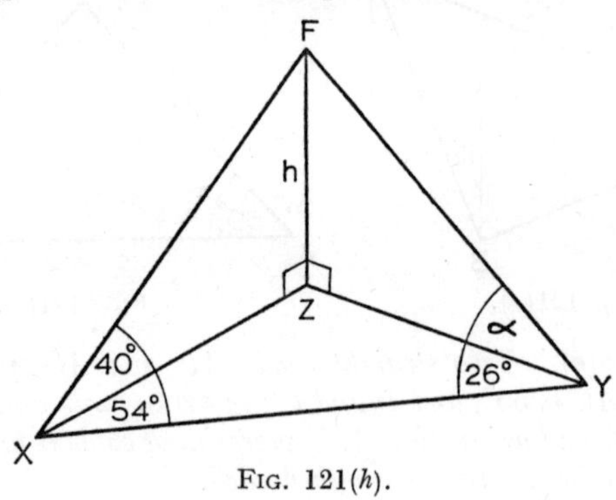

FIG. 121(h).

Thus in the (scalene) base triangle XYZ in the horizontal plane, writing $XZ = y$ and $XY = z$, the Sine Rule gives

$$\frac{y}{\sin Y} = \frac{z}{\sin Z}$$

i.e.,
$$y = \frac{z \sin Y}{\sin z} = \frac{600 \sin 26°}{\sin 100°}$$

But
$$\sin 100° = \sin (180 - 100)° = \sin 80°$$

$$\therefore y = \frac{600 \sin 26°}{\sin 80°}$$

So the height h in the right-angled triangle FZX is given by

$$h = y \tan 40^\circ$$
$$= \frac{600 \sin 26^\circ \tan 40^\circ}{\sin 80^\circ}$$
$$= 224 \text{ feet to the nearest foot}$$

log
2·7782
$\bar{1}$·6418
$\bar{1}$·9238
2·3438
$\bar{1}$·9934
2·3504

Once this height is known, the angle α which the mast subtends at Y is given in the right-angled triangle FZY by

$$\tan \alpha = \frac{h}{x} = \frac{224 \sin 80^\circ}{600 \sin 54^\circ}$$

Whence $\alpha = 24^\circ\ 27'$

using the Sine Rule again, where $YZ = x$.

EXERCISE 8(c)

Use of the Sine Rule, Cosine Rule, and Area of a Triangle

1. If $BC = 3{\cdot}2$ cm., $A = 42^\circ$, and $B = 56^\circ$, calculate the sides AB and AC of triangle ABC, and its area.

2. If $a = 2{\cdot}4$, $C = 113^\circ$, $A = 29^\circ$, find the other three elements of the triangle.

3. A triangle PQR is stated to have $PQ = 3{\cdot}8$, $QR = 2{\cdot}9$ and angle $P = 37^\circ$. Show there are TWO possible triangles, and find the other two angles in each case.

4. If $BC = 2{\cdot}7$, $CA = 3{\cdot}3$, and $B = 49^\circ\ 27'$, find the other elements of the triangle (or triangles) which fit these data.

5. Calculate PR, given that $Q = 62^\circ$, $PQ = 4{\cdot}2$, $PR = 3{\cdot}8$.

6. If $a = b = 4{\cdot}2$ and $C = 82^\circ$, calculate side c and verify by another method.

7. If in a triangle ABC, a, b, c are known, find a formula for calculating angle A.

   If the sides are 48, 36, 14, find the three angles of the triangle.

8. Given $a = 2{\cdot}1$, $b = 3{\cdot}7$, $C = 113^\circ$, calculate c and then use the Sine Rule to find A and B. Check the angle-sum.

9. Sketch the triangle ABC, where $a = 2{\cdot}9$, $b = 4{\cdot}8$, $c = 2{\cdot}4$. Which is the best angle to calculate first by the Cosine Rule? Complete the solution of this triangle, using the Sine Rule as well.

10. A factory roof has a roof truss in the shape of a triangle ABC with a horizontal base BC of length 16 ft., and with angles 32° and 75° at B and C. Calculate the length of the beams AB, AC and the height of A above BC.

11. A triangular building site PQR has frontages QP and QR on two

intersecting roads. If $PQ = 350$ ft., $QR = 270$ ft,. and $PR = 230$ ft., find the angle at which the roads intersect and the area of the site.

12. A level straight road from A to B is 3240 yd. in length in the direction S.E. The bearings of a radar station from A, B are 83° and 24° respectively.

Find the distance of the station from A and also its shortest distance from the road.

13. A stay-wire is attached to a telegraph pole at a height of 28 ft. from the ground and is anchored 8 ft. uphill from the base of the pole. If the hill slopes at 12° with the horizontal, calculate the length of stay-wire required allowing 4 ft. for making off each end.

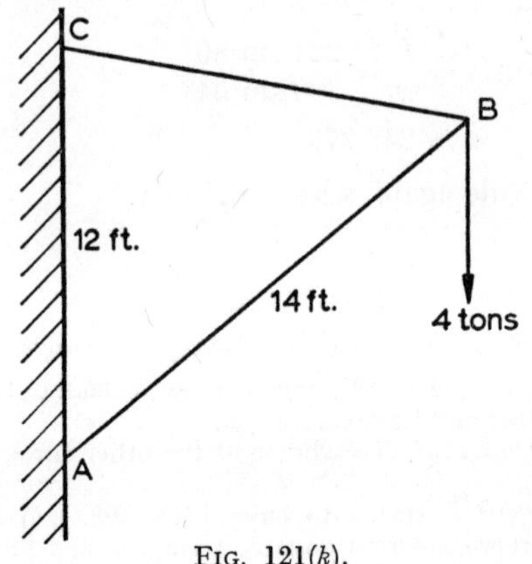

Fig. 121(k).

14. With the same situation as in Question 13, calculate the minimum length of stay-wire needed if the direction of the stay must not be less than 40° with the vertical.

15. In Fig. 121(k) the line AB represents the jib of a simple hoist, of length 14 ft., hinged at A to a vertical wall. A load of 4 tons hangs from B, and a steel rope connects B to a point C on the wall vertically above A.

ABC may be used as a triangle of forces for the three forces acting at B. This means that the sides AB, BC, CA are proportional to the thrust on the jib at C, the tension in the rope, and the load.

If the tension must never exceed $2\frac{1}{2}$ tons, show that the maximum inclination of the jib to the vertical is α, given by $\cos \alpha = \frac{5}{7}$.

How does the thrust in AB vary with different lengths of BC?

16. An angle pole on an overhead cable route has a single stay, anchored in level ground 16 ft. from the foot of the pole. The pole is set with its top towards the stay at an angle of 5° with the vertical.

If the stay is attached to the pole at a point 26 ft. up the pole, calculate the length of stay required. Allow 4 ft. for making off.

CHAPTER 9

AN INTRODUCTION TO MECHANICS

9.1. The A.B.C. of Vector and Scalar Quantities

List X.	*List Y.*
Attraction of a needle by a magnet.	**A**ge, next birthday: 34.
Boat's speed in knots.	**B**lood temperature: 98° F.
Combined pull of two tugs.	**C**ar rating: 14 h.p.
Deceleration of a car.	**D**iet: 2300 calories daily.
Electrical field strength.	**E**lectric power consumption (2 kilowatts).
. . . .	. . . .

What other quantities could you add to each list? How does List X differ fundamentally from List Y?

Notice that all the quantities in List X have *direction* as well as size. For instance, in manœvring an ocean liner it is the *direction* of the combined pull of the two tugs which is more vital than the actual amount of the pull in tons weight. Similarly, in applying a crowbar to shift a heavy weight, it is important to apply the muscular effort at the right point of the bar and in the best direction. Also, in studying the alternating-current generator, we find that the direction of the magnetic field cut by the moving coils is as important as its intensity. (Note para. 8.14 of the previous chapter.)

List Y, on the other hand, contains a few of the large number of quantities for which direction can have no meaning, *i.e.*, they can be expressed completely as a *number of units*. Some may have sign (+ or —), such as " below zero " in temperature, but in all instances they may be represented by points on a *line scale*, as in the base line of a graph. Such quantities are called *Scalar*, and the normal processes of arithmetic and algebra suffice for their study. But what kind of " arithmetic " can be devised for quantities which involve both number and direction? First let us give such quantities a *name*!

These List X items, which have direction as well as magnitude, are termed *Vector* quantities. Their study must inevitably deal with relationships in space as well as number, involving geometrical concepts and argument, and the technique of trigonometry for their calculation. (A special " vector " algebra has been developed, of great importance to the engineer, for calculating with vectors. This we shall begin to discuss in Volume II.)

The vectors we most commonly meet are forces, speeds, and accelerations, which are involved when we study matter in motion or in equilibrium. This study is called " *Mechanics* ".

9.2. Adding Vector Quantities (*e.g.*, Speeds)

(a) *In one Dimension*, i.e., *Speeds in a Straight Line*

Trains A and B (Fig. 122) are travelling in parallel tracks at the speeds shown. The driver of A takes a " pot-shot " at the

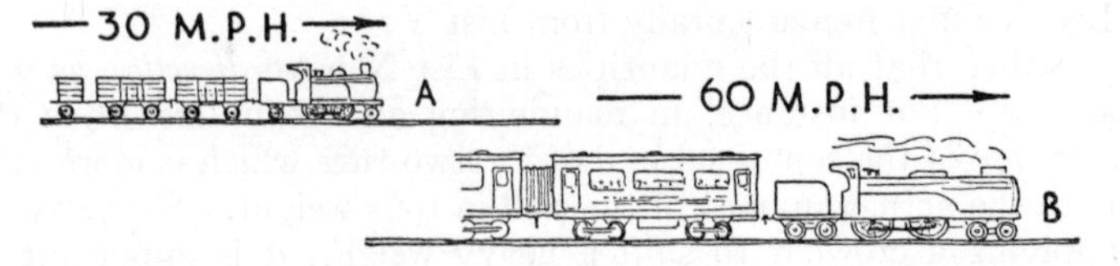

Fig. 122.—Motion in One Dimension.

guard of B. If the muzzle velocity of the gun is 1500 f.p.s., at what speed does the bullet travel in the air as it leaves the gun?

At the moment of firing, the gun shares the velocity of the train, *i.e.*, 30 m.p.h. or 44 f.p.s. The bullet leaves the barrel at a speed of 1500 f.p.s. *relative to the barrel*, so its total speed in the air (relative to the *ground*) will be 1500 + 44, *i.e.*, 1544 f.p.s.

Also the speed at which the bullet approaches Train B (moving at 88 f.p.s.) will be 1544 − 88, *i.e.*, 1456 f.p.s.

We see then that *vectors in the same direction may be added or subtracted just like scalar quantities.*

(b) *Speeds in Two Dimensions*

As a contrast to the above, let us consider how to add speeds in *two* dimensions. In Fig. 123(*a*) we have such a situation; the

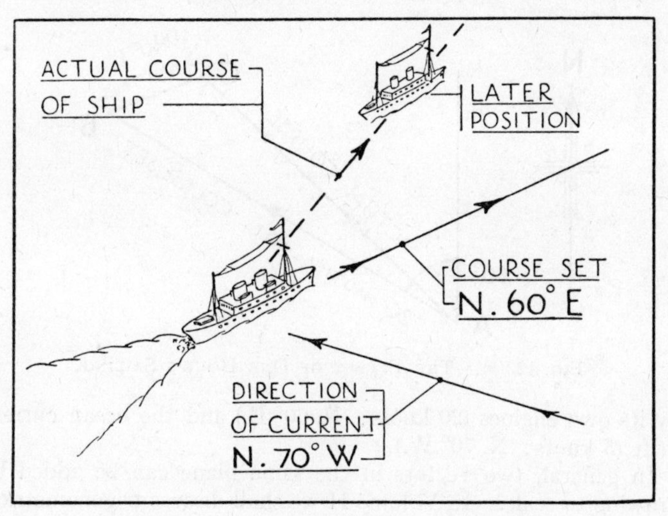

FIG. 123(*a*).—MOVEMENT IN TWO DIMENSIONS.

liner is cruising on a course N. 60° E. in the path of an ocean current running at 5 knots in a direction N. 70° W. If the cruising speed of the ship in still water is 20 knots, what is the actual (total) speed of the ship in magnitude and direction?

Fig. 123(*b*) shows in plan (drawn to scale) the combined effect of engine speed and current drift at the end of 1 hr. under these conditions. *A* is the position of the vessel at 2 p.m., shall we say. Then at 3 p.m. the vessel would have been at *B*, 20 miles N. 60° E. from *A*, if the current had not caused a drift of 5 miles to *C*, N. 70° W. from *B*. As the drift is a continuous effect (assuming the current remains at 5 knots) the actual course of the vessel will be *AC*. By measurement in our scale drawing $AC = 17{\cdot}2$ nautical miles, and the bearing of *C* from *A* is N. 47° E.

Expressing our result in another way, the ship travels 17·2 nautical miles in 1 hr.—which means an actual speed of 17·2 knots in a direction N. 47° E. This last statement is the *vector sum* of the two "component" speeds given to the vessel

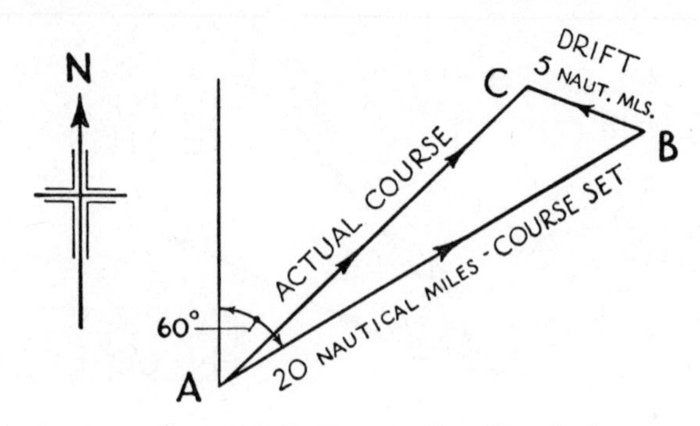

FIG. 123(*b*).—THE EFFECT OF ONE HOUR'S SAILING.

by its own engines (20 knots; N. 60° E.) and the ocean current drift (5 knots; N. 70° W.).

In general, two vectors in the same plane can be added by drawing to scale. In Volume II we shall discuss trigonometrical techniques for doing the job by calculation.

(c) *Speeds in Three Dimensions*

A train is constrained to move in one direction in the *line* of the rails; a ship is free to move in two dimensions on the *surface* of the sea; but an aircraft is free to move in *space, i.e.*, in *three* dimensions.

Consider a helicopter in flight (Fig. 124). Its airscrew gives it a forward speed of 280 f.p.s., its horizontal (rotor) blades impart a vertical lift equivalent to 60 f.p.s., and there is an effective cross-wind (at right angles to the course set) of 40 f.p.s.

Then in 1 sec. the aircraft will travel from O to Q, where $OM = 280$ ft. (forward movement), $MP = 60$ ft. (upward movement), and $PQ = 40$ ft. (wind drift).

In this instance we may *calculate* the actual speed of the aircraft.

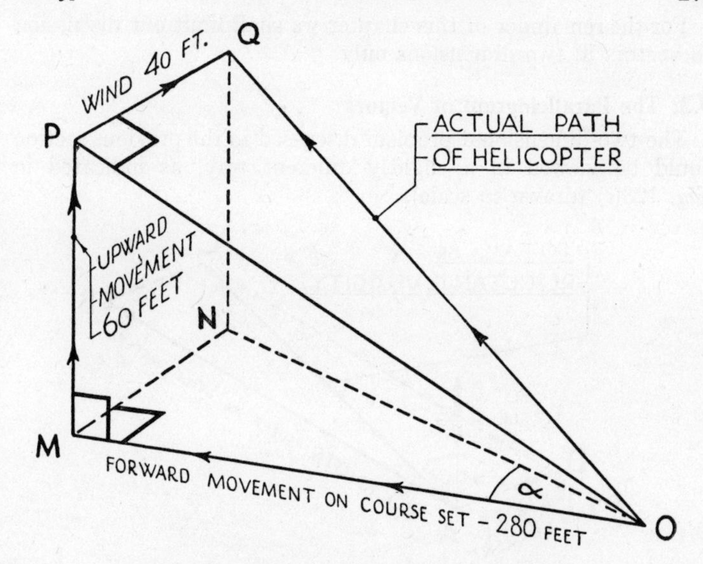

FIG. 124.—MOTION IN THREE DIMENSIONS.

For by Pythagoras' Theorem,

$$OP^2 = 60^2 + 280^2 \text{ (right-angled } \Delta OMP)$$

also

$$OQ^2 = 40^2 + OP^2 \text{ (right-angled } \Delta OPQ)$$

$$\therefore \quad OQ^2 = 40^2 + (60^2 + 280^2)$$

$$= 1600 + 3600 + 78400$$

$$= 83600$$

$$\therefore \quad OQ = 289 \text{ (approx.)}$$

So the actual speed of the aircraft is 289 f.p.s.

Also the aircraft is blown out of its course an angle α, given by

$$\tan \alpha = \frac{40}{280} \text{ (in horizontal right-angled } \Delta OMN)$$

$$= 0{\cdot}1429$$

$$\therefore \quad \alpha = 8^\circ \, 8'$$

So the aircraft is flying 8° off its set course.

Note : All bearings or courses set are in a *horizontal* plane.

For the remainder of this chapter we shall limit our discussion to vectors in two dimensions only.

9.3. The Parallelogram of Vectors

The two-dimensional problem discussed in the previous section could be viewed in a slightly different way, as indicated in Fig. 123(*c*) (drawn to scale).

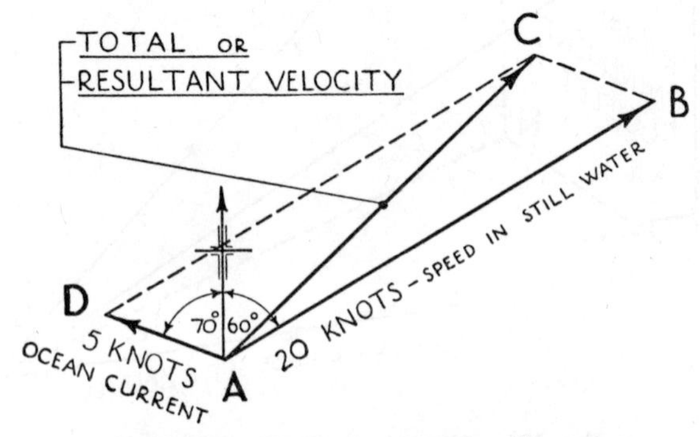

FIG. 123(*c*).—THE PARALLELOGRAM OF VELOCITIES.

At any *instant* the vessel can be said to have two *component* velocities: 20 knots in a direction N. 60° E. (represented by AB) due to its own engines and helmsman, and 5 knots in a direction N. 70° W. (represented by AD) due to the ocean current then running. From Fig. 123(*b*) we know that the *resultant* speed of the vessel is given in magnitude and direction by the closing line AC of the $\triangle ABC$. But this line AC is also the diagonal of the complete parallelogram $ABCD$.

So we see that the resultant of the two component speeds $\overrightarrow{AB}$ and $\overrightarrow{AD}$ is just $\overrightarrow{AC}$; *i.e.*, it is represented in magnitude and direction by the diagonal of the parallelogram $ABCD$.

Observe the *notation*, which is sometimes useful. AB just means the "length of the line AB", whereas $\overrightarrow{AB}$ represents the *vector* which the line AB represents in *direction* as well as in magnitude (*i.e.*, from A to B).

So far we have discussed only velocities, but the same ideas are equally useful in dealing with forces and accelerations, indeed any vector quantities.

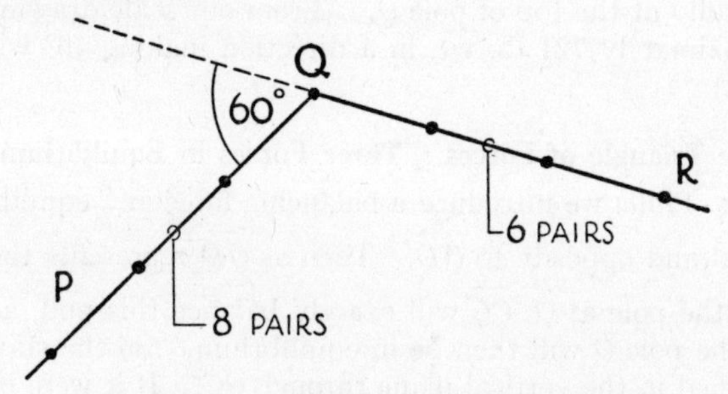

FIG. 125(*a*).—O/H ROUTE DIAGRAM (PLAN).

In Fig. 125(*a*) *PQR* represents in plan an overhead route which changes direction at *Q* by 60°. *Q* is a terminal pole, and if it were not stayed the effective pull due to the tensioned wires

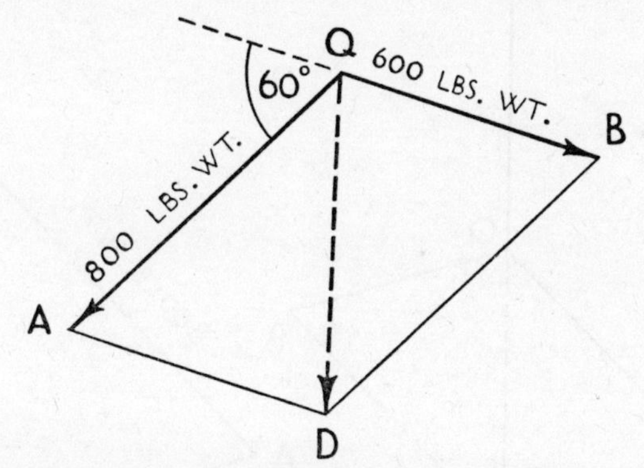

FIG. 125(*b*).—THE RESULTANT PULL ON THE ANGLE-POLE.

would tend to pull it over. If we assume each wire to be tensioned to 50 lb. wt. (*e.g.*, 40-lb. cadmium–copper conductors), then in the direction *QP* there is a total pull (assumed horizontal) of 800 lb. wt. ($\overrightarrow{QA}$), and along *QR* a combined pull of 600 lb. wt.

$(\overrightarrow{QB})$ as suggested to scale in Fig. 125(*b*). Then, completing the parallelogram $QADB$, $\overrightarrow{QD}$ represents the ***resultant pull*** exerted horizontally at the top of pole Q. (From our scale drawing, this is approximately 721 lb. wt. in a direction making 46° with the line PQ.)

9.4. The Triangle of Forces : Three Forces in Equilibrium

In Fig. 125(*c*) we introduce a balancing force or "equilibrant" $\overrightarrow{QC}$ equal and opposite to $\overrightarrow{QD}$. Then as $\overrightarrow{QD}$ represents the total pull on the pole at Q, $\overrightarrow{QC}$ will exactly balance this pull, and the top of the pole Q will then be in equilibrium. So the stay must be fastened in the vertical plane through QC. If it were possible for the stay to be horizontal (*e.g.*, attached to a building), then the tension in the stay, if correctly aligned, would be 721 lb. wt.

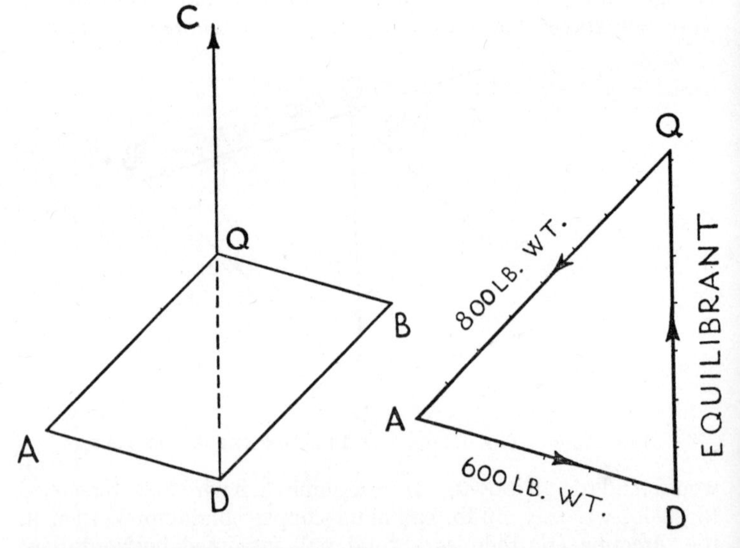

FIG. 125(*c*).—THE EQUILIBRANT.

FIG. 125(*d*).—THE TRIANGLE OF FORCES.

In Fig. 125(d) we have drawn the $\triangle QAD$. Note that $\overrightarrow{AD} = \overrightarrow{QB}$ and $\overrightarrow{DQ} = \overrightarrow{QC}$; in words, $\overrightarrow{AD}$ represents the 600-lb. wt. force along QR in magnitude and direction, and $\overrightarrow{DQ}$ similarly represents completely the " equilibrant " 721-lb. wt. force.

So the three forces in equilibrium acting at Q are in fact represented in magnitude and in direction by the sides of $\triangle$QAD *taken in order.*

Such a triangle as QAD is called a *Triangle of Forces.* Any three forces in equilibrium must be representable by the three sides of a triangle in order—" tail-chasing ".

This principle gives us a simple geometrical construction for finding unknown forces, where a system is in equilibrium. If extended to a system of more than three forces, it involves a " Polygon of Forces ".

9.5. Experimental Justification for the Triangle of Forces Principle

(a) *Using Three Spring-balances*

On a horizontal drawing-board the fixed ends of three spring-balances A, B, and C are secured, as shown in Fig. 126(a). To

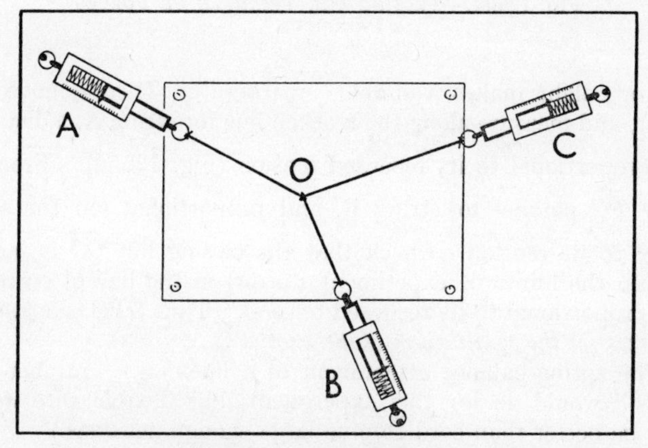

FIG. 126(a).—THE SPRING-BALANCE EXPERIMENT—HORIZONTAL FORCES IN EQUILIBRIUM.

the moving (spring) end of each balance a thin silk cord or linen thread is attached, and the three cords knotted together at O so that all three are well tensioned. The tension in each of the three cords can then easily be read on the spring-balances. Note these tensions.

Mark the " lie " of each cord on the paper (its shadow under

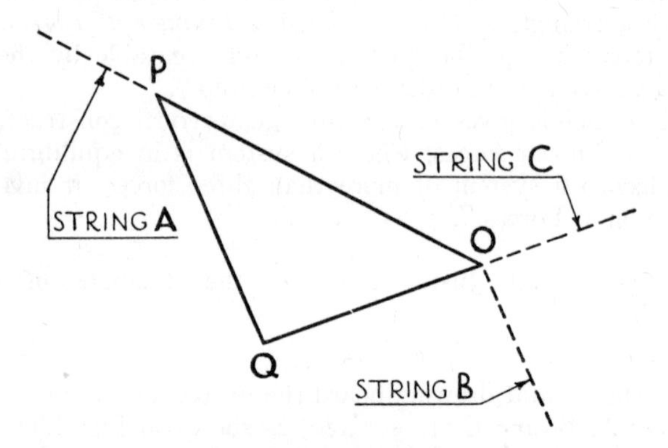

FIG. 126(*b*).—TESTING THE TRIANGLE OF FORCES PRINCIPLE.

a strong light makes a simple " marker "). Then remove the paper, and measure along the marked line for string A, a distance $\overrightarrow{OP}$ proportional to its recorded tension (Fig. 126(*b*)). From P draw $\overrightarrow{PQ}$ parallel to string B, and proportional (on the same scale) to its tension. Check that the closing line $\overrightarrow{QO}$ is indeed (within the limits of experimental error) in the line of string C, and proportional to its recorded tension. Thus OPQ is a *triangle of forces for the equilibrium of the knot at* O.

(The spring-balance attachment of a lineman's " ratchet and tongs " would do for this experiment, but flexible wire would then be better than cord to give large enough tensions.)

(b) *Using Pulleys and Weights*

A similar experiment to the above may be performed, using two smooth-running pulleys mounted on a vertical board, as suggested in Fig. 127.

The effect is that the tensions in the three strings at the knot are measured by the weights (10 lb., 6 lb., and 8 lb. in our diagram) which cause these tensions. Each pulley is thus a

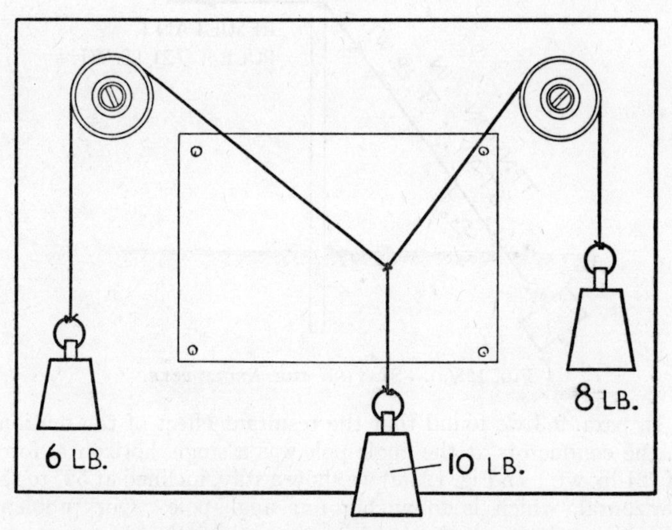

FIG. 127.—THE PULLEYS AND WEIGHTS EXPERIMENT—FORCES IN EQUILIBRIUM IN A VERTICAL PLANE.

device for altering the *direction* of a force without affecting its magnitude. (To do this effectively, they must be as nearly frictionless as possible.)

The " lie " of each string at the knot is marked on the drawing-paper, and the principle of the " triangle of forces " tested as in the previous experiment.

Whichever method is adopted, the experiment should be repeated with different positions of the spring-balances in (*a*) or with different weights in (*b*).

9.6. The Component Effects of a Vector

(a) *Forces.* Just as two forces may be considered usefully as having a *resultant* effect (by the parallelogram of forces) it is frequently convenient to study the *component* effects of a single force in two directions. Very often the components are considered horizontally and vertically.

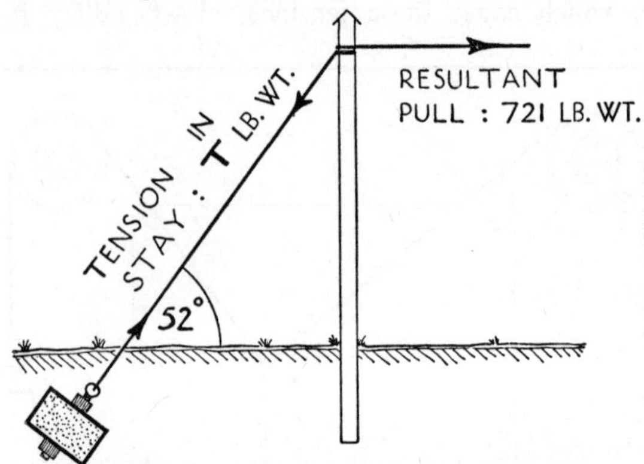

FIG. 128(*a*).—STAYING THE ANGLE-POLE.

In para. 9.3 we found that the resultant effect of the tensions in the conductors at the angle-pole was a single horizontal force of 721 lb. wt. In Fig. 128(*a*) we show a stay, inclined at 52° to the horizontal, which is to anchor our angle-pole. Our problem, then, is to estimate the resulting tension in the stay-wire—for this fact will govern the gauge of wire to be chosen.

In Fig. 128(*b*) the single tension T lb. wt. pulling on the pole at O in a downward direction at 52° with the horizontal is shown "resolved" into two components at right angles—X lb. wt. horizontally to the left, and Y lb. wt. vertically downwards. By the Parallelogram of Forces principle the combined effect of the two perpendicular forces (proportional to the two adjacent sides $\overrightarrow{OQ}$, $\overrightarrow{OR}$ of the rectangle $OQPR$) will be exactly equal to the effect of the single force T lb. wt. (proportional to the diagonal OP of the same rectangle).

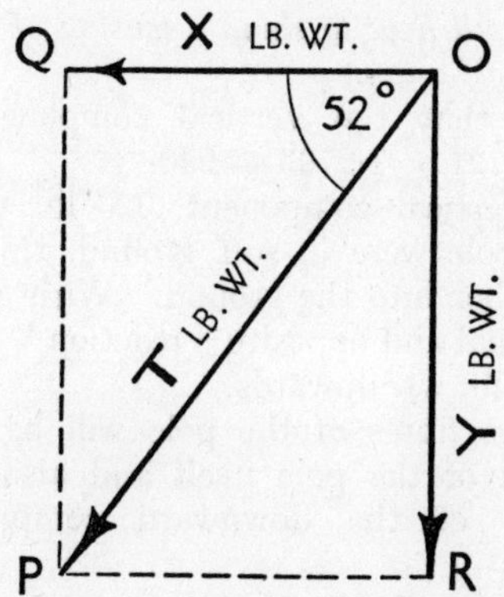

FIG. 128(*b*).—COMPONENT EFFECTS OF THE TENSION IN THE STAY-WIRE.

Symbolically $$\vec{T} = \vec{X} + \vec{Y}$$

or $$\vec{OP} = \vec{OQ} + \vec{OR}$$

The horizontal component X lb. wt. will have to balance an equal and opposite horizontal force of 721 lb. wt.—the sum effect of the tensions in the line wires. Clearly the vertical component Y lb. wt. cannot help in this task.

So $$X = 721$$

We now know X, so we could draw Fig. 128(*b*) to scale, and measure the tension T.

Alternatively, we may use trigonometrical argument. For in the right-angled ΔOQP,

$$\cos 52° = \frac{X}{T}$$

$$\therefore \quad T \cos 52° = X$$

$$= 721$$

i.e., $$T = \frac{721}{\cos 52°}$$

$$= \frac{721}{0{\cdot}6157}$$

$$\simeq 1180 \text{ (slide-rule)}$$

So the stay-wire will have to stand a tension of 1180 *lb. wt.* (i.e., *the weight of over half a ton.*)

Similarly, note that the vertical component is given by $Y = X \tan 52° = 721 \times 1{\cdot}2799 \simeq 920$.

What of this vertical component, 920 lb. wt. of the total tension? If the pole were in soft ground, this would tend to drive the pole farther into the ground. With good firm ground there will be an equal and opposite " reaction " from the ground, *i.e.*, a push of 920 lb. wt. upwards.

Furthermore, the fibres of the pole will be in compression, due to the weight of the pole itself and also to the addition (below the stay) of the downward compressive force of 920 lb. wt.

As far as the safety of the pole is concerned, *the main effect of the staying is to replace a harmful bending effect by a harmless vertical compressive stress.* (See para. 9.10 below.)

(b) *Velocities.* Where a body may have velocities in two or more dimensions (such as a cricket ball in flight) it is only possible to study its motion mathematically by resolving its velocity at any moment into component velocities.

An electrical example of this is the generation of A.C. voltages, discussed descriptively in the previous chapter (para. 8.14).

In Fig. 129(a) PQ represents the cross-section of the moving

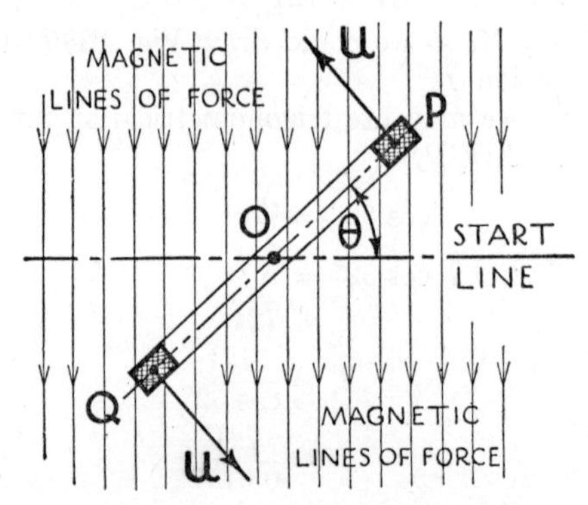

FIG. 129(a).—THE A.C. GENERATOR.

coil rotating at constant speed (shall we say u f.p.s.) in a constant magnetic field. Now the amount of voltage induced in the windings at any instant is proportional to the speed at which the coil is cutting the lines of magnetic force.

So we are interested in the speed at which the coil is moving perpendicularly to the lines of magnetic force, *i.e.*, in the component of the speed u f.p.s. parallel to the start line.

In Fig. 129(*b*) we have resolved the velocity u, at the instant when OP makes an angle θ with the start line, into components x f.p.s. parallel to the start line, and y f.p.s. at right angles to this direction.

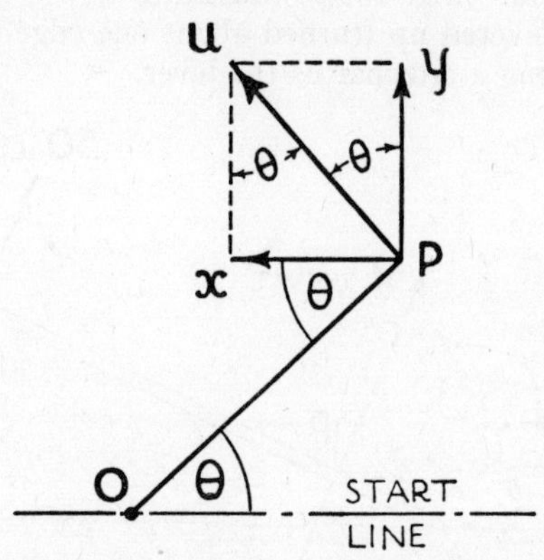

FIG. 129(*b*).—THE COMPONENTS OF A VELOCITY.

From the trigonometry of this " rectangle of velocities ",*

$$\frac{x}{u} = \sin\theta \quad \text{(opposite side : hypotenuse)}$$

$$\therefore \quad x = u \sin\theta$$

which is the speed in f.p.s. at which the coil is cutting the lines of force.

So that the induced voltage at this instant is proportional to sin θ.

In general, if a coil is rotating at constant speed in a uniform magnetic field the voltage generated at any position will be of

* Note in Fig. 129(*b*) the pairs of *alternate angles*, each equal to θ.

the form $V \sin \theta$, where V is the "amplitude" of the induced voltage.

As θ will be proportional to the time t sec., one may also write the voltage v volts at any instant as $v = V \sin kt$ (substituting kt for θ).

9.7. The Turning Effect of a Force—Moments

In everyday experience force is often exerted to produce a *turning* effect—in using a crowbar, turning a key in a lock, undoing a nut with a wheel-brace, prising open a manhole cover, and so on. Fig. 130(*a*) illustrates how a heavy block of stone may be levered up (turned about one edge which rests on the ground) using a crowbar as the lever.

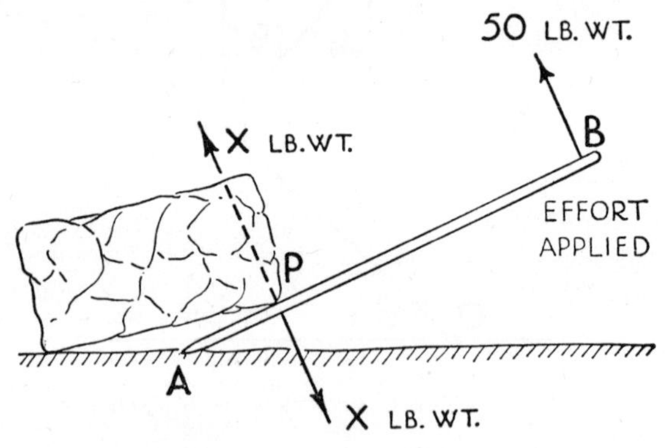

FIG. 130(*a*).—LEVERAGE—THE CROWBAR.

A muscular effort, 50 lb. wt., is exerted at the end B of the crowbar, which is 8 ft. long. The other end A of this crowbar is resting on the rough ground, and we call this point A, about which the bar turns, its *fulcrum*.

We know from experience that to get maximum leverage we need to apply the effort as far from the fulcrum as possible, and at right angles to the crowbar. As the result of experiment, it is found that the turning effect of a force which we call its *moment* is conveniently measured by using the formula

$$\text{Moment of Force} = \frac{\text{Amount of Force} \times \textit{Perpendicular} \text{ Distance}}{\text{of its Line of Action from the Fulcrum}}$$

The turning effect of the 50 lb. wt. effort is therefore $50 \times 8 = 400$ lb. ft. units. If the load rests on the point P of the bar, where $AP = 1$ ft., then the block will exert a downward force on the bar at P (at right angles to the bar) of X lb. wt.

If the bar is resting in equilibrium the turning effect of the 50-lb. wt. force (anti-clockwise) about the fulcrum A will exactly balance that of the X lb. wt. force (clockwise) about the same fulcrum. In other words

Taking moments about the fulcrum A

$$50 \times 8 = X \times 1$$

So that $X = 400$ (lb. wt.)

The bar will press upwards against the block at P with an equal but opposite force X lb. wt. (dotted in the diagram).

So the effect of using the crowbar is to magnify a muscular effort of 50 lb. wt. into an effective lifting force of 400 lb. wt. on the block.

It is worth noting that the end B of the bar must be raised 8 in. for every inch the point P is raised. (The same ratio 8 : 1 as the magnifying " leverage " obtained.)

There is one other force acting on the crowbar which is not shown in the diagram. This is the reaction from the ground, and it is not difficult to see that it must be equal to 400 – 50, *i.e.*, 350 lb. wt. perpendicular to the bar. As this force acts through the fulcrum A it can have no turning effect about A.

(This discussion has assumed a rigid crowbar and neglects friction between the bar and the block at P.)

Fig. 130(*b*) represents another example of the same type of leverage—a guillotine used to trim the end of a piece of wood or card. If the arm AB is 18 in. long, and the blade is cutting the material at P, 4 in. from the fulcrum A, what force (Y lb. wt.) is exerted on the material when 40 lb. wt. applied at B is just sufficient to cut the material?

In this case the " dotted " force represents the action of the blade on the material, the forces marked with continuous lines being those acting on the guillotine.

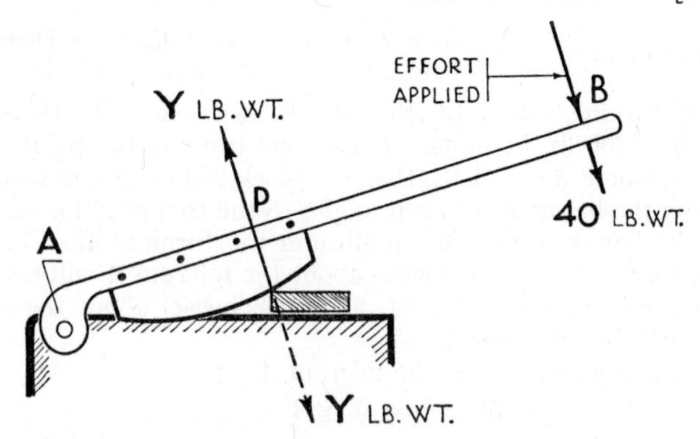

FIG. 130(*b*).—LEVERAGE—THE GUILLOTINE.

Taking moments about A

$$Y \times 4 = 40 \times 18$$
$$\therefore \quad Y = 180 \text{ lb. wt.}$$

So the blade must exert a force of at least 180 lb. wt. in order to cut the material. (In practice, much more than the minimum is exerted, in order to give a quick, clean cut.)

Look out for other instances of levers (the action of an old-fashioned self-filling fountain pen is another example). In any type of lever action there is a *load* to be supported or moved, a *fulcrum,* and an applied *effort.* An intriguing problem in leverage is the action of an oarsman—does the rowlock act as the fulcrum, or the tip of the blade? Or is it neither of these?

In passing, it is worth noting that many levers are bent or curved. A very common type is the " bell-crank " lever which is bent at right angles at the fulcrum. This not only provides leverage, but also changes the direction (by 90°) of the applied effort. A relay armature normally operates as a bell-crank lever, the attraction of the energised relay core acting as the " effort " and the resistance of the spring-set as the load. Bell-crank levers are important components of teleprinter mechanisms, too.

9.8. Parallel Forces in Equilibrium

Systems of parallel forces are very common in practical loading problems. Thus, on a windless day the external loads of a building are nearly always gravitational ones, *i.e.*, vertical forces due to the attraction of the Earth. Clearly one cannot draw triangles or parallelograms of forces when all the forces concerned are parallel. So a fresh approach is necessary.

Fig. **131** represents a light pole 8 ft. long with loads of 30 and 20 lb. attached at the ends. If we wish to carry this loaded pole with a minimum of discomfort, where should we grip the pole? Intuitively we would feel for the " point of balance "—somewhere nearer to the 30 lb. end A than to the 20 lb. at B. But how near to A?

We need to grip the pole at the fulcrum P so that the moment (clockwise in our diagram) of the 20 lb. wt. about P is counterbalanced exactly by the moment (anti-clockwise) of the 30 lb.wt. about the same point.

If $AP = x$ ft. we have:

$30 \times x$	$=$	$20 \times (8 - x)$
The moment of the 30 lb. wt. about P (anti-clockwise)		The moment of the 20 lb. wt. about P (clockwise)
i.e., $30x$	$=$	$20\ (8 - x)$
or $3x$	$=$	$2\ (8 - x)$
$\therefore\ 3x$	$=$	$16 - 2x$
$\therefore\ 5x$	$=$	16
i.e., $\underline{x}$	$\underline{=}$	$\underline{3{\cdot}2}$

So the point of balance of this loaded pole is 3·2 ft. from A.

Does it mean that when we hold the pole at P there is no force to overcome? Clearly not; we must exert an upward force of 50 lb. wt. to counterbalance the total load on the pole. But it *does* mean that there is no tendency for the pole to rotate, and the load may be carried with the least inconvenience.

As the combined load is now counterbalanced entirely by one upward force through P, it also follows that the *resultant* of the

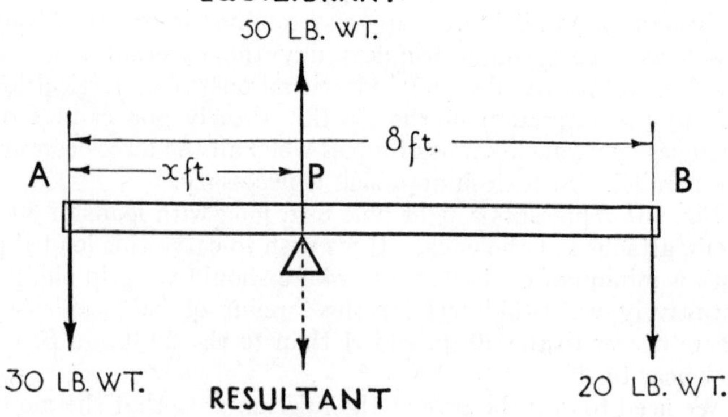

FIG. 131.—THE RESULTANT OF PARALLEL FORCES.

30 lb. weight at A and the 20 lb. weight at B is a simple downward force of 50 lb. wt. acting through the particular point P (shown dotted in Fig. 131).

This principle of " Taking Moments " about a convenient point is a very useful one in Mechanics. And not merely for parallel forces; it must always be used in calculations in which the forces involved do not act through one point.

Many practical weighing-machines employ this principle. The butcher's steelyard (Fig. 132) is a homely example of this. (Steelyards of this type were used by the Romans.)

AB is the " beam " of the steelyard, and C the " fulcrum " or point of balance. With a load L hung below A (the side of mutton at the butcher's) the " rider " R is moved along the beam until the latter just balances in a horizontal position.

Then the turning effect of weight R (clockwise) will exactly balance the turning effect (anti-clockwise) of the load L about the fulcrum C.

As the position of A is fixed during the weighing, the right-hand side of the beam is graduated in pounds or stones to read directly the weight L. In our illustration, if the markings denote pounds, then the load weighs 6 lb. and the rider must weigh 1 lb. only.

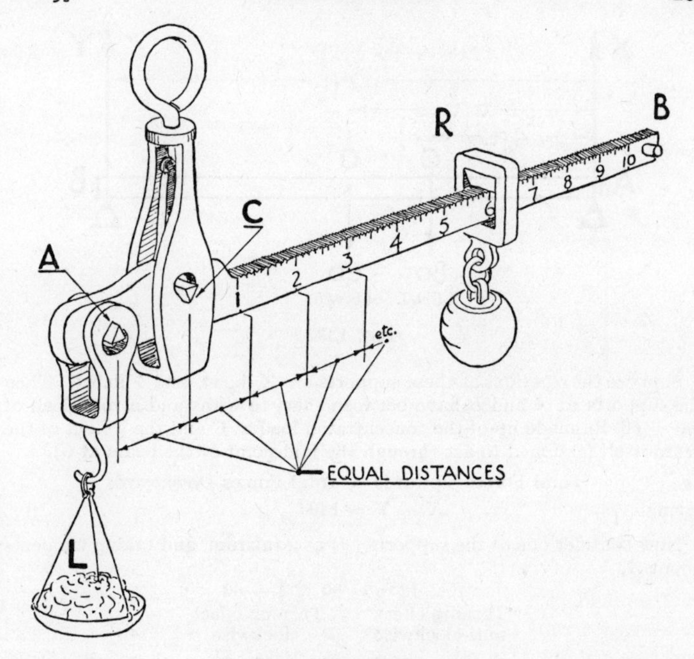

FIG. 132.—THE STEELYARD.

(In practice, the centre of gravity of the steelyard is not at the fulcrum, but on the load side. See Exercise 9, Question 26).

Practical steelyards are often modified to give a wider range of weighing—either by providing alternative positions to hang the load or by supplying alternative riders. Thus the rider of 10 times the weight of R would multiply the scale readings by 10.

You will discover many more examples of similar devices, and even though they be heavily encased in chromium plate and glass the underlying principle is unaltered.

Example 9(*a*). *A uniform horizontal beam* 12 *ft. long, weighing* 60 *lb., is supported at its ends* A *and* B, *and carries a concentrated load of* 80 *lb. at a point* C, 4 *ft. from* A (Fig. 133). *Calculate the reactions at the supports* A *and* B.

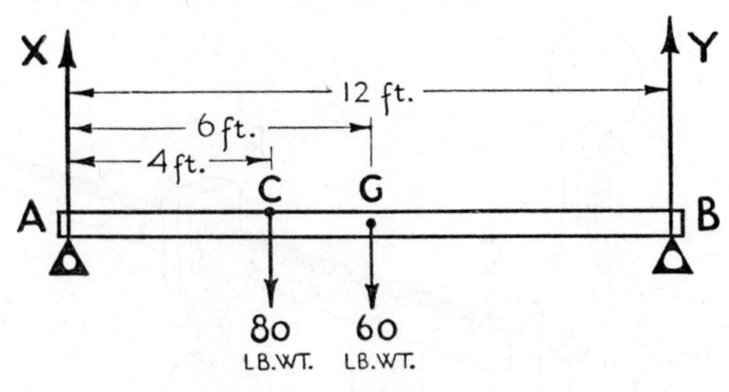

FIG. 133.

Suppose the reactions at these supports are X lb. wt. and Y lb. wt. Then the supports at A and B have between them to withstand a total load of (80 + 60) lb. made up of the concentrated load at C and the weight of the beam itself (assumed to act through the mid-point of the beam at G).

i.e., Total Forces Upwards = Total Forces Downwards

giving $$X + Y = 140 \qquad \text{(i)}$$

Now consider one of the supports (A) as a fulcrum, and taking moments about A,

$$\underset{\text{Turning effect anti-clockwise}}{Y \times 12} = \underset{\text{Turning effect clockwise}}{80 \times 4 + 60 \times 6}$$

i.e., $$12Y = 320 + 360$$
$$= 680$$
$$\therefore\ Y = 56\tfrac{2}{3} \qquad \text{(ii)}$$

Substitute for Y in equation (i), and we get

$$\underline{X = 83\tfrac{1}{3}}$$

So the reactions at A *and* B *are* $83\frac{1}{3}$ *and* $56\frac{2}{3}$ *lb. wt. respectively.*

In the above solution we took moments about A; note that we could also have taken moments about B, for

$$\underset{\text{(clockwise)}}{X \times 12} = \underset{\text{(anti-clockwise)}}{80 \times 8 + 60 \times 6}$$
$$= 1000$$
$$\therefore\ \underline{X = 83\tfrac{1}{3}}$$

Check for yourselves that you can take moments about *any* point (*e.g.*, a point in the beam 9 ft. from A) and still the total moments clockwise balance the total moments anti-clockwise.

It is quite an easy matter to verify the results calculated above by *experiment*, using spring-balances to support the beam at A and B.

We can generalise the ideas used in the discussion above in mathematical language as follows:

A system of forces in a plane is in equilibrium if

(*a*) the components of these forces in any direction balance (sum algebraically zero); and

(*b*) the moments of these forces about any point in the plane balance (sum algebraically zero).

*9.9. The Effects of Forces on Materials

The telecommunications engineer, like any other engineer, is very much concerned with the way in which his materials are affected by forces, such as wind pressures on pole-lines and radio towers, or the weight of heavy vehicles passing over manholes, or the bending of springs in switches, mechanisms, and relays. He needs to know not only that such effects exist but also how to calculate their magnitude and direction. In some cases he has to make sure that his structures are sufficiently strong for their purposes and yet use no more material than is necessary; in others he must ensure that they have a sufficiently long life and do not fail after a limited number of operations.

Stress

When a material is subjected to the action of a force it becomes *stressed*. With a large force acting on a small piece of material the latter is heavily stressed; with a small force acting on a large piece it is only lightly stressed. So a measure of stress is the ratio of the force acting to the dimensions of the material withstanding it. If a force P lb. wt. is transmitted through a part of the material having a cross-section a sq. in., then the average intensity of the stress is said to be p lb. per sq. in., where

$$p = \frac{P}{a}$$

Example (*i*). A stay wire exerts a pull of 1500 lb. wt. on the stay-rod which anchors it to the ground. If the rod has a diameter of 0·50 in., then the rod is subjected to a stress of

$$\frac{1500}{\frac{\pi}{4}(0\cdot5)^2} = \frac{24}{\pi} \times 10^3$$

$$= 7\cdot6 \times 10^3 \text{ lb. per sq. in.}$$

Example (ii). The insulating block on which a radio mast rests measures 8 by 8 by 6 in. high. The total load due to the weight of the mast, together with the vertical components of the tensions in the stays, is estimated to be 10 tons. So the stress in the block, assuming it to be uniformly distributed, is

$$\frac{10 \times 2240}{8 \times 8} = 350 \text{ lb. per sq. in.}$$

It is important to remember that the stress calculated in this way is only the *average* intensity of stress, and in using any figure so calculated we are assuming that the stress is uniformly distributed over the cross-section. Irregularities in shape or finish may produce a non-uniform distribution, with stresses greater than the average at some points.

Changes of Shape and Dimensions

Whenever a piece of material is subject to stress it changes its shape and dimensions. The stay-rod in Example (i) above stretches and becomes a little longer under the pull of the stay —and incidentally, a little thinner too. The block in Example (ii) loses a little height and expands slightly horizontally. Such changes in shape and dimension are easily recognisable in a material like rubber, but they occur in all materials. Although they may not be large enough to be visible, these changes may be measured with special instruments.

Strain

Under the same tension, a long rod will obviously stretch more than a short one. If the rod be of uniform cross-section throughout, each unit of length will stretch by the same amount. Similarly, under compression each unit of length will *shorten* by the same amount (assuming a uniform stress distribution).

The change in dimension per unit length in the line of action of the force applied is called the *strain*. Thus, if a dimension L in. changes by an amount l in. the strain q is given by the ratio

$$q = \frac{l}{L}$$

Remember that, while in conversation we speak of a rope " taking the strain ", in engineering " strain " is the measure of the proportionate change in dimension due to a " stress ".

*9.10. Different Kinds of Stress

When a tie-bar or a stay-wire is resisting a pull tending to lengthen it, it is said to be *in tension*, and the stress which exists across a section at right angles to the direction of the pull is a *tensile* stress (Fig. 134(*a*)). Thus, a wire by which a weight hangs is under a tensile stress.

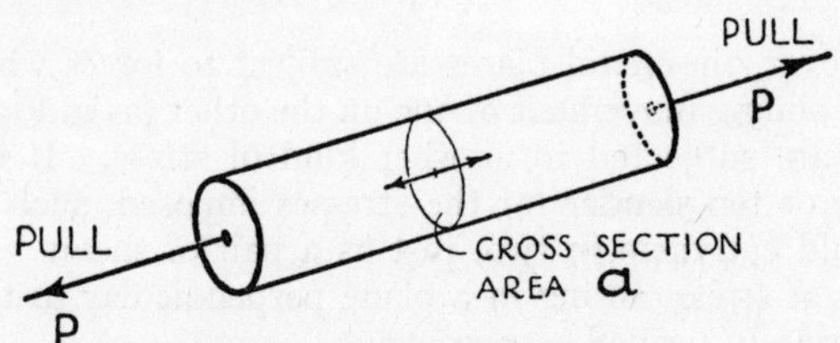

FIG. 134(*a*).—TENSILE STRESS.

If on the other hand, a piece of material is resisting a push tending to shorten it, it is said to be in *compression* or under a *compressive* stress (Fig. 134(*b*)). This occurs, for example, in a block of material on which a machine or other heavy structure stands.

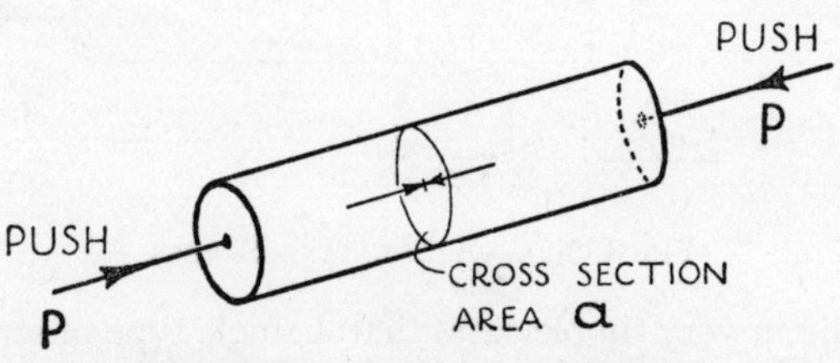

FIG. 134(*b*).—COMPRESSIVE STRESS.

When bolts and nuts are used to clamp together two pieces of metal, the bolts are in tension and the metal clamped is in compression. The strain effects in the material are suggested diagrammatically in Fig. 135(*a*).

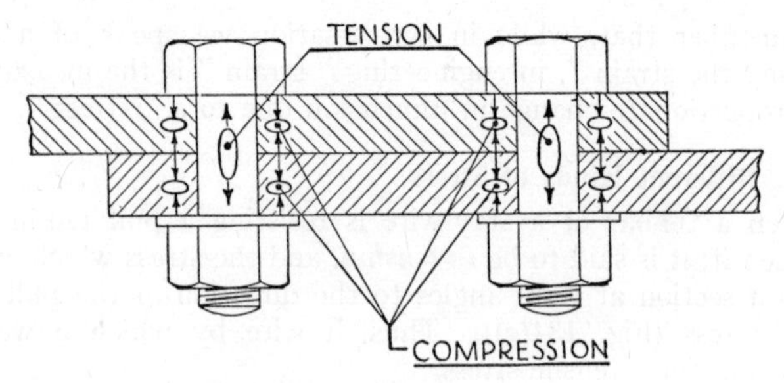

FIG. 135(*a*).—STRAIN EFFECTS.

If, however, the metal plates are subject to forces which tend to cause a sliding movement of one on the other (as in Fig. 135(*b*)) the bolts are subjected to another kind of stress. If the bolts were soft, or too slender for the stresses imposed, such a movement would cut them in two, just as a pair of shears would do. This kind of stress, acting in a plane perpendicular to the bolts, is appropriately termed a *shear* stress.

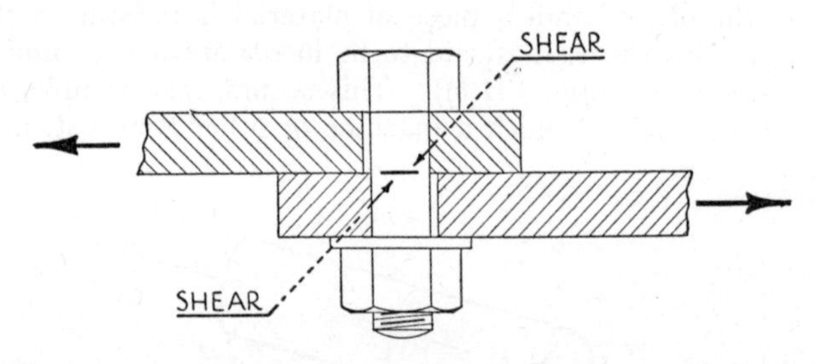

FIG. 135(*b*).—A SHEARING STRESS.

It is only in very simple cases that a single type of stress exists in the material of a member of a framework or a piece of mechanism. Stresses of different kinds are normally present in different parts of it. *Bending* is a very common example of such compound stressing in a structure. Consider, for instance, the bending of a flat steel spring, such as a relay spring or the type often used as the restoring spring of a selector magnet armature. Under a load (generally a pull or push at its free end) such a spring bends, becoming convex on (say) its upper side and concave underneath. The effect is to lengthen the upper side and shorten the lower. Hence the upper surface is in *tension* and the lower in *compression*. The same effects occur in a horizontal beam supporting a load.

When a screw is being used to draw two metal plates together, its shank is subjected to *tensile* stress. But if the screw be tight and excessive force be used to turn it further, it is possible that the shank may *shear*, the head parting company with the threaded end. So the shank is then subject to both tensile and shearing forces.

Again, imagine that a flat spring or beam is supported near one end, and a load is applied very near to the support. Then there will be little bending, but if the load be increased sufficiently it may ultimately shear the spring or beam. So there are present in the material of a spring shearing stresses as well as the tensile and compressive stresses already discussed.

*9.11. Relation between Stress and Strain

Elasticity

As we have seen in para. 9.9, all materials undergo a change in shape and dimensions as an effect of an imposed stress. In many cases, when the force causing the stress is removed, the material returns precisely to its original (unstressed) shape and size, or, if the force is only partially removed, to a shape and size corresponding to the new value of the stress. In these cases there is no permanent distortion, and the material is said to behave *elastically*. But if the stress exceeds a certain value, which differs greatly for different materials, the material does not recover its original shape on the removal of the force, it has *yielded* and is permanently distorted.

Hooke's Law

At first we will confine our attention to the cases in which the material behaves elastically. For practically all materials and for all kinds of stress it may be shown experimentally that the strain ($q = l/L$) produced by any force is directly proportional

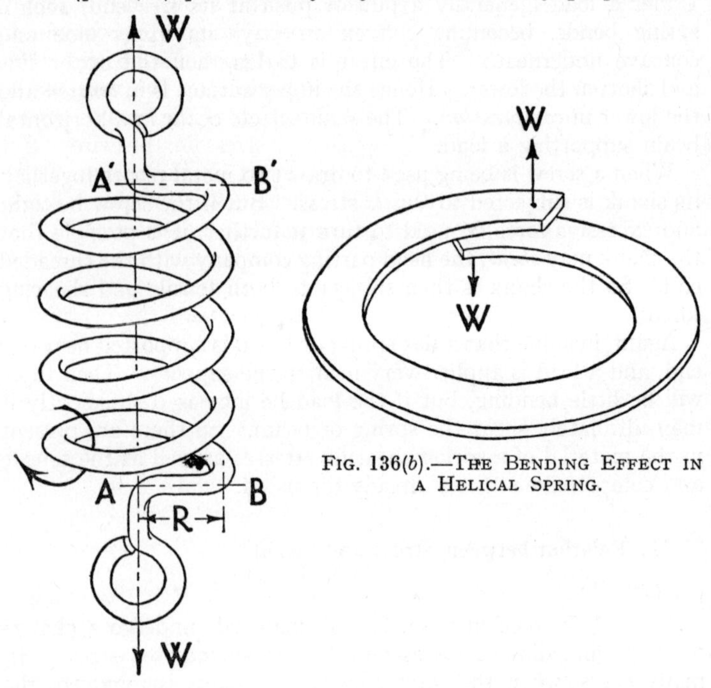

FIG. 136(*a*).—THE EFFECT OF LOADING A HELICAL SPRING.

FIG. 136(*b*).—THE BENDING EFFECT IN A HELICAL SPRING.

to the stress ($p = P/a$) induced by that force, provided that the limits within which the material behaves elastically are not exceeded. This fundamental linear relation is known as *Hooke's Law*. It is of great importance, for it enables the engineer to calculate the extent to which his structures will change shape or dimension or to estimate the forces operating in them by measuring their deformation.

In the case of all simple structures Hooke's Law may be simply expressed as " the change in dimension is proportional to the force producing it ".

Example (*iii*). Consider the helical spring shown in Fig. 136(*a*), in which the two ends A and A' have been brought to a position in line with the axis of the spring by radial portions AB and $A'B'$. Then the load W exercises a turning effect on the wire of the spring, tending to twist it in the direction of the curved arrow. There is thus a *torsional* (" twisting ") stress in the wire. But the load W also exercises a bending effect on the spring, which is more clearly seen in Fig. 136(*b*), where we consider a single turn of the spring, treated as of rectangular cross-section.

Both for the torsional stress and for the bending stress, so long as we keep within the elastic limit of the material, Hooke's Law holds, and the strain is proportional to the stress. So the extension of the spring which results from the combined effect of the torsional and the bending strains is proportional to the load W. This we verified in Experiment C (Chapter 4).

The effect of this last conclusion is to be seen on every spring balance, such as the ordinary domestic variety; *the divisions on its dial are equally spaced throughout its range.*

Example (*iv*). In his switches and relays the telecommunications engineer uses innumerable flat springs of the type shown in Figs. 137(*a*) and 137(*b*), in which a force W applied near the free end exerts a bending moment Wl on the spring, producing a deflection δ. Imagine the same arrangement magnified some 50 times, so that the length becomes say 10 ft. and the cross-section correspondingly greater, and we have what the mechanical engineer describes as a *cantilever beam,* as, for instance, a simple jib for hoisting equipment into a building. The same theory as he uses for calculating the stresses and deflections in his cantilever can be used with equal ease in determining the stresses and deflections of flat springs. A detailed treatment of this theory of beams is beyond the scope of this book.

Considering the relation between load and deflection, as long as the elastic limit is not exceeded the deflection δ is proportional to the load W. Since the deflections with which we are concerned in practice are small, of the order of a millimetre or less, it is difficult to measure them directly. It is more convenient

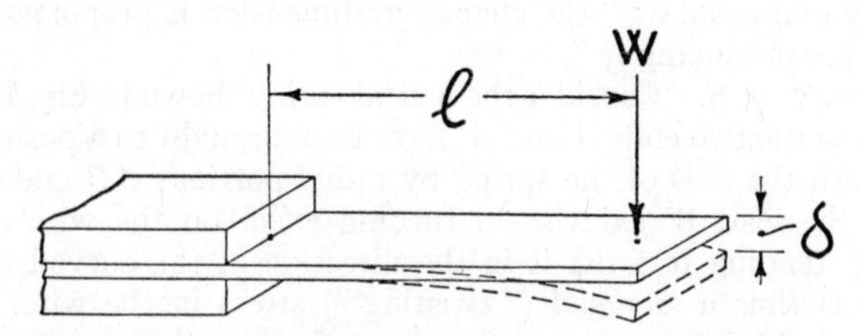

FIG. 137(*a*).—THE DEFLECTION OF A FLAT SPRING.

to reverse the process (as indicated in Fig. 137(*b*)) and produce a known deflection—then measuring the *reaction* by the spring.

The spring is mounted in a rigid clamp, furnished with a screw S, whose pitch is accurately known. This screw is first adjusted so that it is only just touching the spring, *i.e.*, the slightest force

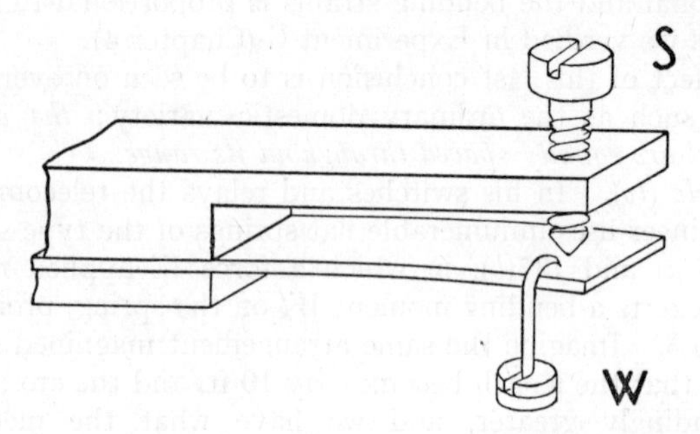

FIG. 137(*b*).—MEASURING THE STIFFNESS OF A FLAT SPRING.

applied to the spring will separate them. This corresponds to the condition of zero stress and zero deflection. Then the screw is advanced by one turn, creating a deflection equal to the pitch of the screw. The force exerted by the spring against the screw can then be tested by hanging weights on the spring until it is again just, but only just, touching the screw. The screw is then advanced one turn at a time and the force similarly measured for each deflection. If we plot the weights required against the deflections we should expect a straight line through the origin—the graph of direct proportion.

The following are the results of an actual test, in which a No. 13 B.A. screw having 4 threads to the millimetre (102 to the inch) was used :

No. of turns of screw . .	0	3	6	10	15
Deflection in mm. . .	0	0·75	1·5	2·5	3·75
Weight in gm. to balance upthrust of spring . .	0	30	70	112	176

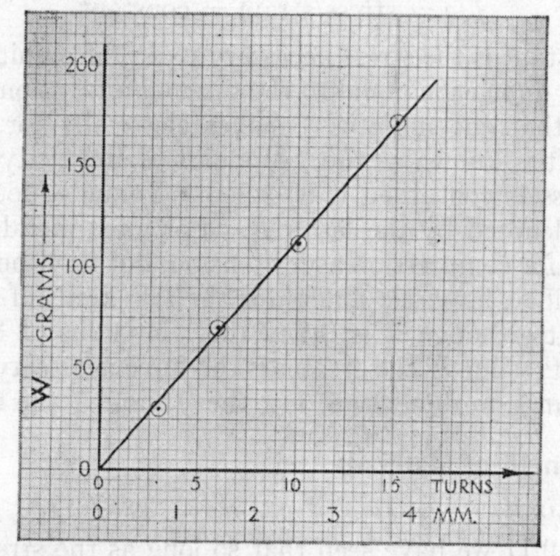

FIG. 137(c).—THE STIFFNESS OF A FLAT SPRING.

These results, plotted in Fig. 137(c), fall on a straight line (within the limits of experimental error) having a slope of 46 gm. per mm. The stiffness of the spring could thus be expressed as 460 gm. per cm.

Compliance and Stiffness

If a force F gm. wt. produces a change x cm. in the dimension in line with that force, then the ratio x/F is a constant for that particular structure (within the limits of elastic behaviour), and this ratio is called the *compliance* of the structure in the direction of that force. The compliance is therefore the change in dimension (or the " displacement ") produced by unit force ($F = 1$): for example, 0·5 mm. per kg. The reciprocal ratio F/x is called the *stiffness* of the structure in the direction considered. Thus " stiffness " is the force required to produce unit change in dimension ($x = 1$), in this case 2 kg. per mm.

Young's Modulus

We have seen that a constant ratio exists between the stress (p) induced in a material and the strain (q) accompanying it, *i.e.*,

$$p/q = \text{stress/strain} = \text{constant}$$

This constant is the stress necessary to produce unit strain, *i.e.*, a change in dimension equal to the original dimension (assuming, of course, that the material behaves elastically for so great a change). It is known as the " Modulus of Elasticity ". Where tensile stress is concerned it is called " Young's Modulus " and is usually denoted by the letter E. For most metals it has the same value for compression as for tension, but for other materials, such as timber, its value differs for the two kinds of stress.

Note that although it is called the " Modulus of Elasticity ", the larger its value is, the greater is the force necessary to produce a given change in dimension, *i.e.*, the " stiffer " the material is.

*9.12. Strength of Materials

Elastic Limit

In para. 9.11 we have seen that so long as the stress is within a certain limit the change in dimension is proportional to the stress.

But if the stress exceeds this limit, proportionality no longer holds; the change in dimension increases more rapidly than the stress. The minimum stress at which this effect is noticeable is called the *Elastic Limit*. Obviously in order to ensure dimensional stability in the structures or apparatus he designs, the engineer must see that the stresses occurring in them under the most adverse conditions of operation are well below the elastic limit of the materials used.

Yield Point, Ultimate Strength

If a piece of material has been stressed up to the elastic limit and the stress is still further increased, change of dimension will occur rapidly for very little change of stress; the material yields to the force and acquires a *permanent set*; it will no longer return to its original shape and dimension when the force is removed. The stress at which this effect occurs is known as the *Yield Point.*

If now the force stressing the material is still further increased, the material after yielding responds to the stress at first somewhat as it did below the elastic limit, but soon change of dimension

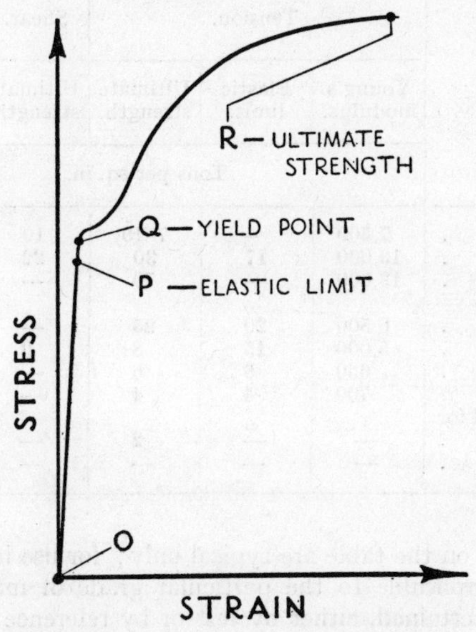

FIG. 138.—A "TEST TO DESTRUCTION".

occurs at an ever-increasing rate compared with the increase of stress, until at some value of the stress the material breaks down altogether. This value of the stress is called ths *Ultimate Strength.*

Fig. 138 shows how a piece of steel rod behaves under an increasing tensile stress. At first (portion *OP* on the graph) the extension is proportional to the load, *i.e.*, stress/strain is constant. Beyond *P* this proportionality no longer holds; we have reached the " elastic limit ". At *Q* the material begins to " yield "; finally at *R* the " ultimate strength " is reached, and the rod breaks.

The following table gives values for Young's Modulus and for the strengths of some common materials:

Elasticity and Strength of Common Materials

Material.	Tension.			Shear.	Compression.
	Young's modulus.	Elastic limit.	Ultimate strength.	Ultimate strength.	Ultimate strength.
	Tons per sq. in.				
Cast iron . .	7,500	—	7–10	10	40
Mild steel . .	13,000	17	30	22	—
Steel wire . .	13,000	—	70	—	—
Copper—hard drawn . .	6,500	20	25	—	—
Brass . .	5,000	15	8	9	5
Oak . . .	650	6	6	—	3
Red pine . .	700	4	4	—	4
Paper (as used for cables) . .	—	—	2	—	—
Concrete . .	—	—	—	—	1

The values on the table are typical only; for use in designing, the values applicable to the particular grade of material used should be ascertained, either by test or by reference to relevant published data.

Factors of Safety

A great deal of data is available for the properties (*e.g.*, the ultimate strength) of a very large range of materials. But the engineer cannot be certain that the material he is using is precisely similar to that on which the data was obtained. Moreover, its properties may be dependent to a certain extent on the methods of fabrication and treatment which the parts have undergone. More important still, conditions may occur occasionally during the operational life of the structure or apparatus concerned which impose considerably higher stresses than those calculated, even although the most adverse conditions expected have been assumed in the calculation. From experience derived from failures in service in cases for which the stresses calculated as in normal design are known, engineers have established " factors of safety " for use in design. A factor of safety is the ratio between the ultimate strength of the material employed (*i.e.*, the stress at which it breaks, as determined from tables of data or from test pieces) and the calculated maximum stress under foreseeable conditions of use. It is a safeguard against failure due to chance variations or other uncontrolled effects.

For example, if the ultimate strength of a steel rod is 30 tons per sq. in. and we decide to apply a factor of safety of 5, then we select its cross-section so that the calculated stress under the most adverse conditions cannot exceed $30/5 = 6$ tons per sq. in.

Effect of Reversals and Variations of Stress : Fatigue

If we pull on a piece of wire with sufficient force, so as to stress it beyond its ultimate strength, it will break. But if we wish to break a length of wire with our hands we do not normally try to do it in that way; we bend the wire to and fro, taking it beyond its " yield point " in alternate directions. Doing it this way involves a succession of bends, *i.e.*, a number of reversals of stress. The less the range of stress to which we subject the wire, the greater the number of reversals which it will stand before breaking. In breaking a wire in this way we take the stress beyond the yield point, but a similar relation exists between the range of stress and the number of reversals at which breakage occurs, even if the stress does not exceed the elastic limit. There is a critical range of stress within which the stress may be reversed an unlimited number of times without risk of breakage. The relations connecting the safe limit with the size and variations of stress are complex; as a rough and ready way of dealing with the problem the engineer increases his " factor of safety " in all cases where the stress is not steady.

This lowering of the ultimate strength of a material due to repeated variations of stress is usually referred to as " Fatigue " —a picturesque term which, of course, does not convey any explanation of the effect. It is due to slight but cumulative changes in molecular structure which take place as the result of repeated fluctuations in stress.

Some commonly used Factors of Safety are given below, distinguishing between those appropriate to a steady (" dead ") load, and higher values applicable in cases where the stresses are varied or reversed.

Material.	Factor of Safety for Stress which is:		
	Steady.	Variable, without reversals.	Reversed.
Cast iron . . .	4	6	10
Steel . . .	3	5	8
Timber . . .	7	10	15

EXERCISE 9

A.

1. Decide which of the following quantities are scalar and which vector: the frequency of a pendulum, the shadow of the gnomon of a sundial, the speed of an electron in a cathode-ray tube, the amplitude of an alternating current in a wire, the intensity of illumination due to a filament lamp, the total power radiated from a wireless transmitter.

2. A boy cycles 12 miles due W. and then 8 miles N.N.E. on straight roads. (*a*) How far now has he cycled? (*b*) How far is he from the starting point? Are these answers scalar or vector quantities? Which is easier to measure?

3. (*a*) Why is it easier to loosen the wheel-nut of a car with a wheel-brace than with a box-spanner?

(*b*) Why is it easier to lift a weight using a rope over an overhead pulley than by direct lifting by the same rope?

4. Find graphically the resultant (in magnitude and direction) of: (*a*) Forces 8 lb. wt. and 14 lb. wt., at an angle of 84°. (*b*) Velocities of 30 f.p.s. and 50 f.p.s. whose directions differ by 127°.

5. A man swims a river, which is 120 yd. wide, always heading directly towards the opposite bank. If his speed in still water is 45 yd. per min. and the river is flowing at 20 yd. per min., calculate where he will land on the opposite bank, and find his actual speed relative to the banks. Would he have crossed more quickly if there had been no current?

6. A ship sails at 16 knots, on a bearing of 127°. At the end of 1 hour's sailing how far is it: (*a*) East; (*b*) North of its starting point? (Draw to scale, and verify by calculation.)

7. An aircraft whose cruising speed is 270 m.p.h. is set on a course due West, but a cross-wind causes a drift Northwards of 64 m.p.h. Calculate the actual speed of the aircraft in magnitude and direction. In what direction should it be steered under these wind conditions to maintain a westward course?

8. A ferry-boat has to cross a river between two landing-stages A and B. If A is directly opposite B and a current of 3 knots is running, find the direction in which the ferry-boat must be steered, if its speed in still water is 8 knots.

9. With the same data as in Question 8, but with B downstream from A, such that AB makes an angle 71° with the direction of the stream, find by scale drawing the direction in which the boat should be steered, and the actual speed of the boat from A to B, and also on the return journey from B to A.

10. A yacht is sailing at 16 knots in a direction N.73° E. A nearby tramp steamer is steaming at 12 knots in a direction N.17° W. What meaning in words can you give to the DIFFERENCE of their speeds? Find this difference by scale drawing, and check by calculation. (It helps to consider first the meaning of the difference in speed of two vehicles moving on the same road.)

11. At the terminal pole of an 8-pr. overhead route, the stay-rod enters the ground 7 ft. 6 in. from the base of the pole, and the stay-wire is attached to the pole 18 ft. above the ground level. Calculate the tension in the stay if each conductor is assumed horizontal and tensioned to 120 lb. wt.

12. With the remaining data as in Question 11, how near (x ft.) to the pole may the stay-rod enter the ground if the tension in the stay may not exceed 4 tons wt.? What is the angle of inclination (α) of the stay in this critical position? (It is easier to calculate α first!)

B.

13. A 14-pr. overhead route changes direction by 24° at an angle pole. Find the correct direction of a horizontal stay-wire and calculate the tension in it. Assume 100 lb. wt. tension in each wire. If the horizontal

stay were replaced by one inclined at 56° with the horizontal, calculate the tension in it.

14. An 8-pr. route changes direction by 18° at a certain pole on the inner curve of a canal bank. To avoid a fly-over stay, a strut is used instead of an angle-stay. Calculate the compressive force in this strut if it enters the ground 4 ft. from the base of the pole and is attached to the pole 14 ft. above ground-level. (Allow 100 lb. wt. tension per wire.)

15. 10 pairs of wires arrive at a certain pole of a rural overhead route. 6 pairs continue straight on and the remaining 4 pairs leave on a branch route making an angle of 70° with the forward direction of the original route. Find the best position for a stay at this pole, and calculate its tension, if it is inclined at 54° with the horizontal. (Each wire tensioned at 100 lb. wt.)

16. An underground cable is terminated at a certain overhead distribution pole, and 6 pairs leave this pole on a bearing of 83°, and 4 pairs on a bearing of 192°. Find by drawing to scale the direction of the stay required.

17. A machine weighing 2 tons is supported by two chains attached to the same point on the machine. One chain goes to a hook in a roof-beam and is inclined at 60° to the vertical; the other chain goes to a pulley and is inclined at 45° to the vertical. Find by scale drawing the tension in each chain.

18. In a simple jib-crane the jib is inclined at an angle of 70° with the horizontal, and the tie-rod at an angle of 40°. If a load of 4 tons is suspended from the crane-head, find by scale drawing the forces acting in the jib and in the tie-rod, stating in each case the nature of this force. Could the tie-rod be replaced by a steel hawser?

19. The vertical crane-post of a jib-crane is 16 ft. high, the jib is 26 ft., and the tie 14 ft. long. Find graphically the forces in the jib and tie-rod when a load of 3 tons is supported at the crane-head. Can these be calculated?

20. A mercury-arc lamp weighing 24 lb. is suspended above a city street by means of two equal steel wires which are inclined at 6° with the horizontal. Calculate the tension in each wire. Why is a scale-drawing solution unsatisfactory in this problem? Try it!

C.

21. Sketch the following tools in action: side-cutting pliers, tin-snips, spade (turning a sod), claw-hammer (extracting a nail), tweezers, nut-crackers.

Label on each diagram: (*a*) the "fulcrum"; (*b*) the "load" or resistance to be overcome; (*c*) the applied "effort".

22. A steel bar 4 ft. long, weighing 7 lb., is placed on a bench so that 3 in. overhang. What is the greatest weight that may be hung on the overhanging end before the rod is pulled over?

23. A uniform girder weighing 10 cwt. is 24 ft. long and rests on supports

2 ft. from each end. What weight is suspended 8 ft. from one end, if the thrust on one support is double that on the other?

24. The arm AB of a guillotine (see Fig. 130(*b*) above) is 3 ft. 6 in. long, and the blade is in contact with the edge of a piece of wood at a distance 4 in. from the hinge at A. If 20 lb. wt. is exerted at the end B, what force is exerted by the blade (both forces perpendicular to AB)? In cutting through a $\frac{1}{2}$-in. thickness of wood how far does the end B move? What is the force on the hinge at A, perpendicular to AB?

25. A rough spring-board is made by fixing an 8-ft. plank at its end A and at a point B 3 ft. from A. A man weighing 12 st. 12 lb. stands on the free end C. Calculate the forces acting on the supports at A and B if the plank itself weighs 60 lb.

26. A practical steelyard (without the rider) weighs 12 lb., and the load is suspended from a knife-edge support $2\frac{1}{2}$ in. from the fulcrum. The centre of gravity of the steelyard when the beam is horizontal is $1\frac{1}{2}$ in. to the load side of the fulcrum. A 10-lb. rider is used. Find the distances from the fulcrum of graduations corresponding to loads of 10 lb., 40 lb., W lb. Where should the rider rest when no load is attached?

27. The safety valve of Fig. 139 has an aperture 2 in. in diameter, and is just on the point of blowing off steam. The lever is pivoted at A, and its weight 4 lb. may be considered to act through the centre of gravity G, where $AG = 7$ in. If the rider at B is 50 lb., what is the pressure in the boiler? $AC = 3$ in., $AB = 32$ in.

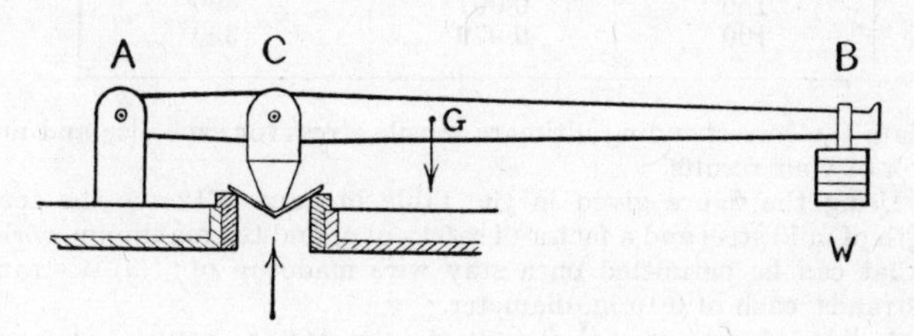

FIG. 139.

28. A lever safety valve with the same physical dimensions as in Question 27 is required to " blow " when the boiler pressure exceeds 120 lb. per sq. in. What weight W lb. is required? What is the force acting on the hinge joint at A when the valve " blows "?

29. The framework of a dockside crane is bolted to the quay at four points A, B, C, D forming a rectangle. AB is nearer to the edge than CD and parallel to it. A load of 8 tons is suspended above the water at a horizontal distance of 20 ft. from AB. If $BC = 12$ ft., calculate the vertical components of the forces on the bolts at A, B, C, D due to the load.

30. With the data of Question 29, find the *total* vertical force acting on the crane at each point A, B, C, D if the crane weighs 24 tons and its centre of gravity is equidistant from A and B and 8 ft. horizontally shorewards from AB.

D.

31. Allowing a factor of safety of 5, what is the greatest tension permissible in " 100-lb." copper wire (diameter 0·08 in. approx.) assuming the ultimate (tensile) strength of hard-drawn copper is 25 tons per sq. in. ?

32. A steel wire stretches 0·029 in. under a tension of 500 lb. wt. Find the ratio of the extension to the load. How much would a 2-ton weight stretch the wire, assuming the elastic limit is not exceeded ?

33. A typical specification requires that hard-drawn copper wire as used for telephone and telegraph lines shall not break at less than the following loads :

Weight per mile, lb.	Diameter, in.	Minimum breaking load, lb.
800	0·224	2400
600	0·194	1800
400	0·158	1250
300	0·137	950
200	0·112	650
150	0·097	490
100	0·079	350

Calculate the corresponding ultimate tensile stress for each size and make a graph of your results.

34. Using the figure given in the table in para. 9.12 for the tensile strength of mild steel and a factor of safety of 4, find the maximum working load that can be permitted on a stay wire made up of : (*a*) 4 strands; (*b*) 7 strands, each of 0·16 in. diameter.

35. A line of ground poles supports ten 200-lb. copper wires each tensioned to 160 lb. wt. A pole at an angle in the route is 10 yd. out of line with its immediate neighbours, and the length of span on each side of the angle pole is 50 yd. If the angle stay is attached to the pole 30 ft. from the ground, and the stay enters the ground 10 ft. from the base of the pole, would a 4-strand or a 7-strand stay wire be used ? (Use the results of Question 34 and assume the wires to be horizontal, at an average height of 30 ft. above the ground.)

36. A contact bank for a selector is made up of arcs of contact segments, insulating spacers, and metal spacers, all held between two clamping plates and tightened together by six steel screws passing through from one clamping plate to nuts on the other. It is found that for secure fixing of the contact segments the total tension of the screws must be 4800 lb. wt.

Using the ultimate strength given in the table on p. 304 and a factor of safety for a steady load, what is the minimum diameter of screw which is suitable? What assumption have you made in arriving at the tension in each screw?

37. What is the extension of the screws in the above case (*i.e.*, the length by which they are stretched) assuming their unstretched length is 2 in., and Young's modulus is 13,000 tons per sq. in.?

38. A machine weighing 2000 lb. is supported on four blocks of rubber each measuring (unstressed) 3 by 3 by 3 in. Calculate the average compressive stress in the rubber, and taking Young's modulus as 20,000 lb. per sq. in., find the loss of height of the blocks when loaded.

39. Using the figure for ultimate tensile stress in the table in para. 9.12, say what load should be supported by a test strip of insulating paper 0·0025 in. thick and 1 in. wide.

40. The armature of an electromagnet is attached to a straight bar which is pivoted on a knife edge at P, 2·8 cm. distant from the centre of the pole face of the magnet. At a distance 1·2 cm. on the opposite side of the knife edge P a helical spring is fixed, tending to hold the armature in the unattracted position, up against a stop S, which is fixed 6 cm. from P, beyond the magnet. The spring is adjusted to exercise a pull of 500 gm. wt. when the armature is resting against the stop S, and its stiffness is 100 gm. per mm. (*a*) Calculate the thrust against the stop S in the non-operated condition. (*b*) Make a graph and derive a formula showing how the minimum pull X gm. wt. which the magnet must exert varies with the travel d mm. of the armature at the pole face when the magnet is energised. A total travel of 4 mm. can be allowed.

REVISION PAPERS C

Use sketch diagrams freely. C1–4 are mainly concerned with Chapters 7–9.

C1.

1. ABC is an equilateral triangle. A straight line, parallel to AC, cuts BA, BC at P, Q. AC is produced to R so that $BQ = CR$. Prove that PR bisects CQ. (State carefully the case of congruency you use.)

2. An isosceles triangle has a base 4·8 in. and sides 6·4 in. long. Find, by calculation, the angles and the area of this triangle.

3. The scale of a map is $\frac{1}{2}$ in. to the mile. Two places are on 250-ft. and 700-ft. contours and are 2·35 in. apart on the map. Find the average slope of a straight road joining them.

4. A vessel is sailing along a course N. 32° W. when an iceberg is observed in a direction N. 4° E. After the ship has sailed $2\frac{1}{2}$ miles, the iceberg then bears N. 22° E. If the ship continues on the same course, how near the iceberg will she pass? (Assume the iceberg is stationary.)

5. A cable-ship is laying a submarine cable at a steady 3 knots on a bearing of 126°. There is a tide running which is causing a drift of 1·2 knots in a direction N. 53° E. On what course must the vessel be steered, and what would be its speed in still water?

6. Sketch: (*a*) a wheelbarrow; (*b*) a nail extractor in action; and label on each sketch the effort E applied, the fulcrum F, and the load L.

How do these two examples of leverage differ in principle?

C2.

1. If a $2\frac{3}{4}$-in.-diameter shaft 3 ft. 6 in. long weighs 70·7 lb., determine the weight of a $4\frac{1}{2}$-in.-diameter shaft of the same material 6 ft. 9 in. long.

2. $ABCD$ is a parallelogram. P is any point on BD. Prove that $\triangle PAB$ equals $\triangle PBC$ in area.

3. The angle between two tangents drawn to a circle of radius 7·2 cm. is 47°. What are the lengths of the tangents and of the chord joining their points of contact?

4. From a boat at sea the angle of elevation of a mountain peak is 21° 14′. The peak is 10,500 ft. above sea-level, and it is 2 miles horizontally from the shore. How far is the boat from the shore? (Assume the nearest point of the shore from the boat is in line with the mountain peak.)

5. A barge is being towed along the centre of a canal by a rope 75 ft. long, the pull in the rope being 60 lb. wt. If the canal is 30 ft. wide, find the effective force pulling the barge along, and the (component) force urging it towards the bank.

6. A 9-ft. oar is in contact with the rowlock 3 ft. from the handle end.

What is the propelling force on the boat when the oarsman exerts a pull of 24 lb. wt.? (Assume the fulcrum is in the water at the tip of the blade, which is at right angles to the side of the boat.)

C3.

1. AB (12 ft.) is the jib of a light crane, anchored at A, and BC (8 ft.) is its tie-rod. If C is fixed in masonry 6 ft. vertically above A, find by scale drawing or calculation the tension or compression in AB and BC when a load of 6 cwt. is suspended from B.

2. A 1-in.-diameter bolt has 8 threads to the inch. Find the angle of lead of the thread.

3. The vertical angle of an isosceles triangle is 45°, and the base is 6 in. long. Calculate the area of the triangle.

If an isosceles triangle has a vertical angle a and base b in. obtain a formula for its area A sq. in.

4. Two cubes, of total volume 87·5 cu. in., have sides in the ratio of 2 to 3. Find the volume of each and the side of the larger cube.

5. An observer is in a plane at 25,000 ft. and finds that the angle of depression of a building is 43°, and $\frac{1}{2}$ min. later it is 59°. How fast is the plane flying, if it flies directly over the building?

6. The safe load L tons (uniformly distributed) which can be taken by 15-in. by 6-in. steel girders of span S ft. is shown by the accompanying table:

S . .	8	10	14	18	22
L . .	44·7	37·3	26·7	20·7	17

Plot a graph of L against S. What is the safe load to apply to a 20-ft. span of this type of girder?

If 32 tons is the maximum loading of a steel joist of the same cross sectional dimensions, what is the greatest span permissible?

C4.

1. The development of the curved surface of a solid cone is a quadrant of a circle of radius 8 in. Find the height of the cone, the angle at its vertex, and its volume.

2. A lineman loses his footing whilst suspending cable, and grabs the steel suspension wire in the middle of a 40-yd. bay. If the wire sags by 2 ft. calculate the additional tension in the wire due to the man's weight of 12 st. 4 lb.

3. Loads up to $\frac{1}{2}$ cwt. are to be raised by winding a rope on an axle, which is to be turned by a handle at 12 in. radius. Find the diameter of

the axle if the effort is not to exceed 14 lb. wt. With this size axle how many times must the handle be turned to lift the load 132 ft.?

4. From a church tower the elevation of the top of an aerial mast at a wireless-telegraphy installation is 14° and of the base is 11°. If the distance from the church tower to the mast is 1·7 in. on the 1 : 25,000 survey map, calculate the height of the mast.

5. In Question 4 above, if the angles of elevation were α, β, and the distance on the map d in., obtain a formula for the height h ft. of the mast in terms of α, β, and d.

6. A steel spring l in. long, b in. broad, and d in. thick is fixed at one end (*i.e.*, it is a "cantilever"). If a weight W lb. is hung on the free end, the deflection δ in. is determined by the formula $\delta = \frac{4Wl^3}{Ebd^3}$, where $E = 30 \times 10^6$ lb. per sq. in. (Young's modulus for steel). Calculate the deflection for a load of 2 oz., if $l = 2{\cdot}6$ in., $b = 0{\cdot}2$ in., $d = 0{\cdot}014$ in.

Express E in terms of the other variables. How could you find E experimentally with a flat spring?

C5.

1. The addition of a lubricant to the petrol improves a car's consumption for long journeys from 17 to 20 m.p.g. If petrol costs 2*s*. 8*d*. per gall., and the lubricant increases fuel costs by 3*s*. 4*d*. per 100 gall., find the percentage saving in running costs, using the lubricant.

2. The angles of elevation 42°, 28° of a tall wireless mast were noted from two points A and B, 150 ft. apart on level ground, in line with the base of the mast and on the same side of it. Find the height of the mast.

3. The velocity v cm. per sec. of an electron reaching the anode of a cathode-ray tube is given by an approximate formula $v = 6 \times 10^7 \times \sqrt{V}$, where V volts is the anode voltage. What is the electron velocity when 4000 volts are applied to the anode? Express this also in m.p.h.

4. A steelyard weighs 9 kg., and its weight acts 4 cm. to the load side of the fulcrum (beam horizontal). The movable rider weighs 12 kg., and the graduation corresponding to 80 kg. is at a distance 60 cm. from the fulcrum. How far is the load (horizontally) from the fulcrum? Where will the 100-kg. graduation mark be?

5. The velocity V f.p.s. of sound in a solid is given by the formula $V = \sqrt{\frac{gE}{\rho}}$, where E is Young's modulus in lb. per sq. ft., ρ the density of the solid (lb. per cu. ft.). Calculate the speed of sound in a copper wire, if $E = 15 \times 10^6$ lb. per sq. in. and the density of copper is 559 lb. per cu. ft. ($g = 32{\cdot}2$ ft./sec./sec.)

6. The permissible current for rubber-insulated cables having stranded copper conductors (2 cables running side by side) is given by the following figures:

Conductor, standard size	3/0·029	3/0·036	7/0·029	7/0·036	7/0·044
Cross-sectional area (sq. in.)	0·0020	0·0030	0·0045	0·0070	0·0100
Current rating (amperes)	5	10	15	29	38

Conductor standard size	7/0·052	7/0·064	19/0·052	19/0·064
Cross-sectional area (sq. in.)	0·0145	0·0225	0·040	0·060
Current rating (amperes)	45	56	78	102

Plot the cross-sectional area of the cables against the current rating, to show a range 0 to 100 amperes. Mark the standard sizes at the points plotted. What is the likely current rating of 19/0·044 cable, of total cross-sectional area 0·03 sq. in.?

*C*6.

1. The jib of a crane will bear a maximum safe compressive force of 6 tons wt. If it is inclined at 60° to the horizontal and has to deal with loads up to $2\frac{1}{2}$ tons, find by scale drawing the steepest inclination the tie-rod may have, and the tension in the tie-rod in this case, when the load is $2\frac{1}{2}$ tons.

2. Calculate the length of the chord of a circle (radius 16·7 cm.), if it subtends an angle of 117° 42′ at the circumference.

3. APQ is a secant and AT a tangent to the circle PQT. Prove $AT^2 = AP \,.\, AQ$.
If $AT = 6$ cm., and the chord $PQ = 9$ cm., calculate AP (x cm.)

4. A piece of round rod, 0·20 in. in diameter, is to have flats milled on it to form a key which may be either (*a*) square or (*b*) triangular (equilateral) in cross-section. What is the length of side of (*a*) the square and (*b*) the triangle, assuming that the minimum amount of material is removed?

5. The speed v f.p.s. of a body falling under gravity is proportional to the square root of the distance fallen (s ft.) If v is 24 when it has fallen 9 ft., express s in terms of v. Plot the graph s against v for speeds up to 80 f.p.s. How fast will a body be moving when it strikes the ground, if knocked off a window-ledge 42 ft. up?

6. A sq. ft. is the area of cross-section of a 20-ft. log, measured x ft from one end.

A	4·8	3·6	2·9	2·3	1·8	1·6
x	0	4	8	12	16	20

Use the trapezoidal rule to calculate the average cross-sectional area of this log, and thence obtain its volume in cu. ft.

*C*7.

1. For a certain type of telephone equipment A, it is calculated that the annual charges will be at the following rates :

Interest	3% p.a.
Depreciation . . .	5% p.a.
Maintenance . . .	4% p.a.

An alternative type B costs 15% more in the first place, but it is more durable, so that it is estimated that the depreciation can be reckoned at 20% less than for type A, and also more reliable, so that maintenance can be reckoned at 25% less than for type A. Calculate the total annual charges in each case as a percentage of the first cost, and say which type will be the more economical.

2. AB is a chord 4·2 in. long in a circle 7·2 in. in diameter. If the tangents at A and B meet at T, calculate $\angle ATB$ and the length AT.

3. The heat H calories generated by the passage of a current through a resistor varies directly as the time t sec. during which the current is passing, directly as the square of the voltage (V volts) across the resistor, and inversely as the resistance R ohms. If $V = 50$, $R = 100$, the heat generated is 6 calories per sec. How many calories will be generated in 8 min., if $V = 40$, $R = 5$?

4. Water to a depth of 2·5 in. flows from a V-shaped gutter, in which the sides meet at right angles, into a cylindrical drum of diameter 2 ft. 6 in. If the flow is at the rate of 2 f.p.s. past a fixed point in the gutter, how long will it take to fill the drum to a depth of 3 ft.? (Take $\pi = 3{\cdot}142$.)

5. A "tank circuit" composed of a capacitor (capacitance C farads) and an inductor (inductance L henrys) will oscillate at a frequency f cycles per second, where $f = \dfrac{1}{2\pi\sqrt{LC}}$. Calculate this frequency for a 2-microfarad capacitor and a 10-millihenry inductor. Express L in terms of f and C.

6. In a compression test on a specimen of mild steel the compression x in. was measured for different values of the load W tons :

W . .	10	12	13·2	14·5	15·6	16
x . .	0·004	0·008	0·0105	0·014	0·023	0·036

Comment on the graph you obtain from these figures. What is the likely load for a compression of 0·02 in.?

*C*8.

1. The force to stop a train travelling at V m.p.h. in d ft. is $0{\cdot}69\ V^2/d$ tons wt. What force will stop it in 87 yd. when moving at 23 m.p.h.?

The maximum compression of the buffers at the terminus is 4 ft. What is the average resistance of the buffers, if they can just stop a train travelling at 3 m.p.h.?

2. A heap of earth has a horizontal rectangular base of 15 by 25 ft. It is levelled at the top to form a rectangle 7 by 17 ft. The sides, which are smooth and straight, each slope at an angle of 30° to the horizontal. Calculate its volume.

3. The depth d ft. of a river is measured at distances x ft. from one bank, as under:

x	0	5	10	15	20	25	30	35
d	0	4·1	9·4	12·0	14·2	6·5	2·1	0

Find the average depth of water, and the rate of flow in gallons per hour, if the current is 2 m.p.h.

4. A hollow metal pipe 0·08 in. thick is used as a tie-bar and has an external diameter of 1·36 in. Find its cross-sectional area. What is the greatest safe tension it may bear, at $2\frac{1}{2}$ tons per sq. in.?

5. A block of flats is 36 ft. in height to the eaves, and the ridge of the roof (which is symmetrical) is parallel to the street and 10 ft. above the eaves. The width from eave to eave is 40 ft. The landlord agrees that a tenant may erect a television aerial at the eave farthest from the street, provided it is not visible to an eye 6 ft. above ground level on the opposite side of the street. How high may it be if the width of the street is: (*a*) 75 ft.; (*b*) 50 ft.?

(*Hint:* Find the "line of sight" from eave to ridge of the roof first.)

6. A new class of telephone service is introduced on 1st January, and orders are received at the uniform rate of 10 per week, but owing to the delays in the supply of material it is impossible to begin to execute them until 5th March. Make graphs showing: (*a*) the number of orders received up to any date in the year, and (*b*) the rate at which they must be completed if by the end of the year the completion time is to be reduced to 2 weeks. If from 2nd July till 10th September, the rate of completion falls to one-half the planned rate, to what must it be increased in order to achieve a completion time of 2 weeks by the end of the year?

ANSWERS TO THE EXERCISES

EXERCISE 1(*a*) (p. 4)

1. $3 \times 7 \times 11$. 2. 3^5. 3. $2^3 \times 3^3$.
4. $2 \times 3 \times 7^2$. 5. $3^2 \times 11 \times 17$. 6. $2 \times 5^3 \times 13$.
7. $2^6 \times 5$ pints. 8. $2^7 \times 3^2 \times 5 \times 11$ in. 9. $2^8 \times 3 \times 5$ acres.
10. $2^4 \times 3^8$ cu. in. 11. 3888. 12. 25,725.
13. 247,808. 14. 504 ($2^3 \times 3^2 \times 7$); 6 (2×3).
15. 288 ($2^5 \times 3^2$); 4 (2^2). 16. 1152 ($2^7 \times 3^2$); 16 (2^4).
17. 504 ($2^3 \times 3^2 \times 7$); 7. 18. 3234 ($2 \times 3 \times 7^2 \times 11$); 21 ($3 \times 7$).
19. 792 ($2^3 \times 3^2 \times 11$); 3. 20. 1344 ($2^6 \times 3 \times 7$); 16 (2^4).
21. $\frac{25}{84}$. 22. $\frac{59}{84}$. 23. $9\frac{41}{48}$. 24. $8\frac{29}{90}$. 25. $1\frac{23}{35}$. 26. $2\frac{23}{180}$.
27. $\frac{9}{20}, \frac{5}{12}, \frac{6}{15}, \frac{7}{18}$ $\left(\frac{81, 75, 72, 70}{180}\right)$. 28. $\frac{17}{36}, \frac{15}{32}, \frac{19}{42}$ $\left(\frac{952, 945, 912}{2016}\right)$.
29. $\frac{18}{49}, \frac{22}{63}, \frac{19}{56}$ $\left(\frac{1296, 1232, 1197}{3528}\right)$.
30. $2^7 \times 5 \times 7$ lb. : 2^7 or 128 lengths.
31. 15 in. square (H.C.F. of measurements in in.).
32. 13 ft. 9 in. (L.C.M. of 33, 55, 66 " half-inches ").
33. 9, 4 revs. respectively (*i.e.*, 144 " contacts ").
34. Conductance $\frac{19}{72}$ mhos, resistance $\frac{72}{19}$ or $3\frac{15}{19}$ ohms.
35. (The L.C.M. of 4, 6, 10 is 60). 36. 14 hr.

EXERCISE 1(*b*) (p. 20)

1. ". . . dividing the distance in miles by 15." $P = \frac{m}{15}$.
2. ". . . cubing the number of feet in the side of the cube." $V = a^3$.
3. ". . . dividing the total weight in pounds by the length in feet." $w = \frac{W}{l}$.
4. ". . . multiplying the voltage by the current in amps, and then dividing this product by 1000." $P = \frac{VI}{1000}$.
5. $l = 2p + 2q$ or $l = 2(p + q)$. 6. $V = \frac{s}{t}$.
7. The volume of a pyramid in cu. yd. is one-third the product of its base area in sq. yd. and its height in yd.
8. The quantity of electricity in coulombs held by a capacitor is found by multiplying the potential in volts by the capacitance in farads.
9. The distance in feet fallen from rest by a stone is sixteen times the square of the time in secs.
10. The current flowing in amperes is found by dividing the voltage by the resistance in ohms.
11. It is another way of presenting Ohm's Law—instead of the more usual formula $V = IR$.

12. The surface area of a sphere is four times the product of π and the square of the radius. Alternatively, it is four times the area of a circle of the same radius.
13. $5a - 2b$. 14. a^4b^6. 15. $12p^2q$. 16. $\frac{8p^3}{3q}$.
17. $21a^4b^2$. 18. $6\pi r^2 l$. 19. $5{\cdot}19 \times 10^2$. 20. $2{\cdot}4 \times 10^5$.
21. $1{\cdot}8 \times 10^{-4}$. 22. $5{\cdot}4 \times 10^8$. 23. $8{\cdot}04 \times 10^{-7}$.
24. $3{\cdot}313 \times 10^4$ cm. per sec. 25. $3{\cdot}4 \times 10^7$ ohms. 26. $4{\cdot}2 \times 10^{-5}$ henrys.
27. 3×10^{-3} amperes. 28. $1{\cdot}2 \times 10^{-2}$ in. 29. $6{\cdot}336 \times 10^4$ in.
30. $2{\cdot}78 \times 10^{-4}$ hr. 31. 299,100 km. per sec. 32. 239,000 miles.
33. 0·0000014 mhos.
34. 0·0000000000000000000000000000911 gm. (" point 27 noughts ").
35. 27·9. 36. 47,900. 37. 0·0917. 38. 103,000.
39. 0·0908. 40. 0·700. 41. 1,611,000; $5{\cdot}88 \times 10^8$ miles.
42. (*a*) $2{\cdot}4 \times 10^{-6}$ sec. (2·4 microsec.); (*b*) 1·3 sec. 43. 3100 ohms.
44. $1{\cdot}7 \times 10^{-3}$ ohms (1700 microhms).
45. 3 ft. 4 in. (nearest inch); 5500 microhms.
46. 1900 tons. 47. $2{\cdot}25 \times 10^{10}$ cm. per sec.
48. $5{\cdot}9 \times 10^{12}$ miles. 49. 5·9 cu. ft.
50. Instrument readings are accurate to 1 in 100; *i.e.*, to 2 significant figures at best. This restricts the accuracy of our answer: 2·17227 ohms is absurd; 2·2 ohms is a sensible answer.
51. 22·9 ohms (the copper wire makes no sensible difference).
52. 1·49 amperes (the voltmeter makes no difference for practical purposes).

EXERCISE 2(*a*) (p. 31)

3. Average height of blocks 196·4 units; annual bill £2 9*s*.
4. 2*s*. 3*d*. in the £; total area 100% (unit width per block, and " surplus " included).
7. 15 out of 193, *i.e.*, 7·8% rejects. The process does not appear to be working well: an investigation justified.
8. 35% survive 10,000 hr.
9. Gives at-a-glance information of the " spread " of adjustments; 13 out of 134 (9·7%).
10. Lopsided! Capacitance tends to be too high rather than too low; 4·4% rejects.
11. Broken line—to suggest a trend—intermediate points have no meaning.
12. No. Dotted line to show trend. 2030 to 2040 probably.
13. 134 yd.; 52 m.p.h. 14. 6.48 p.m.; 19th April.
15. 25 sec.; at the start (the higher above room temperature, the faster it cools).
16. The beeswax begins to solidify at about 72 sec. from the start; " latent heat " is given out during this change of state from liquid to solid; melting point 61° C. = 142° F.
17. Just over 8 hr. 18. 12.55 p.m. 19. Approx. 21 lb. per mile.
20. (*a*) 11·6; (*b*) 8·5 amperes.

EXERCISE 2(*b*) (p. 42)

5. 357 cu. ft.
7. 11th year; $S = 360 + 15n$.
8. $R = 0{\cdot}042T + 10{\cdot}8$.
9. $Q = 15 + \dfrac{x}{16}$.
10. Needs 7 guests to clear £10 in one week, an average profit of £10 per week requires 6·3 guests weekly.
11. 16·9 mils; 0·047 in. (No. 13 S.W.G.).
12. 223 lb. per mile; approx. S.W.G. No. 6, diameter 0·192 in.
13. (*a*) 65 gall. per min.; (*b*) 9·3 in.
14. 6·64 hr. or 6 hr. 38 min.; 1·3 min.
15. About 300 ohms—release and operate times equal; 2·3 millisec. too long; 16·8 millisec. too short.

EXERCISE 3(*a*) (p. 53)

1. 1·15, 3·63, 42·2, 31·6 (3 significant figures).
2. 1·9, 2·2, 4·2, 9·3.

(thus: $1{\cdot}9 \times 2{\cdot}2 = 4{\cdot}2$ approx., etc.).

3. $10^{0{\cdot}8633}$, $10^{0{\cdot}6893}$, $10^{0{\cdot}7024}$, $10^{0{\cdot}8025}$, $10^{0{\cdot}2797}$, $10^{0{\cdot}8456}$.
4. 7·52, 4·045, 1·022, 9·836, 2·027.
5. 8·372.
6. 1·593.
7. 6·683.
8. 3·619.
9. 1·273.
10. 1·512.
11. 9·077.
12. 1·197.
13. (*a*) 5·561 cm.; (*b*) 2·866 in.
14. 4117 sq. in.
15. 5·25 ohms.
16. 8·056 cu. in.
17. 3·174 kg.
18. $\dfrac{8}{3{\cdot}282} = 2{\cdot}438$, $\dfrac{3{\cdot}5}{3{\cdot}282} = 1{\cdot}067$.

The bay is 2·44 metres high, 1·07 metres wide.

19. 3·11 miles (3 significant figures).
20. 5·44 kW.

EXERCISE 3(*b*) (p. 57)

1. $2{\cdot}92 \times 10$.
2. $5{\cdot}003 \times 10^3$.
3. $4{\cdot}7 \times 10^{-2}$.
4. $2{\cdot}37 \times 10^6$.
5. $8{\cdot}18 \times 10^{-7}$.
6. 11·66.
7. 1510.
8. 3·976.
9. 0·0263.
10. 0·1538.
11. 0·1393.
12. $1{\cdot}633 \times 10^{-6}$.
13. 145 litres (Check $\frac{3200}{22} = \frac{1600}{11} = 145\frac{5}{11}$).
14. (*a*) 4167 watt-hours; (*b*) 1422 B.Th.U.
15. 259 hectares.
16. 14·93 sq. in.
17. 96·32 sq. cm.
18. $r^2 = \dfrac{261}{\pi}$; $r = 9{\cdot}11$ ft. or 9 ft. 1 in.
19. $\dfrac{39{\cdot}7}{0{\cdot}7854(6{\cdot}2)^2} \simeq 13$ mm.
20. $\frac{4}{3}\pi(0{\cdot}21)^3 \times 11{\cdot}37 \simeq 0{\cdot}441$ gm.

EXERCISE 3(*c*) (p. 67)

1. 2·20 lb.
2. 22·9 kg.
3. 778 ft.-lb.
4. $\dfrac{200}{2{\cdot}205} \times \dfrac{0{\cdot}62}{1} \simeq 56$ kg. per km. Draw a straight-line graph through the origin and the point (200, 56).

5. (*a*) 0·094 litres per km.; (*b*) 18·8 m.p.g. 6. 588 ohms.

7. 20; 0·914 mm.; $\frac{11\cdot77 \times 1\cdot094}{2\cdot205} = 5\cdot84$ kg. per km. 26; 0·457; 1·46. 36; 0·193; 0·260.

8. $\frac{4 \times 5280 \times 12}{100{,}000} \simeq 2\cdot5$ km. to the cm.

9. $\frac{63{,}360}{25{,}000} \simeq 2\cdot6$ in. to the mile. 10. 4·17 cm. ≡ 1 metre.

11. 1 ft. 10 in., 6 ft. 7 in., 1 ft. 2 in. 12. 2·22 kg.

13. 680. 14. 4·35 years. 15. 6·2 days.

16. (i) 22·3 lb. per mile, 41·3 ohms per mile; (ii) 38·8 lb., 22·3 ohms; (iii) 10·4 lb., 84 ohms. 17. 40 tons approx.

EXERCISE 4(*a*) (p. 92)

1. $x(3x + y)$. 2. $2p(3p - 2q)$. 3. $\pi(a + b)(a - b)$.
4. $(r + 2s)(r - 2s)$. 5. $3q(p + 3q)(p - 3q)$.
6. $7x^2(3x + 1)(3x - 1)$. 7. $\pi t(a - b)$. 8. $\pi a(2a + b)$.
9. $(4pq + r^2)(4pq - r^2)$. 10. $2lm(lm + 2)(lm - 2)$.
16. $x^2 - 3x + 2$. 17. $2p^2 + 7p - 4$. 18. $y^2 + 2y + 1$.
19. $p^2 - 2pq + q^2$. 20. $9x^2 - 24x + 16$. 21. $x^2 + 6x + 9$.
22. $x^2 - 14x + 49$. 23. $x^2 + 10x + 5^2 = (x + 5)^2$.
24. $y^2 - 8y + 16 = (y - 4)^2$. 25. $p^2 - 6p + 3^2 = (p - 3)^2$.
26. $x^2 + 7x + (\frac{7}{2})^2 = (x + \frac{7}{2})^2$. 27. $x^2 - 14x + 49 = (x - 7)^2$.
28. $r^2 - 2\cdot4r + (1\cdot2)^2 = (r - 1\cdot2)^2$. 29. $x^2 - 2px + p^2 = (x - p)^2$.
30. $x^2 + qx + \left(\frac{q}{2}\right)^2 = \left(x + \frac{q}{2}\right)^2$. 31. $A = \pi t((2a + t)$; 17·1 lb.
32. 21·1 sq. ft.; 47·9 cu. in. per 1000; 22·7% saved.
33. $V = \frac{\pi r^2}{3}(3l + 2h)$; 105 gm.
34. $\frac{1}{R} = \frac{1}{R_1} + \frac{1}{R_2} + \frac{1}{R_3}$; 5·54 ohms. 35. 8·8 sq. in.
36. (*a*) $V = \frac{\pi r^2}{3}(h + 2r)$; (*b*) $A = \pi r(l + 2r)$; 50·7 cu. in.; 68·8 sq. in.
37. 9·42 sq. in.; 2·23 cu. in.; $V = \frac{\pi r^2}{3}(3l - r)$.
38. 28·9 cu. in. 39. 19·3 cwt. 40. 78·0 kg.
48. $(x + 3)(x - 2)$. 49. $(x - 3)(x - 4)$. 50. $(x + 4)(x + 1)$.
51. $(3x + 1)(x - 1)$. 52. $(2y - 1)(y - 2)$. 53. $(4p - 3)(p + 1)$.
54. No rational factors. 55. $(2x - 1)(x + 4)$.

EXERCISE 4(*b*) (p. 104)

1. 7·73. 2. 0·9045. 3. $3\cdot49 \times 10^{-5}$. 4. 0·299. 5. 0·217.
6. 0·540. 7. 0·585 cm. 8. 1950. 9. 1·05 sec. 10. 48·8 f.p.s.
11. $A = \frac{c^2}{4\pi}$; 17·4 in. 12. $l = \frac{A}{2\pi r} - r$; 9·91 in. 13. 5·08 in.

14. $L = 1/(4\pi^2 f^2 C)$; 48·1 mH. required.
15. $r = \dfrac{V}{24\pi t} - \dfrac{t}{2}$; diameter 4·15 in. 16. $S = \dfrac{8d^2}{3l} + l$; 0·038%.
17. (a) $\rho = \dfrac{\pi R d^2}{4l}$; (b) $d = \sqrt{\dfrac{4\rho l}{\pi R}}$; 1·68 microhms per cm. cube; 1·46 mm.
18. 123 tons. 19. 2·0 kg. per sq. cm. 20. 49 watts.
21. 0·9 in. 22. 153 lb. lead; 1000 lb. copper; 13·3% lead.
23. $B = \sqrt{(8\pi g P/A)}$; 3390 lines per sq. cm.
24. 95 ft. per min.; 510 hr. 25. 131 volts.
26. 11·4 sq. cm. 27. 10 cm.
28. $3{\cdot}0 \times 10^{10}$ cm. per sec.; 0·91; 0·20, 1·22; 0·65, 1·55 Mc/s.
29. A.P. 6·67, 7·2 metres; S.C. 4·9, 5·1 metres.
30. 123, 3·4 $\mu\mu$F.

REVISION PAPERS A (p. 107)

*A*1. 1. $\frac{19}{42} = 0{\cdot}452$, $\frac{4}{9} = 0{\cdot}444$, $\frac{3}{7} = 0{\cdot}429$, $\frac{5}{16} = 0{\cdot}3125$.
2. (a) 32, 4480; (b) $6ab^2$, $252a^4b^4c^4$. 3. 1·57 cm.
4. $\frac{224}{30} = 7{\cdot}47$; £56. 5. 147 above, 133 below; + 7·4%, − 6·7%.
6. 1·5 probably.

*A*2. 1. (a) 3·71 ft.; (b) 2 lb. 14 oz.
2. Easier to compare with another town.
3. 72,090 ($\pi = 22/7$). 4. 41%.
5. 8*d.*; (a) $\dfrac{75p}{112}$ shillings; (b) $\dfrac{Bp}{112}$ shillings; £1 8*s.* 4*d.*
6. Insulation rapidly deteriorating, *i.e.*, cable or joint faulty! A graph suggests continuity of insulation figures from day to day.

*A*3. 1. £4·758; 13·32 dollars.
2. Extra length per bay to allow for sag; 90 ft. 2 in.
3. 2·3 cu. ft. 4. 1024; $\log_{10} 2 \simeq 0{\cdot}3$. 5. 7·7%.
6. Little variation; gas used for cooking only? Average: 34·5 therms.

*A*4. 1. Multiply by 0·400; 325 dollars per short ton.
2. 5532; 7810; ratio 2·16 : 1.
3. $h^2 = \dfrac{A^2}{\pi^2 r^2} - r^2$; 2·81 cm. (Use difference of 2 squares.)
4. 10·8 lb. 5. 40 ohms. 6. 49%; unsatisfactory.

*A*5. 1. 3·80. 2. 41·4. 3. 24·5%.
4. 199 gm.; 5620 lines per sq. cm. 5. 60. 6. 20*s.* 10½*d.* Dotted.

*A*6. 1. 5·67 gm. 2. 2 ft. 6½ in. square; 131. 3. 5·7%.
4. $T = \dfrac{pv}{17{\cdot}8} - 273$; 58·7° C. 5. 83%; 3·7%.
6. Heat is absorbed: (a) in turning water into steam (at 100° C.); (b) in turning aniline liquid into aniline vapour (at 184° C.). About 120° C.

EXERCISE 5(*a*) (p. 122)

1. 3. 2. $\frac{1}{3}$. 3. 12. 4. 7. 5. 6·5 6. $1\frac{1}{3}$. 7. 4.
8. 3. 9. 15. 10. $\frac{3}{8}$. 11. 5. 12. 9. 13. $\frac{3}{5}$. 14. 7.
15. NOT an equation : this is an identity; true for any value of P!
16. 8. 17. 4. 18. 11. 19. $\frac{1}{3}$. 20. 1. 21. 12.
22. $1\frac{2}{3}$. 23. 9. 24. $2\frac{5}{7}$. 25. -4.
26. Extra voltage $1{\cdot}5n$; $\therefore 1{\cdot}5n + 50 = 80$, $\therefore n = 20$.
27. 2·5 metres.
28. Total distance travelled $12(t + \frac{1}{4}) + 10t = 69$; at 5 p.m.
29. 30 ohms. 30. 37·5 in. of mercury. 31. 0·2 ohms.
32. 4 tons. 33. 2*s*. 6*d*. 34. 18·5 miles. 35. 24 holes.

EXERCISE 5(*b*) (p. 132)

1. 3, 5. 2. 7, 2. 3. 4, 9. 4. 6, 1. 5. 0·02, 1·06.
6. 8, − 6. 7. 6·2, 1·24. 8. 1·8, − 2·4. 9. 5·6, 4·2.
10. 3·85, − 1·35. 11. 3 tons. 12. 182 men. 13. 8, 7.
14. 9 : 1. 15. $I_1 = 4\frac{2}{3}$, $I_2 = 3\frac{1}{3}$. 16. £10, £5.
17. $C = 10p + 5w$: £220.
18. Mains 3·28 amperes; battery 0·63 amperes : 7·35 ohms.
19. (*a*) 17 sec. : (*b*) 30 sec. approx. 20. 1·3 volts (lead–acid 2·0 volts).
21. £6 10*s*. per line; 0·8*d*. per call: $C = 6{\cdot}5e + \frac{t}{300}$: £9500.
22. x lb. lead per barrel, y lb. copper : Then $5x + 3y = 810$ and $2x + 7y = 730$. Whence $x = 120$, $y = 70$.
If n barrels contain lead, $(10 - n)$ contain copper.
$\therefore 120n + 70(10 - n) = 900$, etc., giving $n = 4$.
$\therefore$ There were 4 barrels of lead and 6 of copper.
23. $a = 0{\cdot}293$, $b = 70{\cdot}7 = R_0$; $a = 0{\cdot}0041$. 24. 3*s*. 3*d*.; 2*s*. 9*d*. each.

EXERCISE 6(*a*) (p. 145)

1. $y = -3x + 16$, $a = -3$, $b = 16$. 2. 7, 4. 3. − 1, 72.
4. − 0·4, 2·2. 5. 0·25, − 2.
6. − 1·5, 3 (2 and 3 are the intercepts on the two axes).
7. 0·75, − 1·75. 8. $-\frac{p}{q}, -\frac{r}{q}$.
9. $-\frac{n}{m}, n$. 10. − 0·53, − 1·1.
11. $a = 64$, $b = -322$: 314 lb. 12. $a = 712$, $b = 58{\cdot}8$; 0·35 in.
13. $a = 0{\cdot}06$, $b = 25$; 25 in.; 417 lb. approx.
14. $a = 0{\cdot}49$, $b = 0$, *i.e.*, R is directly proportional to l; 52 ohms for S.W.G. No. 21.
15. $R_0 = 100$ ohms is resistance at 0° C. $R_0 a = \text{slope} \simeq 0{\cdot}04$.
$\therefore a = 0{\cdot}0004$.
16. $x = 4y = -2$. 17. 0, − 1·5. 18. 1·2(5), 2·3(3).

19. 1·3(3), −0·25. 20. − 3, 4. 21. − 4·2, − 2·4.

Nos. 22 to 24 give a straight-line graph through the origin, No. 25 gives an intercept 1*s*. on the " cost " axis.

26. Carbon, 740 hr.; metal, 1460 hr. 97%.

27. Carbon filament : $P = 106 - 0{\cdot}035t$; metal filament : $P = 103 - 0{\cdot}013t$.

28. A " family " of straight lines through the origin.

29. (*a*) 273 ohms; (*b*) 400-lb. conductor just makes the grade.

30. $S = 0{\cdot}005t + 1{\cdot}167$ between $t = 3$ and $t = 8$.
(Scale for S. 1 in. ≡ 0·01, *i.e.*, between 1·170 and 1·220.)

EXERCISE 6(*b*) (p. 161)

1. 4·6, 0·56. 2. 3·5, − 11. 3. 1, − $1\frac{1}{3}$.
4. 2·3, 0·22. 5. 0·6, − 1. 6. 0·27, − 6·3.
7. 7·80, − 1·80. 8. − 0·69, − 4·30. 9. 3, $\frac{1}{3}$.
10. 0·712, − 4·21. 11. $a = 774$; 88 × 49·1 = 4320 lb.

12. $A = \frac{\pi d^2}{4} \times (2{\cdot}54)^2 = 5{\cdot}07d^2$; 3·98 in.

13. (*a*) 683 ft. (This would have to be a shore station !); (*b*) 16·7 miles.

14. $2\frac{1}{4}$ by $2\frac{1}{4}$ in. (nearest $\frac{1}{4}$ in.). 15. $a = 0{\cdot}03$, $b = 12$; 30·6 m.p.h.

16. Clearance 30 ft.; height 40 ft. above ground; 62 ft. approx.

17. In the middle ($M = 1080$). 18. At the clamped end ($x = 0$).

20. $y = 0{\cdot}000357x^2 + 10$. 21. 2 ft. $5\frac{1}{2}$ in. 22. 0·88 in.

23. Equation $x(14 - 2x) = 24$; 4 by 6 in. or 3 by 8 in.

24. $A = \pi r(r + s)$; $r^2 + 5{\cdot}2r = \frac{48}{\pi}$; radius 2·10 cm.

25. 2 or $\frac{1}{3}$; 2·47 or − 0·135.

26. (*a*) 180 lb. per mile; (*b*) 11·7 ohms per (loop) mile.

27. $y = \frac{250p}{3x}$; (*a*) about 20*s*. 10*d*.; (*b*) 33*s*. 3*d*.

28. 3·1 μF.

29. $1{\cdot}09 \times 10^6$ c/s; $6{\cdot}02 \times 10^5$ c/s; $7{\cdot}48 \times 10^5$ c/s; $1{\cdot}73 \times 10^5$ cm. or 1730 metres.

30. 23·8 volts.

REVISION PAPERS B (p. 165)

*B*1. 1. $x = 72$, $y = 108$.
2. Suppose tank is x ft. by $(x - 3)$ ft. Then solve $5x^2 - 12x = 812$; $x = 14$. 13,500 gall. 3. (i) 5; (ii) − 9.
4. 26·9 by 53·8 in.; 100·5 yd. 5. (i) 0·4861; (ii) 9·0.
6. $t = 2$ to $t = 9$. Slope $a = 0{\cdot}0044$. Intercept ($t = 0$) $b = 1{\cdot}169$.
(Calibrate S-axis from 1·16 to 1·22.)

*B*2. 1. $\frac{3}{4}d$. per unit; £23 R.V. 2. (*a*) $x = -1$; (*b*) $y = -0{\cdot}0028$.
3. £70; £6 10*s*.
4. (*a*) + 0·36 or − 2·11; (*b*) + 1 or − 2·75; $4x^2 + 7x - 11 = 0$

5. 35 ft. (nearest ft.).
6. Linear variation for the range $V_g = -6$ to $V_g = +2$; $I = 1{\cdot}1V_g + 9{\cdot}3$.

*B*3. 1. 24 cm. (approx.). 2. $4{\cdot}4 \times 10^3$ to 1.
3. (*a*) $y = 17$; (*b*) $x = 3$.
4. Quadratic $x^2 - 100x + 1275 = 0$. £85 (or is it £15?). Either answer fits the problem!
5. $-\frac{5}{365} \times 100 = -1{\cdot}4\%$; $P = \frac{3p}{2}$.
6. $W = 120 + 0{\cdot}5x$; $Q = 4 + 0{\cdot}01x$.

*B*4. 1. (*a*) Together $\left(\frac{1}{x} + \frac{1}{y}\right)$ per day; $\frac{xy}{x+y}$ days.
(*b*) $S = MR/(R - \pi TM)$.
2. (i) $x = 2{\cdot}32$, $y = -0{\cdot}30$; (ii) 2·59 or $-0{\cdot}26$.
3. $s = \frac{1}{2}(u + v)t$. Distance travelled = average speed × time.
4. 4*s*. 7*d*., postage 3*d*. 5. 36 years.
6. $E = 3{\cdot}8W + 7{\cdot}7$.

*B*5. 1. $a = 570$, $b = 0{\cdot}90$; 43 m.p.h. 2. 34 sq. cm.
3. Quadratic $2r^2 + 9r = 84/\pi = 26{\cdot}8$; radius 2·05 in.
4. $17\frac{1}{3}\%$; $\frac{3}{4}(100 + x) = 88$. 5. $l = \frac{gt^2}{4\pi^2} = 24{\cdot}85t^2$; 2·01 sec.
6. (*a*) 6 ft.; (*b*) 1·75 in.

*B*6. 1. Eqn. $x^2 + 10x = 336$; 24, 14 m.p.h.
2. $B = \sqrt{8\pi gP/A}$; 4170 lines per sq. cm.
3. 5 yd. per min. 4. 364 lb. lead; 275 lb. copper.
5. $1{\cdot}7 \times 10^{26}$.
6. $V = 0{\cdot}11T + 29$. Substituting in the formula, the volume is doubled at 290° C. (2 significant figures).

EXERCISE 7(*a*) (p. 189)

1. (*a*) 30°, (*b*) 150°, (*c*) $30x°$.
2. (*a*) 110°, (*b*) 130°; a.m., p.m. are irrelevant.
3. 45°, 225°, 270°, $157\frac{1}{2}$°, 195°.
4. (*a*) 11° 27′ West; (*b*) 116° 33′ (in 1951); (*c*) A.D. 2037.
5. (*a*) 600°, (*b*) 10°.
6. (*a*) 18°, (*b*) 56° 23′, (*c*) $(90 - y)°$. 7. 49°.
8. 100°. 9. 28°; $\left(90 - \frac{x}{2}\right)$ degrees.
10. 29°, 69°. 11. 11·8 miles, 213°.
12. 1020 yd., S. $32\frac{1}{2}$° E.; 4 rt. ∠s.
13. 81 miles; $74\frac{1}{2}$°. 14. 136 or 280 m.p.h.
15. (*a*) 6 a.m., 2 p.m., 9.20 p.m., 2.08 a.m., $L/15$ p.m.; (*b*) 90° E.; $67\frac{1}{2}$° W.; $L = 15t°$ E.; (*c*) the effectiveness of radio propagation varies with the time of day.
16. (*a*) 14 rt. ∠s; (*b*) 10 rt. ∠s.

17. 72°, 108°; 45°, 135°; 36°, 54°. 18. 36°.
19. $157\frac{1}{2}$°. 20. 9; $\frac{360}{180 - x}$ sides. 21. $2n - 4 = 24$; $n = 14$.
22. Yes (20); no; yes (80); yes (144). 23. $112\frac{1}{2}$°.
24. $x = 18$; angles 18°, 72°, 108°, 162°; adjacent angles $72° + 108° = 180°$, ∴ 2 sides parallel. 25. 4; 14°.
26. 24 poles; regular polygon with 24 sides.
27. (*a*) Straight line parallel to groove and $\frac{5}{8}$ in. above it; (*b*) circle; (*c*) a small (minor) arc of a circle. 28. $(14 + \pi)$ in.
29. Two positions: a circle radius 25 feet from tree base; a straight line parallel to the track. 30. 4 (ignoring width of track).
31. 8 miles (Pythagoras; 3 : 4 : 5 △). 32. 9·89 miles.

EXERCISE 7(*b*) (p. 206)

1. $A = 46°$, $B = 80°$, $C = 54°$. 2. Impossible.
3. Impossible. 4. $A = 98°$, $C = 50°$, $AC = 2·84$ cm.
5. $C = 100°$, $AB = 5·56$ in., $AC = 2·99$ in. 6. Ambiguous.
7. Gives shape, but not size; only 2 independent pieces of information given. Must know ONE side at least to fix scale.
8. $A = 50°$, $B = 40°$, $AC = 4·12$ cm. 9. Impossible; $AB // AC$!
10. Impossible for 3 angles not to add up to 180°. 11. 642 yd.
12. 1860 yd. bearing 295°.
13. 3° bearing; 16,800 yd. distance. 14. 34·8 miles S. $31\frac{1}{2}$° E.
15. Aerial cable; 10,000 yd.; U.G. cable; 5400 yd. Aerial cable cheaper by $\frac{800}{108} = 7·4\%$. 16. 354 ft.
17. 3·43 naut. miles N. 12° E.; 10·3 knots. 18. 6 in.; 7 ft. 9 in.
24. AC. 32. $A = 56°$, $C = 62°$. 33. $Q = R = 36°$.
34. $L = (180 - 2x)°$, $N = x°$. 35. $D = E = 37°$, $F = 106°$.
36. $Z = X = \left(90 - \frac{y}{2}\right)^{\circ}$.
37. $(180 - y)°$ at U and V; $(2y - 180)°$ at W. 38. $A = 72°$, $C = 18°$.
39. $C = (90 - x)°$. 41. $x = 2(180 - y)$. 43. 60°.
48. 6°, 24°. 50. 6. 52. 4; 110, 155, 255, 360 yd.
53. Intersection of perpendicular bisectors of any two sides of $\triangle PQR$; 2·8 in.
54. On the bisectors of the angles between the given lines. 1·35 in.
55. 3. 56. 89 yd.
58. An ellipse; foci (singular: focus)! $PA + PB$ is constant.

EXERCISE 7(*c*) (p. 234)

1. 10 cm. (3 : 4 : 5 △).
2. Join A to centre O and produce to D. Use Pythagoras in △ OBD. Radius is 7·04 cm. approx.
3. $PQ = 3·12$ in.; $\sqrt{8} = 2·83$ in.
4. 1·5 in.; on part of the surface of a concentric sphere of diameter 3 in.

5. (*a*) 4; (*b*) 5; (*c*) 1; (*d*) 3; (*e*) 1; (*f*) 6.
7. Construct adjacent ∠s of 64° and 212° at the centre.
8. Subtract instead of add in the last stage.
9. $QPR = QSR = 37°$ (use "angle in semicircle" and "same arc" properties). 10. 107°.
14. Use a large set-square, and check its right angle remains firmly in contact with the inner curve in all positions (angle in semicircle a right angle).
16. Join P, R to centre: use cyclic quad. $PORS$: $\angle PSR = 54°$.
17. 66°. 20. $L = 62°$, $M = 66°$, $N = 52°$. 21. 6·69 cm.
22. (*a*) 44 in.; (*b*) 47·5 in. 23. (*a*) 10·95; (*b*) 8·49 in. 24. 10·6 in.
27. 1020 acres. 28. (i) 8 times; (ii) halved.
29. 2 ft. 8 in., 2 ft. 4 in.; area reduced in ratio 4 : 9; 32 lb.
30. 125%; 237·5% (increases).
31. Similar △s; $BN^2 = NM . NA$; ∴ BN is a tangent to ⊙ABM; 4 more. 32. Magnification (*a*) 2, (*b*) 1·2.
33. No real image; "virtual" image 24 cm. from lens, same side as object; lens acting as a magnifying glass, magnification 2.
34. $V = -24$. 35. 10 in.; 96 sq. in. 36. 96 ft.
38. Pythagoras: $r^2 = (r - 2)^2 + 36$; 10 ft. radius.
39. 11 ft. 10 in.; $4\frac{1}{4}$ in. approx.
40. $0{\cdot}293r = 2$. ∴ $r = 6{\cdot}8$ (mm.). Normal separation 9·6 mm.; 45°.
41. 8 cm. 42. 7100 cu. yd. (2 significant figures).
43. 323,000 gall. 44. 8·7 acres or 42,000 sq. yd.
47. 17·5 m.p.h.; 2·34 miles.
49. Average ordinate 24·71; area under curve 2966.
50. Average depth 2·8 ft.; 42·4 cu. yd.

EXERCISE 8(*a*) (p. 253)

1. 0·8942, 0·9088, 0·8656, 2·0338. 2. 29° 26′, 11° 29′, 62° 14′.
3. (*a*) $6 \sin 18° = 1{\cdot}85$ cm.; (*b*) $5{\cdot}3 \tan 32° 17' = 3{\cdot}36$ cm.; (*c*) $\dfrac{4{\cdot}8}{\sin 64° 26'} = 5{\cdot}31$ cm.
4. (*a*) $31\frac{1}{2}°$, $58\frac{1}{2}°$; (*b*) 43° 40′, 46° 20′.
5. $PQ = l \sin (90° - \alpha)$; $RQ = l \sin \alpha$. 6. 3·63 in.; 3·56 in.; 6·47 sq. in.
7. $\sin \alpha = \frac{11}{14}$; 52° (nearest degree). 8. 72°; 15 ft. 2 in.
9. 34° 42′. 10. 4° 46′.
11. 10° 29′; − 11′, about − 2% error. 12. 654 ft.
13. 408 ft.; 1 in 3 approx. 14. 409 yd.
15. 39° 48′. 16. 0° 48′. 17. $\alpha = 7\frac{1}{2}°$, $\beta = 31°$.
18. 10 ft. 5 in. 19. 18 ft. 0 in.; $70\frac{1}{2}°$.
20. 89·7 ft., 3040 sq. ft. and 11,300 cu. yd. 21. (*b*) 11·2 cm.
22. $0{\cdot}868r$ cm. 23. $4 \tan 36° = 2{\cdot}91$ in.
24. $2r \tan 20°$. 25. 1·46 in.
26. (*a*) $l = 2r \sin \left(\dfrac{180}{n}\right)^\circ$; (*b*) $l = 2r \tan \left(\dfrac{180}{n}\right)^\circ$. 27. $9\frac{1}{2}°$ approx.

28. 14·6 ft.; 7·66 in. 29. 1° 12′ either way.
30. Curved part of string subtends 221° at the centre; $\frac{221}{360}$ of the circumference = 10·8 cm.; total length 25·8 cm.
31. (*a*) 14° 22′ or 14·4°; (*b*) 31·8 in.; angles at centre 165·6°, 194·4°.
32. (*a*) 77° 22′; (*b*) 35·0 in.

EXERCISE 8(*b*) (p. 271)

1. 0·6115, 0·6710, 0·1285. 2. 70° 18′, 27° 45′, 85° 33′.
3. 0·6820, − 0·3057, − 0·7314, − 0·9135, − 2·3559, − 0·3584.
4. 17° 26′ *or* 162° 34′; 180° − 53° 8′ = 126° 52′.
5. (*a*) 208° 52′ *or* 331° 8′; (*b*) 77° 10′ *or* 282° 50′.
6. $x = 6 \sin 42° = 4{\cdot}01$ in.; $y = x \cos 42° = 2{\cdot}99$ in.
7. $h \cos \alpha$, $h \sin \alpha$. 8. 20 tan 79° = 103 yd.; height 166 ft.
9. 218 ft. 10. 15·3, 12·9 in.
11. Either (*a*) 121°, 75°, 105°, 59°; (*b*) 3·11 in.; or (*a*) 59°, 133°, 47°, 121°; (*b*) 4·11 in.
12. 7·84 cm.; ratio 9 : 10 very nearly. 13. 35° 32′, 72° 14′, 72° 14′.
14. 74°. 15. 5 ft. 7 in.; 13 ft. 3 in.
16. $AB = AC$ = 14 ft. 9 in.; $PQ = 8\sqrt{2}$ ft. = 11 ft. 4 in.; $\therefore PR = QS$ = 3 ft. 1 in.; A is 5 ft. 8 in. above BC.
17. 6·30 cm., 6·25 cm.
18. $\angle ABD = 90°$ (in semi-⨀); $AB = 30 \cos 30° = 26{\cdot}0$ in.; shorten BC by 2·47 in. 19. $a = 20°$, $d = 42{\cdot}3$ in., $w = 19{\cdot}8$ in.
20. (*a*) 10·67 cm.; (*b*) 5·96 cm.; (*c*) 93° 38′.
21. 4350 yd., when $\theta = 45°$ (sin 90° = 1); 4140 yd.; 17°.
22. Very strong likeness to a " sine wave " (not quite a " pure " one); for short-wave radio the hours of darkness are those favourable to reception. 24. 4·5 millisec.; 222 c/s.
25. $\frac{130}{360} \times \pi(4{\cdot}2)^2 = 20$ sq. in.; 8·25, 11·8 sq. in. 26. 48·8 gall. per min.
27. (*a*)(i) $\frac{3}{4}\pi$; (ii) $\frac{1}{10}\pi$; (iii) $0{\cdot}462\pi$; (iv) $\frac{9}{4}\pi$; (v) $\dfrac{\pi}{10800}$ or $0{\cdot}0000976\pi$; (vi) $\frac{4}{5}\pi$; (vii) $\frac{37}{90}\pi$ or $0{\cdot}411\pi$; (viii) $1{\cdot}095\pi$.
 (*b*)(i) 2·36; (ii) 0·314; (iii) 1·45; (iv) 7·07; (v) 0·000291; (vi) 2·51; (vii) 1·29; (viii) 3·44.
 (*c*)(i) 150° 0′; (ii) 67° 30′; (iii) 255° 0′; (iv) 330°; (v) 41° 22′; (vi) 120° 0′; (vii) 270°; (viii) 11° 10′.
28. To 5 millionths of a radian.
29. (i) 0·9511; (ii) 0·500; (iii) 135°; (iv) 0·7660; (v) − 0·6428; (vi) 1·00; (vii) 0·0698; (viii) 0·9976; (ix) 0·0699; (x) 1·00.
30. 314 radians/sec.; 3·18 m.secs.; 25·25 volts.

EXERCISE 8(*c*) (p. 274*e*)

1. 4·74, 3·97, 6·28 sq. units. 2. 38°, 3·05, 4·56.
3. R is either 52° 4′ or 127° 56′, Q = 90° 56′ or 15° 4′, PR is 4·82 or 1·25.
4. A = 38° 27′, so C = 92° 6′, c = 4·35 (one solution only).

5. 4·13.
6. $c = 5{\cdot}51$. Alternatively, use half the isosceles triangle as right-angled triangle.
7. 143° 42′, 26° 21′, 9° 57′. 8. 49·2, 23° 8′, 43° 52′.
9. The *largest, i.e.*, B, opposite the longest side. If B obtuse, cos B will reveal this fact by being negative. Compare sin B, which is positive for both acute and obtuse angles. Here $B = 139° \, 36'$, $A = 27° \, 45'$, $C = 22° \, 29'$. 10. 16 ft. 2 in., 8 ft. 10 in., 8 ft. 7 in.
11. 41° 4′: 3450 sq. yd. 12. 3530, 2780 yd. 13. 31 ft.
14. 35 ft. 15. It remains $4\frac{2}{3}$ tons. 16. 33·3 ft.

EXERCISE 9 (p. 306)

1. S, V, V, S, V, S.
2. (*a*) 20 miles by road; (*b*) 11 miles N. 60° W. as the crow flies. (*a*) is scalar; (*b*) is vector (a " displacement "). The scalar quantity is easily measured with a " cyclometer "; the vector displacement is difficult or impossible to measure directly.
3. (*a*) Because the effort is applied at a greater distance from the fulcrum (axis of nut) and hence turning effect is greater for the same effort. (*b*) The *direction* of the applied effort is easier (downwards) even if the effort itself is slightly greater (due to friction).
4. (*a*) 16·8 lb. at 28° with 14 lb. force; (*b*) 3·98 f.p.s. at 37° with the 50 f.p.s. velocity.
5. Use similar △s: $\dfrac{x}{20} = \dfrac{120}{45}$; 53 yd. 1 ft. downstream; 49·2 yd. per min.; same time.
6. (*a*) 16 sin 53° = 12·8; (*b*) − 16 cos 53° = − 9·6 (naut. miles).
7. 277 m.p.h.; 13° 20′ N. of W.; steer 13° 43′ S. of W.
8. Upstream, at 68° with the stream.
9. A to B; upstream, 88° with stream, 8·5 knots.
   B to A; ,, 50° ,, 6·5 knots.
10. The speed of " separation "; or the speed at which the yacht is gaining on the steamer; 20 knots in direction N. 36° E. (mathematically, velocity of yacht *relative* to steamer).
11. A 5 : 12 : 13 △; 2 tons $4\frac{1}{2}$ cwt. approx.
12. $x = 18/\tan \alpha$; 3 ft. 11 in.; $\alpha = 77\frac{1}{2}°$ ($\cos \alpha = 3/14$).
13. 1160 lb. wt., 2080 lb. wt.
14. 16·3 cwt.
15. Vertical plane through stay at 55° with forward direction; tension 1600 cos 55°/cos 54° = 1560 lb. wt.
16. Bearing 302°.
17. 1·46, 1·78 tons.
18. Draw the △ of forces: *compressive* force in jib 6·2 tons wt.; *tension* in tie-rod 2·8 tons wt.; yes.
19. The △ of forces is similar to a 16 : 26 : 14△. By calculation forces 4·87 tons (compressive) and 2·63 tons (tensile).

20. $\triangle$ of forces is isosceles; tension $12/\cos 84° = 115$ lb. wt. In drawing, a small error in the angle measured makes a large error in the lengths to be measured.
22. 49 lb.
23. If reactions from supports are $2R$ and R cwt., and W cwt. is the unknown weight. $W + 10 = 3R$; $6W + 100 = 20R$; weight 50 cwt., thrusts on supports 40, 20 cwt.
24. 210 lb. wt.; $5\frac{1}{4}$ in.; 190 lb. wt. upwards.
25. Take moments about B. Resolve vertically; reactions from supports A, B are 320 lb. down, 560 lb. up; Forces acting on the supports are $\therefore$ 320 lb. up, 560 lb. down.
26. 4·3, 11·8, $1{\cdot}8 + 0{\cdot}25W$ in.; 1·8 in. from fulcrum (ZERO calibration).
27. Upthrust due to steam = 543 lb. wt.; therefore pressure (above atmospheric) $= \frac{543}{\pi} = 173$ lb./sq. in.
28. $32W + 28 = 1130$; 34·5 lb.; action of end of lever on the joint is 339 lb. wt. *upwards*.
29. Moments about the line AB: at A, B, $10\frac{2}{3}$ tons DOWN; at C, D, $6\frac{2}{3}$ tons UP.
30. At A, B, $14\frac{2}{3}$ tons UP; at C, D, $1\frac{1}{3}$ ton UP.
31. About 57 lb. wt. tension.
32. $5{\cdot}8 \times 10^{-5}$ in./lb.; 0·26 in.
34. (*a*) 0·6 tons; (*b*) 1·05 tons.
35. Tension 2020 lb. wt.; a 7-strand stay is sufficient.
36. Equal tension assumed in each bolt; diameter 0·21 in.
37. $1{\cdot}6 \times 10^{-3}$ in.
38. 55·5 lb. per sq. in.; 0·0083 in.
39. 11·2 lb.
40. (*a*) 100 gm.; (*b*) $X = 214 + 18{\cdot}4d$.

REVISION PAPERS C (p. 316)

*C*1. 1. 2 $\angle$s, corr. side. 2. 44° 4′, 67° 58′ (twice); 14·2 sq. in.
3. 1° 2′ (1 in 55). 4. Scale drawing, 3·9 miles.
5. Scale drawing, 2·9 knots, bearing $149\frac{1}{2}$°.
6. Wheelbarrow; L between F and E (1st order of levers): nail extractor; F between L and E (2nd order of levers).

*C*2. 1. 365 lb.
2. Same base PB. Prove by congruent $\triangle$s they have same perpendicular height. 3. 16·6 cm.; 13·2 cm.
4. 5480 yd. 5. 59 lb. wt.; 12 lb. wt. 6. 36 lb. wt.

*C*3. 1. ABC is $\triangle$ of forces for the 3 forces acting at B; tension in BC = 8 cwt., compression in AB = 12 cwt. 2. 2° 17′.
3. 21·7 sq. in.; $A = \frac{1}{4}b^2 \tan\left(90° - \frac{\alpha}{2}\right)$.
4. 20 cu. in., $67\frac{1}{2}$ cu. in.; 4·06 in.
5. 270 m.p.h. (2 significant figures).
6. Not greater than 18·7 tons; 11 ft. 9 in.

*C*4. 1. Base diameter 4 in.; height 7·75 in.; vertical angle 29°; volume 32·5 cu. in. 2. 1 ton 3 cwt.
3. Use "moments"; diameter 6 in.; 84 turns. 4. 194 ft.
5. $h = 2083d\,(\tan\alpha - \tan\beta)$. 6. 0·534 in.

*C*5. 1. Saving 13·9%. (Consider 340-mile journey.)
2. 195 ft. 3. $3{\cdot}8 \times 10^9$ cm. per sec.; $8{\cdot}5 \times 10^7$ m.p.h.
4. $80x + 36 = 720$; 8·55 cm.: $12y = 855 + 36$; 74·25 cm. from the fulcrum. 5. 11,000 f.p.s. 6. 65 amperes.

*C*6. 1. 42°; 4·03 tons wt. 2. 29·6 cm.
3. Quadratic $x^2 + 9x = 36$; $AP = 3$ cm.
4. (*a*) 0·141 in.; (*b*) 0·173 in. 5. $s = \frac{v^2}{64}$; 52 f.p.s.
6. Average 2·3 sq. ft.; volume 46 cu. ft.

*C*7. 1. Type B; 11·5%, compared with 12% p.a. for type A.
2. 108° 38′; 2·58 in. 3. $3{\cdot}7 \times 10^4$ calories.
4. 2·83 min. 5. 1126 c/s; $L = \frac{1}{4\pi^2 f^2 C}$.
6. 15·4 tons wt.; graph linear for loads up to about 14 tons—the probable yield point.

*C*8. 1. 1·40 tons wt.; 1·55 tons wt.
2. Consider as a rectangular block + 4 prisms + 4 pyramids; 546 cu. ft. 3. 6·04 ft.; 14 million gall. per hr.
4. 0·324 sq. in.; 16 cwt. 5. (*a*) 18·4 ft.; (*b*) 24 ft.
6. 11·6 per week; 14·8 per week.

APPENDIX

PAST PAPERS IN MATHEMATICS FOR TELE-COMMUNICATIONS CITY AND GUILDS OF LONDON INSTITUTE

City and Guilds of London Institute

Telecommunication Technicians' Course Practical Mathematics

1961–62

1. Use tables to calculate:

(*a*) (*i*) $\dfrac{32{\cdot}62 \times 4{\cdot}257 \times 10^{-4}}{\sqrt{23{\cdot}42}}$ (*ii*) $\dfrac{1}{(5{\cdot}6)^2} - \dfrac{1}{35{\cdot}6}$

(*b*) the percentage error (to two significant figures) when $(1{\cdot}08)^5$ is approximated to 1·40.

2. (*a*) The following calculation arises in the solution of a traffic problem. Simplify the result, leaving your answer as a fraction in its lowest terms.

$$\frac{\dfrac{2^3}{3 \times 2}}{1 + 2 + \dfrac{2^2}{2} + \dfrac{2^3}{6}} \times 2$$

(*b*) The length (s), measured from its lowest point, of a heavy cable suspended between two points and the vertical height (y) are connected by the equation

$$(y + c)^2 = s^2 + c^2$$

where c is a constant. If $y = 4$ when $s = 8$, show that $c = 6$. Find s when $y = 10$.

3. (*a*) Find i_1 and i_2 when

$$2{\cdot}5i_1 - 3{\cdot}0i_2 = -0{\cdot}75$$
$$1{\cdot}6i_1 + 0{\cdot}8i_2 = 0{\cdot}88$$

(*b*) Find the Highest Common Factor and the Lowest Common Multiple of

$$24pq^2r, \qquad 16p^2qr, \qquad 6pqr^3.$$

(*c*) The power P watts generated in a circuit is

$$P = \frac{E^2}{R}$$

where E is in volts and R is in ohms. Keeping P in watts, find a formula for P in terms of K and Y where K is in megavolts and Y is in megohms.

4. (a) Divide $x^3 + x^2 + 2x - 4$ by $x - 1$.

(b) Multiply $x^2 + 3x$ by $x + 2$.

(c) If $\log 2 = a$, $\log 3 = b$ find, *without using tables*, log 6, log 12, and log 4·5 in terms of a and b.

5. (a) Simplify $15ab + 3bc - 9ba - 2cb$.

(b) Remove the brackets and simplify

$$4x - 3[x - (5x + 1)]$$

(c) Factorise: (i) $x^2 - x$; (ii) $2x^2y - 2y^3$.

(d) Solve for x the equation

$$\tfrac{3}{5}(2x - 3) + \tfrac{1}{4}(2 - 3x) = \tfrac{1}{8}(x - 1)$$

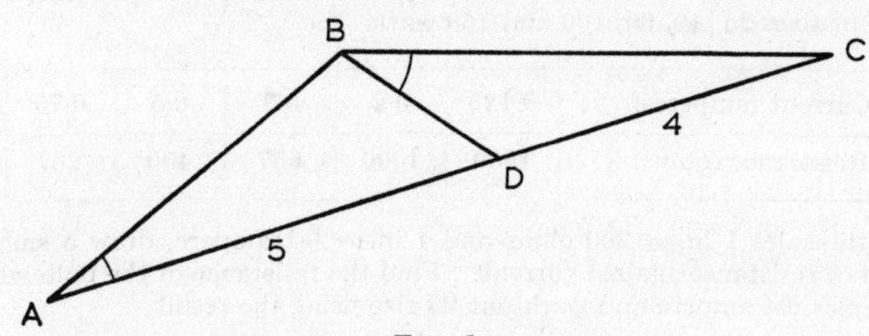

FIG. 1.

6. (a) The sending end resistance (R) of a transmission line is given by

$$R = R_2 + \frac{R_1R_2}{R_1 + R_2}$$

Calculate R when $R_1 = 4270$ ohms. $R_2 = 2320$ ohms.

(b) Make R_1 the subject of the formula in (a).

(c) State the value of R_2 for which R is zero.

7. (a) The power P in a circuit is proportional to the square of the current i. If $P = 36$ when $i = 3$, calculate i when $P = 156$.

(b) When a control switch is moved from one position to the next a variable resistance is increased by the same amount. At position 5 the resistance is 5200 ohms and at position 9 it is 6400 ohms. Find: (i) the resistance at position 17; (ii) the position of the switch for a resistance of 6100 ohms.

8. (a) A triangle ABC is right-angled at B with $AB = 2{\cdot}6$ in. and $BC = 3{\cdot}6$ in. Find the angles and third side of the triangle by graphical construction and verify your answers by calculation.

(b) In Fig. 1 the two marked angles are equal. Name the two similar triangles. If $AD = 5$ cm. and $DC = 4$ cm., calculate BC and state the ratio of BD to AB.

9. Copy and complete the following table:

$x°$	0	20	30	60	80	90
$\cos x$	1·0			0·5		
$\sin x$	0			0·866		
$\tan \frac{1}{2}x$	0			0·577		

On the same axes draw the graphs of $y = \sin x$, $y = \cos x$, $y = \tan \frac{1}{2}x$. Read off from your graphs the values of x at which: (i) $\cos x = \sin x$; (ii) $\sin x = \tan \frac{1}{2}x$; (iii) $\cos x = \tan \frac{1}{2}x$.

10. The table shows the resistance and current consumed by electric light bulbs of sizes 25, 40, 60, 100 and 150 watts.

Current (amperes) .	0·125	0·2	0·3	0·5	0·75
Resistance (ohms) .	1600	1000	667	400	267

With scales 1 in. = 200 ohms and 1 in. = 0·1 ampere, draw a smooth graph of resistance against current. Find the resistance of the bulb which consumes 0·4 ampere and work out its size using the result

$$\text{Watts} = (\text{Amperes})^2 \times \text{Ohms}$$

What size of a bulb would have a resistance of 900 ohms?

1962–63

1. (*a*) Give the exact values of the following: (i) $0{\cdot}000461 \times 1{\cdot}2 \times 10^3$; (ii) $0{\cdot}006858 \div (9 \times 10^{-4})$.

(*b*) A tank is fitted with two taps and an outlet valve. The first tap can fill the tank in 5 min., the second in 4 min., and the valve can empty a full tank in 8 min. When both taps are operating with the valve closed the time to fill the tanks is $\dfrac{1}{\frac{1}{5}+\frac{1}{4}}$ min. Calculate this time. How long will it take both taps to fill the tank when the valve is open?

2. (*a*) If $\frac{1}{5}(2x - 3)$ exceeds $\frac{1}{4}(7x - 3)$ by $\frac{3}{8}$, calculate the value of x.

(*b*) Given that $\dfrac{x}{y} = \dfrac{5}{4}$, evaluate $\dfrac{2x - y}{x + y}$ and state why this is insufficient information to evaluate $\dfrac{x^2 - y}{x + y^2}$.

(*c*) If $a = -2$, $b = 3$, $c = -1$, evaluate: (i) ab^2c; (ii) $ab(b^2 - c^2 + a^2)$.

3. (*a*) R is resistance in ohms, I current in amperes, and V potential difference in volts. Find the unit of measurement for each of the following, assuming 1 volt = 1 ohm × 1 ampere, and 1 watt = 1 volt × 1 ampere:

(i) $\dfrac{V}{R}$ (ii) RI^2 (iii) $\dfrac{V^2}{R}$

(*b*) Factorise:

(i) $a^2 + bc - ca - ab$ (ii) $4x(x - 3y) + y(12x - 9y)$

(*c*) Simplify:

(i) $\dfrac{6a^2b \times 2ab^2c}{(ab)^2 \times 24abc^2}$ (ii) $x^{\frac{1}{2}} \times y^{\frac{1}{3}} \times x^{\frac{3}{2}} \times y^{\frac{5}{3}}$

4. (*a*) Evaluate using logarithms:

$$\frac{(0{\cdot}08624)^2 \times (52{\cdot}61)}{76{\cdot}14 \times 0{\cdot}002861}$$

(*b*) If $\log (xy) = a + b$ and $\log x = a$ explain why $\log y = b$. Find in terms of a and b

(i) $\log \left(\dfrac{x}{y}\right)$ (ii) $\log (x^2)$ (iii) $\log (xy^2)$ (iv) $\log (\sqrt{xy})$

5. (*a*) The potential difference (v) across a capacitor charging through resistance (R) is given in terms of the current i by the equation $v = V - iR$, where V and R are constant. Calculate V and R if $v = 150$ when $i = 2{\cdot}8$ and $v = 200$ when $i = 2{\cdot}3$.

(*b*) Evaluate t using the formula

$$t = \frac{R(1 - p)}{p(R + 1)} \text{ with } R = 68{\cdot}6 \text{ and } p = 0{\cdot}0781.$$

Make p the subject of this formula.

6. (*a*) A is the area of a rectangle with length l; express (i) the perimeter and (ii) the length of the diagonal of the rectangle in terms of A and l.

(*b*) How many odd numbers exist between x and $x + 24$ if x is odd?

(*c*) A car travels 40 miles and consumes 1 gallon of petrol. Taking 1 kilometre = $\frac{5}{8}$ miles and 1 gallon = $4\frac{1}{2}$ litres, express this petrol consumption in litres per 100 kilometres. Give your answer to the nearest litre.

7. (*a*) Using mathematical tables find the value of: (i) tan 42° 31′; (ii) sin 63° 44′; (iii) cos 10° 16′.

(*b*) Using tables of squares or otherwise calculate the third side of all right-angled triangles having two sides of 3·2 in. and 4·3 in.

What is the area of the larger triangle?

8. Draw the graph of $y = x^2 + x - 5$ between the values of $x = -4$ and $x = 3$, with scale for x, 1 unit = 1 in., and scale for y, 1 in. = 2 units.

Use your graph to solve the equations: (i) $x^2 + x - 5 = 0$; (ii) $x^2 + x - 6 = 0$.

Verify the solution of $x^2 + x - 6 = 0$ using the method of factors.

9. A regular five-sided polygon has sides 6 in. long. Calculate the angle

subtended by one side at the centre of the circumscribing circle and show that the radius of the inscribed circle is 4·128 in. Hence deduce the area of the polygon.

Calculate the volume of the pyramid with vertical height 30 in. which has this polygon as base.

10. An alternating voltage (e) is measured at equal intervals of time (t) beginning at zero. The table gives the values of e^2.

e^2	0	1·64	4·24	4·0	3·53	4·0	4·24	1·64	0
t	0	1	2	3	4	5	6	7	8

With vertical axis for e^2 and horizontal axis for t, draw a smooth curve showing the relationship between e^2 and t. Use scale 1 in. = 1 unit for t, and e^2.

Determine the approximate area under the curve using (*i*) the mid-ordinate rule, (ii) the trapezoidal rule and obtain the r.m.s. value of e in each case if $e_{\text{r.m.s.}} = \sqrt{\dfrac{\text{Area}}{\text{Total time}}}$

1963–64

1. (*a*) If $\frac{1}{6}(4x - 2)$ added to $\frac{1}{3}(3 - x)$ equals $\frac{2}{5}$, evaluate x.

(*b*) Solve the equations: (i) $x^2 - x - 20 = 0$; (ii) $4x^2 - x = 0$.

(*c*) Write down the H.C.F. and L.C.M. of $4\pi R_1{}^3$, $4\pi R_2{}^3$.

2. (*a*) If $a = 0$, $b = -4$, $c = 2$ evaluate: (i) a^2bc^2; (ii) $bc(a^2 - b^2 + c^2)$.

(*b*) Solve the simultaneous equations for a and b:

$$\frac{2}{a} - \frac{3}{b} = -8 \qquad \frac{4}{a} + \frac{5}{b} = 50.$$

(*c*) Without the use of mathematical tables, give the exact values of the following: (i) $0·0000823 \times 1·1 \times 10^5$; (ii) $8263·5 \div (7 \times 10^4)$.

3. (*a*) Factorise: (i) $p^2 - qr - q^2 + pr$; (ii) $(5x - 2y)^2 - (4x - 2y)^2$.

(*b*) Express as one fraction:

$$\frac{2}{ab} - \frac{3}{bc} + \frac{1}{ca}.$$

(*c*) Simplify:

$$\text{(i)}\ \frac{2a^3}{3bc} \times \frac{5b^2}{4ca^2} \times \frac{3c^2}{2ab^4} \qquad \text{(ii)}\ \sqrt{(81a^4b^8)}$$

4. (*a*) Evaluate, using logarithms:

$$\frac{(234·2)^2}{634·1 \times 10^2} + \frac{0·003261}{(0·06547)^2}$$

(*b*) If $\log x = a$, write down the expressions for

$$\log(10x) \qquad \log\left(\frac{x}{10}\right) \qquad \log(x^2) \qquad \log(1000x)$$

assuming all logarithms are to the base 10.

5. (*a*) Draw the graph of $y = x^2 - 5x + 7$ between $x = 0$ and $x = 6$.

(*b*) With the same scales and on the same axes draw the line $y = x$.

(*c*) Write down the value of x where (*a*) and (*b*) intersect, and the equation of which these values are the roots.

Repeat (*b*) and (*c*) with the line $y = -2x + 6$ instead of the line $y = x$.

6. Evaluate p, using the formula $p = \sqrt{\dfrac{Rt - qm}{st}}$ when $R = 72{\cdot}1$, $t = 5{\cdot}6$, $s = 0{\cdot}91$, $q = 0{\cdot}54$, $m = 625$.

Make t the subject of this formula.

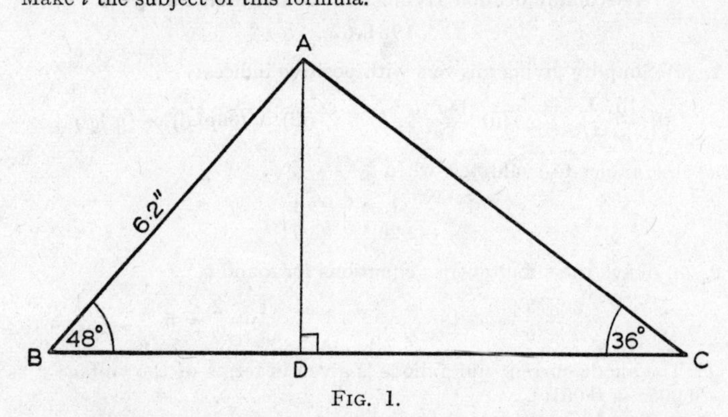

FIG. 1.

7. (*a*) Give a definition of: (i) a quadrilateral; (ii) a parallelogram; (iii) a rhombus.

(*b*) Calculate the shorter side of a parallelogram with the longer side 3·8 in. and the distances between the pairs of sides being 2·3 in. and 3·1 in.

(*c*) Calculate the area of a rhombus with side 7·2 in. and the longer diagonal 10·5 in.

8. (*a*) Express as a single fraction $4 - \dfrac{(\frac{1}{2} + \frac{1}{3})}{(\frac{3}{4} - \frac{1}{7})} \times 2\frac{2}{5}$.

(*b*) An engine gasket is rectangular in shape with dimensions 17·5 in. × 9·2 in. and has four punched holes each of diameter 2·85 in.

Calculate the area of the gasket and the length of lining necessary to reinforce all five edges.

9 (*a*) Use mathematical tables to find: (i) sin 45° 41′; (ii) the angle whose cosine is 0·5876; (iii) the logarithm of the tangent of 32° 17′.

(*b*) Fig. 1 shows a triangle *ABC* in which *AB* = 6·2 in., ∠B = 48°,

$\angle C = =36°$, and AD is the height. Show that $AD = 4{\cdot}608$ in. and calculate the length of BC.

10. Two towns X and Y are 60 miles apart. A motorist A leaves X at 10.0 a.m. travelling at 30 m.p.h. towards Y. At 10.20 a.m. he stops for 15 minutes. He then continues his journey at 40 m.p.h. Motorist B leaves Y at 10.15 a.m. going towards X at a speed of 60 m.p.h. Find graphically (taking horizontal axis 1 in. = 10 minutes and vertical axis 1 in. = 10 miles) the distance from X and the time at which A and B pass one another.

Telecommunication Technicians' Course Mathematics A

1961–62

1. (*a*) Simplify giving answers with positive indices:

(i) $\dfrac{10x^{-\frac{3}{8}}}{5x^{-\frac{3}{4}}}$ (ii) $\dfrac{14a^{-3}b^{-2}}{7a^2b^3}$ (iii) $\sqrt{(36p^{\frac{3}{4}}q^{\frac{1}{2}}) \times (p^{-\frac{1}{4}}q^{\frac{1}{2}})}$

(*b*) Use tables to evaluate z when

$$\frac{1}{z^2} = \frac{1}{(2{\cdot}26)^2} + \frac{1}{(3{\cdot}71)^2}$$

2. (*a*) Solve the simultaneous equations for a and b

$$\frac{2}{a} - \frac{3}{b} = -18 \qquad \frac{1}{a} + \frac{2}{b} = 5$$

(*b*) The anode current i of a diode is given in terms of the voltage v as: $i = 0{\cdot}005v + 0{\cdot}001v^2$.

If $i = 6 \times 10^{-3}$ calculate the values of v.

Comment on your answers.

(*c*) Find the quadratic equation with roots $\frac{3}{2}$ and $-\frac{1}{4}$.

3. (*a*) Solve for x: $\log 2 + \log 4 - \log 12 = \log x$.

(*b*) Evaluate (log 6·32) ÷ (log 3·16).

(*c*) The heating effect H produced in a length of wire during a fixed period is proportional to the applied voltage V and inversely proportional to the square of the resistance R. If V is trebled and R is doubled, by what ratio is H changed?

4. (*a*) In a planar magnetron the cut-off field is given by $V = \dfrac{eH^2d^2}{2mc^2}$ and the transit time by $T = \dfrac{\pi mc}{eH}$.

Use these two expressions to produce a formula for V in terms of T, H, d, and c only.

(*b*) The length and breadth of a rectangular workshop 30 ft. × 20 ft.

are increased by the same amount x ft. If this alteration increases the floor space by 56 per cent of the original area, calculate x.

5. (*a*) Obtain an expression (in a factorised form) for the characteristic resistance R_0 of a single T-section from the result

$$R_0 = R_1 + \frac{R_2(R_1 + R_0)}{R_0 + R_1 + R_2}$$

(*b*) Simplify

$$\frac{2x^2 + x - 3}{x^2 - 1} \times \frac{4x^2 - 12x + 9}{2x^2 - x - 3} \div (4x^2 - 9)$$

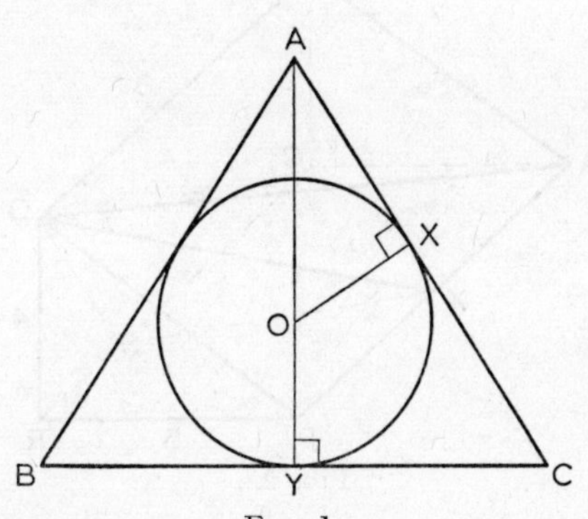

FIG. 1.

6. The discharge current i mA through a circuit is measured after time t sec. at intervals of 0·1 sec.

t sec.	0	0·1	0·2	0·3	0·4	0·5
i mA.	50·0	44·6	40·0	35·7	32·1	28·7

Draw a graph of log i against t and show that $\log i = A + Bt$ is a reasonable assumption where A and B are constants. Estimate from your graph values for A and B.

7. Draw the graphs of $y = 3 \cos x$ and $y = 4 \sin x$ on the same axes at intervals of 30° from $x = 0°$ to $x = 180°$. Use these two graphs to sketch the graph of $y = 3 \cos x + 4 \sin x$ and estimate the value of x to satisfy the equation $3 \cos x + 4 \sin x = 0$.

8. In Fig. 1 ABC is an isosceles triangle with $AB = AC = l$ and $BC = 2R$ with the inscribed circle centre O radius r touching AC in X and BC in Y. Name two similar triangles in the figure. If $AY = h$ using similar triangles or otherwise show that: (*a*) $hR - rR = rl$; (*b*) $lR - R^2 = rh$.

If $R = 2r$ what is the ratio of l to h?

9. Fig. 2 is the diagram of a right triangular prism with cross-section a right-angled triangle. $BQ = 8$ cm., $QR = 4$ cm., $PR = 5$ cm. and X is the mid-point of AP.

(*a*) Prove that $XQ^2 = 57$ cm.^{2} and $AQ^2 = 105$ cm.^{2}.

(*b*) Calculate the angles of the triangle AXQ.

(*c*) What is the least value of the sum $QY + YX$ where Y is a point on RC.

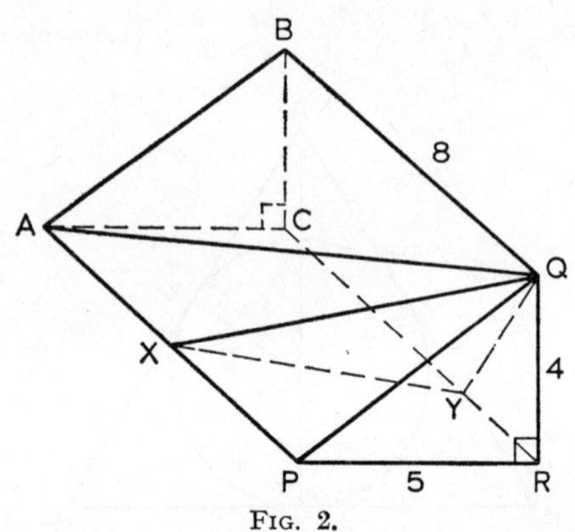

FIG. 2.

10. (*a*) Find the value of θ between 0° and 360° which satisfies both the equations

$$\sin \theta = 0{\cdot}8829, \cos \theta = -0{\cdot}4695.$$

(*b*) Find the values of θ between 0° and 360° which satisfy the equation

$$\sin^2 \theta - 3 \cos^2 \theta = 0.$$

(*c*) By making the substitution $\cos^2 \theta + \sin^2 \theta = 1$ prove that

$$\frac{2 \cos^2 \theta - 1}{1 + 2 \cos \theta \sin \theta} = \frac{\cos \theta - \sin \theta}{\cos \theta + \sin \theta}$$

1962–63

1. (*a*) If $\dfrac{5a - 4b}{2a - b} = \dfrac{3}{2}$ find the value of $\dfrac{a}{b}$.

(*b*) Simplify giving answers with positive indices:

(i) $\dfrac{3x^{-2}y^3}{12xy^{-2}}$ (ii) $\dfrac{27x^{-\frac{1}{2}}y^{-\frac{3}{4}}z^{\frac{1}{2}}}{18x^{\frac{3}{2}}y^{\frac{1}{4}}z^{-\frac{1}{2}}}$ (iii) $(z^{\frac{1}{3}} - z^{-\frac{1}{3}})\,(z^{\frac{2}{3}} + 1 + z^{-\frac{2}{3}})$

(*c*) Find the value of θ between 0° and 360° which satisfies both equations $\tan\theta = -0{\cdot}4663$ and $\cos\theta = -0{\cdot}9063$.

2. (*a*) If $x - \frac{1}{x} = 5$ find the value of $x^2 + \frac{1}{x^2}$ without solving the equation and deduce the value of $\left(x + \frac{1}{x}\right)^2$.

(*b*) Solve, *by completing the square*, the quadratic equation

$$2x^2 - 5x - 4 = 0$$

giving roots correct to 2 decimal places.

3. (*a*) Evaluate using logarithms $(0{\cdot}412)^{2{\cdot}36}$.

(*b*) A triode valve amplification factor μ and anode slope resistance r

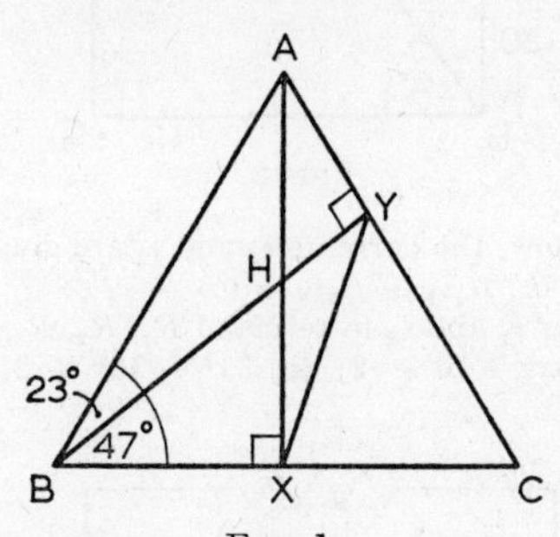

FIG. 1.

is connected as a cathode follower, the value of the cathode resistor is R. The voltage gain G of the stage is given by

$$G = \frac{\mu R}{r + (\mu + 1)R}$$

Make μ the subject of the formula giving the result in a factorised form.

4. In Fig. 1 above a triangle ABC has altitudes AX and BY which intersect in H. $\angle ABH = 23°$, $\angle CBH = 47°$.

Show that $(AX)^2 + (BX)^2 = (AY)^2 + (BY)^2$.

State what is common to the four points A, B, X, Y, and deduce the values of the angles, AXY, XAY, BAX, and BYX. Name the other set of points that all lie on a circle. If $AX = h$, $HX = x$, $BH = k$, express HY in terms of h, k, and x.

5. (*a*) Evaluate tan 157° 13′, cos 256° 27′, sin 325° 15′.

(*b*) In Fig. 2 below, from a point A on the roof of a workshop the angle of elevation of the top of a chimney EC is 32° and from B on the ground 20 ft. vertically below A it is 48°. Show that $\angle AEB$ 16°. Apply the sine rule to triangle AEB to find the distance AE, and hence calculate the height of the chimney EC.

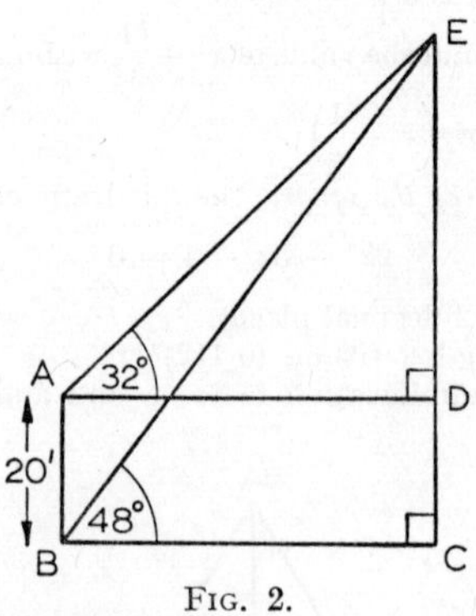

FIG. 2.

6. (*a*) In Fig. 3 below, the currents i_1 and i_2 are given by the equations $R_3(i_1 + i_2) + R_1 i_1 = E$, $R_1 i_1 - R_2 i_2 = 0$.

Find expressions for i_1 and i_2 in terms of R_1, R_2, R_3, and E.

(*b*) Factorise: (i) $3x^2 + 5x - 2$; (ii) $3x^2 - 10x + 3$.

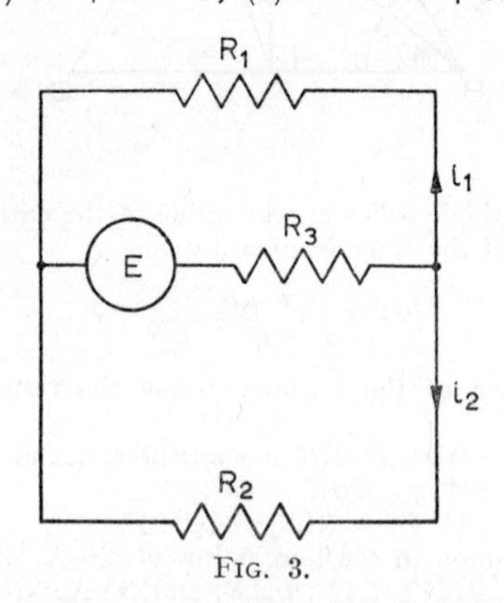

FIG. 3.

Hence reduce to one fraction, with the lowest common denominator

$$\frac{2x - 3}{3x^2 + 5x - 2} - \frac{x - 2}{3x^2 - 10x + 3}$$

7. In Fig. 4 below, C is a point on the circumference of a circle centre O radius 4 in. A circle is drawn with C as centre to cut the first circle in A and

B such that AB passes through O. Show that $AC = 5{\cdot}656$ in. and calculate the area common to both circles.

8. Draw the graphs of $8y = x - 1$ and $y = \frac{1}{x}$ on the same axes between $x = -5$ and $x = -\frac{1}{2}$ and between $x = \frac{1}{2}$ and $x = 5$. Use your graphs to solve the equation $x^2 - x - 8 = 0$. Adopt the same procedure to solve the equation $5x^2 - 2x - 8 = 0$.

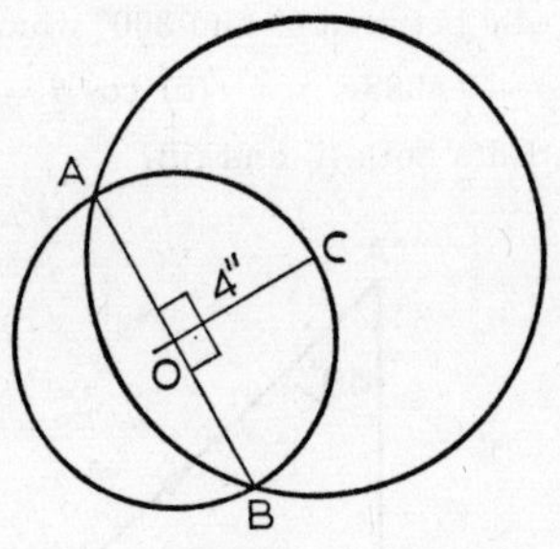

FIG. 4.

9. (*a*) Use the identity $\cos^2 \theta + \sin^2 \theta = 1$ to establish the following results:

(i) $1 + \tan^2 \theta = \sec^2 \theta$ (ii) $\cot^2 \theta + 1 = \operatorname{cosec}^2 \theta$

Prove the identities:

(iii) $(\sec \theta + \tan \theta)(\sec \theta - \tan \theta) = 1$ (iv) $\dfrac{\cos^2 \theta - \cos^4 \theta}{1 - \sin^2 \theta} = \sin^2 \theta$

(*b*) Find the values of θ between 0° and 360° which satisfy the equation $3 \sin \theta = 2 \cos^2 \theta$.

10. A cylinder of radius r, closed at both ends, has a volume 200π. Show that the total surface area S is given by $S = \dfrac{400\pi}{r} + 2\pi r^2$.

What would be the total surface area of a solid cone of the same volume and base radius?

1963–64

1. (*a*) The force F between two magnetic poles varies directly as the strengths m_1 and m_2 and inversely as the square of the distance apart d. Write down the formula with a constant k connecting F, m_1, m_2, and d. If $F = 4$ when $m_1 = 3$, $m_2 = 7$, and $d = 2{\cdot}5$, find F when $m_1 = 4$, $m_2 = 9$, and $d = 3{\cdot}5$.

(*b*) Simplify giving answers with positive indices only

(i) $\dfrac{4e^{-2}f^{-\frac{3}{4}} \times 3g^{\frac{4}{3}}}{18e^{3}f^{\frac{1}{4}}g^{-\frac{1}{3}}}$ (ii) $\dfrac{\sqrt[3]{(27p^{4}q^{5})}}{p^{-\frac{2}{3}}p^{-\frac{1}{3}}}$

2. (*a*) Divide $a^3 - b^3$ by $a - b$ and hence evaluate $x^3 - \dfrac{1}{x^3}$ when $x - \dfrac{1}{x} = 3$.

(*b*) Find the values of θ between 0° and 360° which satisfy:

(i) $\sin \theta = -0{\cdot}8988$ (ii) $\cos \theta = -0{\cdot}4384$

Which value of θ satisfies both (i) and (ii)?

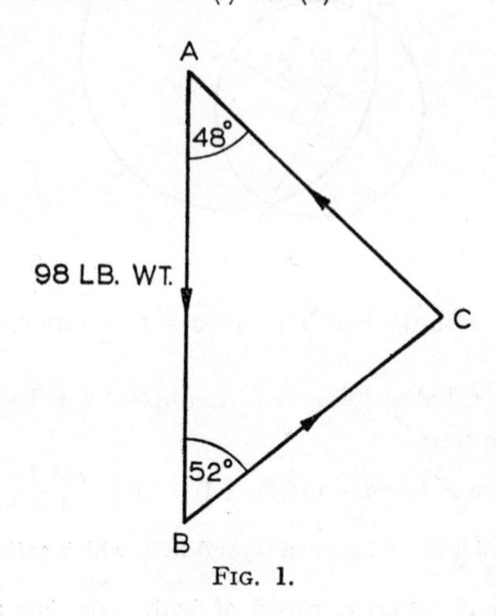

FIG. 1.

3. (*a*) Solve the quadratic equation $4x^2 - 9x + 3 = 0$, giving roots to two decimal places.

(*b*) Rearrange the following formula which occurs in the theory of frequency modulation to make L_1 the subject:

$$f = \frac{1}{2\pi}\sqrt{\frac{L^1 + L^2}{L_1 L_2}}.$$

Write down the formula for L_2.

4. (*a*) Use mathematical tables to find the values of: cot 123° 51′, sec 341° 17′, cosec 263° 26′.

(*b*) A weight of 98 lb. is suspended by two strings, the ends of which are fixed at the same horizontal level. The resulting triangle of forces is shown in Fig. 1. *AB* represents the 98-lb. wt., *BC* the tension T_1 in one string and CA the tension T_2 in the other. The angle B is 52° and the angle A is 48°. Calculate the tensions T_1 and T_2.

5. A train leaves London at 10.00 a.m. and travels at an average speed of 30 m.p.h. before being stopped by signals at 10.20 a.m. After waiting 15 minutes, the train continues its journey to Canterbury at an average speed of 40 m.p.h. A second train leaves Canterbury for London at 10.15 a.m. and averages 60 m.p.h. non-stop. The distance from London to Canterbury is 60 miles. The trains pass one another t minutes after 10.00 a.m. and at a distance of s miles from London. Calculate by the method of simultaneous equations the values of t and s.

6. (*a*) Solve for x:

(i) $\log 3 + \log 4 - \log 6 = \log x$ (ii) $4 \log 2 + \log 4 - \log 8 = x \log 2$

(*b*) Evaluate using logarithms: $(0{\cdot}04625)^{-0{\cdot}361}$.

7. (*a*) Find the values of θ between 0° and 360° which satisfy the equation: $6 \sin^2 \theta - 7 \cos \theta - 8 = 0$.

(*b*) Using Pythagoras' theorem or otherwise, show that, where x is less than 90°, $\cos^2 x + \sin^2 x = 1$.

Establish the following identities:

(i) $\sin x - \sin^3 x = \sin x \cos^2 x$ (ii) $\dfrac{\cos^2 x - \sin^2 x}{\cos x(\cos x - \sin x)} = 1 + \tan x$

8. With the same scales and on the same axes, draw the graphs of $y = \dfrac{1}{x}$, and $3y = 2x - 1$, and find the values of x between -3 and 5 where they intersect. Of what equation are these the roots?

9. (*a*) Calculate the slope and intercept on the y axis of the straight lines given by each of the equations:

(i) $4y = 3x + 2$ (ii) $2y + 5x - 1 = 0$

(*b*) The results in the table below show how the length (x in.) of a given piece of wire varies when subjected to a load (W lb.). Show by drawing a suitable graph that the law $W = ax + b$ is approximately true and establish *from your graph* values for the constants a and b.

x in.	12·51	13·11	13·50	14·13	14·52	14·85
W lb.	80	136	172	235	276	304

10. (*a*) A chord subtends an angle θ at the centre of a circle radius r. Prove that the area of the segment cut off by the chord is given by $\frac{1}{2}r^2(\theta - \sin \theta)$ where θ is measured in radians.

(*b*) Fig. 2 shows a right-angled bend in a road with $AB = BC = a$ ft. (the radius of a circle with centre O). The road is to be modified as shown by the dotted lines to remove the sharp corner. If the width of the road is x ft., show that the saving in road surface area of the modification is given by $(1 - \frac{1}{4}\pi)(2ax - x^2)$ sq. ft.

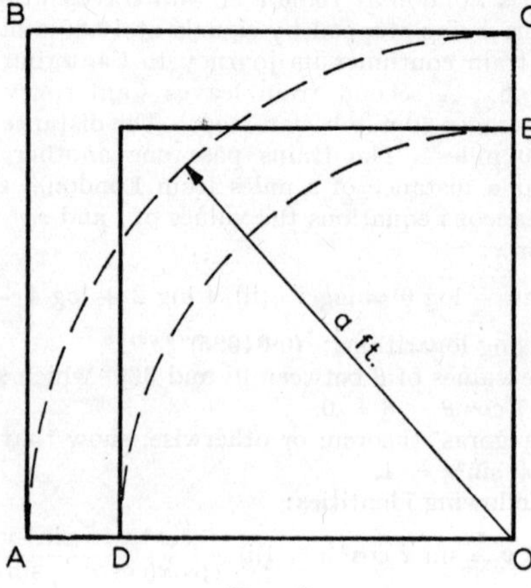

FIG. 2.

ANSWERS TO PAST CITY AND GUILDS PAPERS

PRACTICAL MATHEMATICS 1961–62 (p. 336)

1. 0·00287, 0·0038, 4·7%. 2. $\frac{8}{19}$, 14·8.
3. 0·3, 0·5; $2pqr$, $48p^2q^2r^3$; $\frac{k^2}{y} \times 10^6$.
4. $x^2 + 2x + 4$; $x^3 + 5x^2 + 6x$; $a + b$, $2a + b$, $2b - a$.
5. $b(6a + c)$; $16x + 3$; $x(x - 1)$; $2y(x + y)(x - y)$; $x = \frac{47}{13}$.
6. 3823 ohms; $R_2(R_2 - R)/(R - 2R_2)$; $R_2 = 0$.
7. (*a*) 6·245; (*b*) 8800 ohms; position 8.
8. 4·44, 54° 10′, 35° 50′; *BCD*, *ACB* similar. $BC = 6$ cm. and $BD : AB = 2 : 3$. 9. 45°, 90°, approx. 57°.
10. 500 ohms; 80 watts: approx. 44 watts.

PRACTICAL MATHEMATICS 1962–63 (p. 338)

1. (*a*) 0·5532; 7·62. (*b*) $2\frac{2}{9}$ min; $3\frac{1}{13}$ min.
2. (*a*) 6; (*b*) $\frac{2}{3}$, only the *ratio* of $x : y$ is known; (*c*) 18, −72.
3. (*a*) ampere, watt, watt; (*b*) $(a - b)(a - c)$, $(2x + 3y)(2x - 3y)$. (*c*) $1/2c$, x^2y^2. 4. (*a*) 17·96; (*b*) $a - b$, $2a$, $a + 2b$, $\frac{1}{2}(a + b)$.
5. $R = 100$, $V = 430$; 11·63, $p = R/(R + Rt + t)$.
6. $\frac{2}{l}(l^2 + A)$, $\sqrt{A^2 + l^4}/l$; 13; 7 litres per 100 km.
7. 0·9168, 0·8968, 0·9840; 5·36, 2·87; 6·88.
8. 1·79 or −2·79; +2 or −3. Factors $(x - 2)(x + 3)$.
9. 72°; radius 3 tan 54° = 4·13 in.; 61·9 sq. in. volume $\frac{1}{3}$ base × height = 619 cu. in. approx.
10. (i) 23·6, 1·72; (ii) 23·3, 1·71. (Exact answer for (i) not possible.)

PRACTICAL MATHEMATICS 1963–64 (p. 340)

1. −0·8; $x = 5$ or −4; $x = 0$ or $\frac{1}{4}$; 4π, $4\pi R_1^3 R_2^3$.
2. 0; 96; put $\frac{1}{a} = A$, $\frac{1}{b} = B$ and solve for A, B (5, 6); so $a = \frac{1}{5}$, $b = \frac{1}{6}$; 9·053, 0·11805.
3. $(p - q)(p + q - r)$; $x(9x - 4y)$; $(2c - 3a + b)/abc$; $\frac{5}{4b^3}$; $3ab^2$.
4. 1·626; $(1 + a)$, $(a - 1)$, $2a$, $3 + x$.
5. $x = 7$; $x^2 - 6x + 7 = 0$; $x = 2·62$ and $x = 0·38$; $x^2 - 3x + 1 = 0$.
6. 1·152; $t = qm/(R - p^2)$. 7. (*b*) 2·82 in.; (*c*) 51·8 sq. in.
8. 12/17; 135·5 sq. in.; 89·2 in.
9. 0·7155, 54° 1′, 0·8005; 10·49 in. 10. 22 miles; 10·53 a.m.

MATHEMATICS A 1961–62 (p. 342)

1. $2x^{\frac{3}{4}}$, $2/a^5b^5$, $6p^{\frac{1}{2}}q^{\frac{1}{2}}$; 1·93 (*Hint:* Use tables of Squares and Reciprocals.)
2. (*a*) Solve for $\dfrac{1}{a} = A, \dfrac{1}{b} = B$ first; $a = -\frac{1}{3}, b = -\frac{1}{4}$; (*b*) $v = 1$ (value $v = -6$ has no physical significance in this case); (*c*) $8x^2 - 10x - 3 = 0$.
3. $\frac{2}{3}$; 1·60 (NOT log 2); $H = kV/R^2$; 3 : 4.
4. $V = \pi Hd^2/2c$; 6. 5. $R_0{}^2 = R_1(R_1 + 2R_2)$; $1/(x + 1)^2$.
6. For " best " straight line, slope $= B$, intercept on log i axis is A.
7. 143° approx. 8. AOX, ACY; 5 : 4.
9. 38° 29′, 122° 16′, 19° 15′ (Cosine Rule and Sine Rule); least value $\sqrt{97}$ (see p. 172). 10. 118°; 60°, 120°, 240°, 300°.

MATHEMATICS A 1962–63 (p. 344)

1. $\frac{5}{4}$; $y^5/4x^3$, $3z/2x^2y$, $z - 1/z$; 155°.
2. 27, 29; 3·14 or − 0·64. 3. 0·1115; $\mu = G(r + R)/R(1 - G)$.
4. A, B, X, Y lie in a circle, diameter AB; 23°, 47°, 20°, 20°; H, X, C, Y lie as a circle, diameter HC; $x(h - x)/k$.
5. (*a*) 0·5527, − 0·2343, − 0·5700; (*b*) 48·6, 40·7 ft.
6. $i_1 = kER_2$, $i_2 = kER_1$, where $k = (R_2R_3 + R_3R_1 + R_1R_2)^{-1}$; $(3x - 1)(x + 2)$, $(3x - 1)(x - 3)$, and $(x^2 - 9x + 13)/(3x - 1)(x + 2)(x - 3)$.
7. $AC = 4\sqrt{2}$; angle $BCA = 90°$; area $16(\pi - 1) = 36{\cdot}1$ in.².
8. 3·37, − 2·37; 0·84, − 0·44. 9. (*b*) 30°, 150°.
10. $S = \pi r\sqrt{r^2 + h^2} + \pi r^2$, where $r^2h = 600$.

MATHEMATICS A 1963–64 (p. 347)

1. $k = 0{\cdot}1905$; $F = 3{\cdot}50$; $2g^{\frac{1}{3}}/3e^5f$; $3p^2q^2$.
2. $a^2 + ab + b^2$; 36; 244°, 296°; 244°, 116°.
3. 1·843, 0·407; $L_1 = L_2/(4\pi^2f^2L_2 - 1)$; substitute L_1 for L_2 and vice versa.
4. − 0·6707, 0·3237, − 0·1144; 73·9, 78·3 lb.
5. $t = 53$, $s = 22$.
6. 2, 3; 3·03; suggest use reciprocal tables first and write as $(21{\cdot}63)^{0{\cdot}361}$.
7. (*a*) Put $\sin^2\theta = 1 - \cos^2\theta$; $\cos\theta = -\frac{1}{2}$ or $-\frac{2}{3}$; 120°, 240°, 131° 50′, 228° 10′.
8. $x = 1{\cdot}5, -1$; $2x^2 - x - 3 - 0$.
9. $\frac{3}{4}, \frac{1}{2}$; $-2\frac{1}{2}, \frac{1}{2}$: $a = 98$, $b = -1140$ approx.

MATHEMATICAL TABLES

LOGARITHMS
ANTI-LOGARITHMS
NATURAL SINES
NATURAL COSINES
NATURAL TANGENTS

LOGARITHMS

Proportional Parts

	0	1	2	3	4	5	6	7	8	9	1	2	3	4	5	6	7	8	9
10	0000	0043	0086	0128	0170	0212	0253	0294	0334	0374	4	8	12	17	21	25	29	33	37
11	0414	0453	0492	0531	0569	0607	0645	0682	0719	0755	4	8	11	15	19	23	26	30	34
12	0792	0828	0864	0899	0934	0969	1004	1038	1072	1106	3	7	10	14	17	21	24	28	31
13	1139	1173	1206	1239	1271	1303	1335	1367	1399	1430	3	6	10	13	16	19	23	26	29
14	1461	1492	1523	1553	1584	1614	1644	1673	1703	1732	3	6	9	12	15	18	21	24	27
15	1761	1790	1818	1847	1875	1903	1931	1959	1987	2014	3	6	8	11	14	17	20	22	25
16	2041	2068	2095	2122	2148	2175	2201	2227	2253	2279	3	5	8	11	13	16	18	21	24
17	2304	2330	2355	2380	2405	2430	2455	2480	2504	2529	2	5	7	10	12	15	17	20	22
18	2553	2577	2601	2625	2648	2672	2695	2718	2742	2765	2	5	7	9	12	14	16	19	21
19	2788	2810	2833	2856	2878	2900	2923	2945	2967	2989	2	4	7	9	11	13	16	18	20
20	3010	3032	3054	3075	3096	3118	3139	3160	3181	3201	2	4	6	8	11	13	15	17	19
21	3222	3243	3263	3284	3304	3324	3345	3365	3385	3404	2	4	6	8	10	12	14	16	18
22	3424	3444	3464	3483	3502	3522	3541	3560	3579	3598	2	4	6	8	10	12	14	15	17
23	3617	3636	3655	3674	3692	3711	3729	3747	3766	3784	2	4	6	7	9	11	13	15	17
24	3802	3820	3838	3856	3874	3892	3909	3927	3945	3962	2	4	5	7	9	11	12	14	16
25	3979	3997	4014	4031	4048	4065	4082	4099	4116	4133	2	3	5	7	9	10	12	14	15
26	4150	4166	4183	4200	4216	4232	4249	4265	4281	4298	2	3	5	7	8	10	11	13	15
27	4314	4330	4346	4362	4378	4393	4409	4425	4440	4456	2	3	5	6	8	9	11	13	14
28	4472	4487	4502	4518	4533	4548	4564	4579	4594	4609	2	3	5	6	8	9	11	12	14
29	4624	4639	4654	4669	4683	4698	4713	4728	4742	4757	1	3	4	6	7	9	10	12	13
30	4771	4786	4800	4814	4829	4843	4857	4871	4886	4900	1	3	4	6	7	9	10	11	13
31	4914	4928	4942	4955	4969	4983	4997	5011	5024	5038	1	3	4	5	7	8	10	11	12
32	5051	5065	5079	5092	5105	5119	5132	5145	5159	5172	1	3	4	5	7	8	9	11	12
33	5185	5198	5211	5224	5237	5250	5263	5276	5289	5302	1	3	4	5	6	8	9	10	12
34	5315	5328	5340	5353	5366	5378	5391	5403	5416	5428	1	3	4	5	6	8	9	10	11
35	5441	5453	5465	5478	5490	5502	5514	5527	5539	5551	1	2	4	5	6	7	9	10	11
36	5563	5575	5587	5599	5611	5623	5635	5647	5658	5670	1	2	4	5	6	7	8	10	11
37	5682	5694	5705	5717	5729	5740	5752	5763	5775	5786	1	2	3	5	6	7	8	9	10
38	5798	5809	5821	5832	5843	5855	5866	5877	5888	5899	1	2	3	5	6	7	8	9	10
39	5911	5922	5933	5944	5955	5966	5977	5988	5999	6010	1	2	3	4	5	7	8	9	10
40	6021	6031	6042	6053	6064	6075	6085	6096	6107	6117	1	2	3	4	5	6	7	9	10
41	6128	6138	6149	6160	6170	6180	6191	6201	6212	6222	1	2	3	4	5	6	7	8	9
42	6232	6243	6253	6263	6274	6284	6294	6304	6314	6325	1	2	3	4	5	6	7	8	9
43	6335	6345	6355	6365	6375	6385	6395	6405	6415	6425	1	2	3	4	5	6	7	8	9
44	6435	6444	6454	6464	6474	6484	6493	6503	6513	6522	1	2	3	4	5	6	7	8	9
45	6532	6542	6551	6561	6571	6580	6590	6599	6609	6618	1	2	3	4	5	6	7	8	9
46	6628	6637	6646	6656	6665	6675	6684	6693	6702	6712	1	2	3	4	5	6	7	7	8
47	6721	6730	6739	6749	6758	6767	6776	6785	6794	6803	1	2	3	4	5	5	6	7	8
48	6812	6821	6830	6839	6848	6857	6866	6875	6884	6893	1	2	3	4	4	5	6	7	8
49	6902	6911	6920	6928	6937	6946	6955	6964	6972	6981	1	2	3	4	4	5	6	7	8
50	6990	6998	7007	7016	7024	7033	7042	7050	7059	7067	1	2	3	3	4	5	6	7	8
51	7076	7084	7093	7101	7110	7118	7126	7135	7143	7152	1	2	3	3	4	5	6	7	8
52	7160	7168	7177	7185	7193	7202	7210	7218	7226	7235	1	2	2	3	4	5	6	7	7
53	7243	7251	7259	7267	7275	7284	7292	7300	7308	7316	1	2	2	3	4	5	6	6	7
54	7324	7332	7340	7348	7356	7364	7372	7380	7388	7396	1	2	2	3	4	5	6	6	7
	0	**1**	**2**	**3**	**4**	**5**	**6**	**7**	**8**	**9**	**1·**	**2**	**3**	**4**	**5**	**6**	**7**	**8**	**9**

Proportional Parts

	0	1	2	3	4	5	6	7	8	9	1	2	3	4	5	6	7	8	9
55	7404	7412	7419	7427	7435	7443	7451	7459	7466	7474	1	2	2	3	4	5	5	6	7
56	7482	7490	7497	7505	7513	7520	7528	7536	7543	7551	1	2	2	3	4	5	5	6	7
57	7559	7566	7574	7582	7589	7597	7604	7612	7619	7627	1	2	2	3	4	5	5	6	7
58	7634	7642	7649	7657	7664	7672	7679	7686	7694	7701	1	1	2	3	4	4	5	6	7
59	7709	7716	7723	7731	7738	7745	7752	7760	7767	7774	1	1	2	3	4	4	5	6	7
60	7782	7789	7796	7803	7810	7818	7825	7832	7839	7846	1	1	2	3	4	4	5	6	6
61	7853	7860	7868	7875	7882	7889	7896	7903	7910	7917	1	1	2	3	4	4	5	6	6
62	7924	7931	7938	7945	7952	7959	7966	7973	7980	7987	1	1	2	3	3	4	5	6	6
63	7993	8000	8007	8014	8021	8028	8035	8041	8048	8055	1	1	2	3	3	4	5	6	6
64	8062	8069	8075	8082	8089	8096	8102	8109	8116	8122	1	1	2	3	3	4	5	5	6
65	8129	8136	8142	8149	8156	8162	8169	8176	8182	8189	1	1	2	3	3	4	5	5	6
66	8195	8202	8209	8215	8222	8228	8235	8241	8248	8254	1	1	2	3	3	4	5	5	6
67	8261	8267	8274	8280	8287	8293	8299	8306	8312	8319	1	1	2	3	3	4	4	5	6
68	8325	8331	8338	8344	8351	8357	8363	8370	8376	8382	1	1	2	3	3	4	4	5	6
69	8388	8395	8401	8407	8414	8420	8426	8432	8439	8445	1	1	2	3	3	4	4	5	6
70	8451	8457	8463	8470	8476	8482	8488	8494	8500	8506	1	1	2	2	3	4	4	5	6
71	8513	8519	8525	8531	8537	8543	8549	8555	8561	8567	1	1	2	2	3	4	4	5	5
72	8573	8579	8585	8591	8597	8603	8609	8615	8621	8627	1	1	2	2	3	4	4	5	5
73	8633	8639	8645	8651	8657	8663	8669	8675	8681	8686	1	1	2	2	3	4	4	5	5
74	8692	8698	8704	8710	8716	8722	8727	8733	8739	8745	1	1	2	2	3	4	4	5	5
75	8751	8756	8762	8768	8774	8779	8785	8791	8797	8802	1	1	2	2	3	3	4	5	5
76	8808	8814	8820	8825	8831	8837	8842	8848	8854	8859	1	1	2	2	3	3	4	5	5
77	8865	8871	8876	8882	8887	8893	8899	8904	8910	8915	1	1	2	2	3	3	4	4	5
78	8921	8927	8932	8938	8943	8949	8954	8960	8965	8971	1	1	2	2	3	3	4	4	5
79	8976	8982	8987	8993	8998	9004	9009	9015	9020	9025	1	1	2	2	3	3	4	4	5
80	9031	9036	9042	9047	9053	9058	9063	9069	9074	9079	1	1	2	2	3	3	4	4	5
81	9085	9090	9096	9101	9106	9112	9117	9122	9128	9133	1	1	2	2	3	3	4	4	5
82	9138	9143	9149	9154	9159	9165	9170	9175	9180	9186	1	1	2	2	3	3	4	4	5
83	9191	9196	9201	9206	9212	9217	9222	9227	9232	9238	1	1	2	2	3	3	4	4	5
84	9243	9248	9253	9258	9263	9269	9274	9279	9284	9289	1	1	2	2	3	3	4	4	5
85	9294	9299	9304	9309	9315	9320	9325	9330	9335	9340	1	1	2	2	3	3	4	4	5
86	9345	9350	9355	9360	9365	9370	9375	9380	9385	9390	1	1	2	2	3	3	4	4	5
87	9395	9400	9405	9410	9415	9420	9425	9430	9435	9440	0	1	1	2	2	3	3	4	4
88	9445	9450	9455	9460	9465	9469	9474	9479	9484	9489	0	1	1	2	2	3	3	4	4
89	9494	9499	9504	9509	9513	9518	9523	9528	9533	9538	0	1	1	2	2	3	3	4	4
90	9542	9547	9552	9557	9562	9566	9571	9576	9581	9586	0	1	1	2	2	3	3	4	4
91	9590	9595	9600	9605	9609	9614	9619	9624	9628	9633	0	1	1	2	2	3	3	4	4
92	9638	9643	9647	9652	9657	9661	9666	9671	9675	9680	0	1	1	2	2	3	3	4	4
93	9685	9689	9694	9699	9703	9708	9713	9717	9722	9727	0	1	1	2	2	3	3	4	4
94	9731	9736	9741	9745	9750	9754	9759	9764	9768	9773	0	1	1	2	2	3	3	4	4
95	9777	9782	9786	9791	9795	9800	9805	9809	9814	9818	0	1	1	2	2	3	3	4	4
96	9823	9827	9832	9836	9841	9845	9850	9854	9859	9863	0	1	1	2	2	3	3	4	4
97	9868	9872	9877	9881	9886	9890	9894	9899	9903	9908	0	1	1	2	2	3	3	4	4
98	9912	9917	9921	9926	9930	9934	9939	9943	9948	9952	0	1	1	2	2	3	3	4	4
99	9956	9961	9965	9969	9974	9978	9983	9987	9991	9996	0	1	1	2	2	3	3	4	4
	0	1	2	3	4	5	6	7	8	9	1	2	3	4	5	6	7	8	9

ANTI-LOGARITHMS

Proportional Parts

	0	1	2	3	4	5	6	7	8	9	1	2	3	4	5	6	7	8	9
·00	1000	1002	1005	1007	1009	1012	1014	1016	1019	1021	0	0	1	1	1	1	2	2	2
·01	1023	1026	1028	1030	1033	1035	1038	1040	1042	1045	0	0	1	1	1	1	2	2	2
·02	1047	1050	1052	1054	1057	1059	1062	1064	1067	1069	0	0	1	1	1	1	2	2	2
.03	1072	1074	1076	1079	1081	1084	1086	1089	1091	1094	0	0	1	1	1	1	2	2	2
·04	1096	1099	1102	1104	1107	1109	1112	1114	1117	1119	0	1	1	1	1	2	2	2	2
·05	1122	1125	1127	1130	1132	1135	1138	1140	1143	1146	0	1	1	1	1	2	2	2	2
·06	1148	1151	1153	1156	1159	1161	1164	1167	1169	1172	0	1	1	1	1	2	2	2	2
·07	1175	1178	1180	1183	1186	1189	1191	1194	1197	1199	0	1	1	1	1	2	2	2	2
·08	1202	1205	1208	1211	1213	1216	1219	1222	1225	1227	0	1	1	1	1	2	2	2	3
·09	1230	1233	1236	1239	1242	1245	1247	1250	1253	1256	0	1	1	1	1	2	2	2	3
·10	1259	1262	1265	1268	1271	1274	1276	1279	1282	1285	0	1	1	1	1	2	2	2	3
·11	1288	1291	1294	1297	1300	1303	1306	1309	1312	1315	0	1	1	1	2	2	2	2	3
·12	1318	1321	1324	1327	1330	1334	1337	1340	1343	1346	0	1	1	1	2	2	2	2	3
·13	1349	1352	1355	1358	1361	1365	1368	1371	1374	1377	0	1	1	1	2	2	2	2	3
·14	1380	1384	1387	1390	1393	1396	1400	1403	1406	1409	0	1	1	1	2	2	2	3	3
·15	1413	1416	1419	1422	1426	1429	1432	1435	1439	1442	0	1	1	1	2	2	2	3	3
·16	1445	1449	1452	1455	1459	1462	1466	1469	1472	1476	0	1	1	1	2	2	2	3	3
·17	1479	1483	1486	1489	1493	1496	1500	1503	1507	1510	0	1	1	1	2	2	2	3	3
·18	1514	1517	1521	1524	1528	1531	1535	1538	1542	1545	0	1	1	1	2	2	2	3	3
·19	1549	1552	1556	1560	1563	1567	1570	1574	1578	1581	0	1	1	1	2	2	3	3	3
·20	1585	1589	1592	1596	1600	1603	1607	1611	1614	1618	0	1	1	1	2	2	3	3	3
·21	1622	1626	1629	1633	1637	1641	1644	1648	1652	1656	0	1	1	2	2	2	3	3	3
·22	1660	1663	1667	1671	1675	1679	1683	1687	1690	1694	0	1	1	2	2	2	3	3	3
·23	1698	1702	1706	1710	1714	1718	1722	1726	1730	1734	0	1	1	2	2	2	3	3	4
·24	1738	1742	1746	1750	1754	1758	1762	1766	1770	1774	0	1	1	2	2	2	3	3	4
·25	1778	1782	1786	1791	1795	1799	1803	1807	1811	1816	0	1	1	2	2	3	3	3	4
·26	1820	1824	1828	1832	1837	1841	1845	1849	1854	1858	0	1	1	2	2	3	3	3	4
·27	1862	1866	1871	1875	1879	1884	1888	1892	1897	1901	0	1	1	2	2	3	3	3	4
·28	1905	1910	1914	1919	1923	1928	1932	1936	1941	1945	0	1	1	2	2	3	3	4	4
·29	1950	1954	1959	1963	1968	1972	1977	1982	1986	1991	0	1	1	2	2	3	3	4	4
·30	1995	2000	2004	2009	2014	2018	2023	2028	2032	2037	0	1	1	2	2	3	3	4	4
·31	2042	2046	2051	2056	2061	2065	2070	2075	2080	2084	0	1	1	2	2	3	3	4	4
·32	2089	2094	2099	2104	2109	2113	2118	2123	2128	2133	0	1	1	2	2	3	3	4	4
·33	2138	2143	2148	2153	2158	2163	2168	2173	2178	2183	0	1	1	2	2	3	3	4	4
·34	2188	2193	2198	2203	2208	2213	2218	2223	2228	2234	1	1	2	2	3	3	4	4	5
·35	2239	2244	2249	2254	2259	2265	2270	2275	2280	2286	1	1	2	2	3	3	4	4	5
·36	2291	2296	2301	2307	2312	2317	2323	2328	2333	2339	1	1	2	2	3	3	4	4	5
·37	2344	2350	2355	2360	2366	2371	2377	2382	2388	2393	1	1	2	2	3	3	4	4	5
·38	2399	2404	2410	2415	2421	2427	2432	2438	2443	2449	1	1	2	2	3	3	4	4	5
·39	2455	2460	2466	2472	2477	2483	2489	2495	2500	2506	1	1	2	2	3	3	4	5	5
·40	2512	2518	2523	2529	2535	2541	2547	2553	2559	2564	1	1	2	2	3	3	4	5	5
·41	2570	2576	2582	2588	2594	2600	2606	2612	2618	2624	1	1	2	2	3	4	4	5	5
·42	2630	2636	2642	2648	2655	2661	2667	2673	2679	2685	1	1	2	2	3	4	4	5	6
·43	2692	2698	2704	2710	2716	2723	2729	2735	2742	2748	1	1	2	2	3	4	4	5	6
·44	2754	2761	2767	2773	2780	2786	2793	2799	2805	2812	1	1	2	3	3	4	4	5	6
·45	2818	2825	2831	2838	2844	2851	2858	2864	2871	2877	1	1	2	3	3	4	5	5	6
·46	2884	2891	2897	2904	2911	2917	2924	2931	2938	2944	1	1	2	3	3	4	5	5	6
·47	2951	2958	2965	2972	2979	2985	2992	2999	3006	3013	1	1	2	3	3	4	5	6	6
·48	3020	3027	3034	3041	3048	3055	3062	3069	3076	3083	1	1	2	3	4	4	5	6	6
·49	3090	3097	3105	3112	3119	3126	3133	3141	3148	3155	1	1	2	3	4	4	5	6	7
	0	1	2	3	4	5	6	7	8	9	1	2	3	4	5	6	7	8	9

ANTI-LOGARITHMS

	0	1	2	3	4	5	6	7	8	9	1	2	3	4	5	6	7	8	9
											Proportional Parts								
·50	3162	3170	3177	3184	3192	3199	3206	3214	3221	3228	1	1	2	3	4	4	5	6	7
·51	3236	3243	3251	3258	3266	3273	3281	3289	3296	3304	1	2	2	3	4	5	5	6	7
·52	3311	3319	3327	3334	3342	3350	3357	3365	3373	3381	1	2	2	3	4	5	5	6	7
·53	3388	3396	3404	3412	3420	3428	3436	3443	3451	3459	1	2	2	3	4	5	6	6	7
·54	3467	3475	3483	3491	3499	3508	3516	3524	3532	3540	1	2	2	3	4	5	6	6	7
·55	3548	3556	3565	3573	3581	3589	3597	3606	3614	3622	1	2	2	3	4	5	6	7	7
·56	3631	3639	3648	3656	3664	3673	3681	3690	3698	3707	1	2	3	3	4	5	6	7	8
·57	3715	3724	3733	3741	3750	3758	3767	3776	3784	3793	1	2	3	3	4	5	6	7	8
·58	3802	3811	3819	3828	3837	3846	3855	3864	3873	3882	1	2	3	4	4	5	6	7	8
·59	3890	3899	3908	3917	3926	3936	3945	3954	3963	3972	1	2	3	4	5	5	6	7	8
·60	3981	3990	3999	4009	4018	4027	4036	4046	4055	4064	1	2	3	4	5	6	7	7	8
·61	4074	4083	4093	4102	4111	4121	4130	4140	4150	4159	1	2	3	4	5	6	7	8	9
·62	4169	4178	4188	4198	4207	4217	4227	4236	4246	4256	1	2	3	4	5	6	7	8	9
·63	4266	4276	4285	4295	4305	4315	4325	4335	4345	4355	1	2	3	4	5	6	7	8	9
·64	4365	4375	4385	4395	4406	4416	4426	4436	4446	4457	1	2	3	4	5	6	7	8	9
·65	4467	4477	4487	4498	4508	4519	4529	4539	4550	4560	1	2	3	4	5	6	7	8	9
·66	4571	4581	4592	4603	4613	4624	4634	4645	4656	4667	1	2	3	4	5	6	7	8	10
·67	4677	4688	4699	4710	4721	4732	4742	4753	4764	4775	1	2	3	4	5	7	8	9	10
·68	4786	4797	4808	4819	4831	4842	4853	4864	4875	4887	1	2	3	4	6	7	8	9	10
·69	4898	4909	4920	4932	4943	4955	4966	4977	4989	5000	1	2	3	5	6	7	8	9	10
·70	5012	5023	5035	5047	5058	5070	5082	5093	5105	5117	1	2	4	5	6	7	8	9	11
·71	5129	5140	5152	5164	5176	5188	5200	5212	5224	5236	1	2	4	5	6	7	8	10	11
·72	5248	5260	5272	5284	5297	5309	5321	5333	5346	5358	1	2	4	5	6	7	9	10	11
·73	5370	5383	5395	5408	5420	5433	5445	5458	5470	5483	1	3	4	5	6	8	9	10	11
·74	5495	5508	5521	5534	5546	5559	5572	5585	5598	5610	1	3	4	5	6	8	9	10	12
·75	5623	5636	5649	5662	5675	5689	5702	5715	5728	5741	1	3	4	5	7	8	9	10	12
·76	5754	5768	5781	5794	5808	5821	5834	5848	5861	5875	1	3	4	5	7	8	9	11	12
·77	5888	5902	5916	5929	5943	5957	5970	5984	5998	6012	1	3	4	6	7	8	10	11	12
·78	6026	6039	6053	6067	6081	6095	6109	6124	6138	6152	1	3	4	6	7	8	10	11	13
·79	6166	6180	6194	6209	6223	6237	6252	6266	6281	6295	1	3	4	6	7	9	10	12	13
·80	6310	6324	6339	6353	6368	6383	6397	6412	6427	6442	1	3	4	6	7	9	10	12	13
·81	6457	6471	6486	6501	6516	6531	6546	6561	6577	6592	2	3	5	6	8	9	11	12	14
·82	6607	6622	6637	6653	6668	6683	6699	6714	6730	6745	2	3	5	6	8	9	11	12	14
·83	6761	6776	6792	6808	6823	6839	6855	6871	6887	6902	2	3	5	6	8	9	11	13	14
·84	6918	6934	6950	6966	6982	6998	7015	7031	7047	7063	2	3	5	6	8	10	11	13	14
·85	7079	7096	7112	7129	7145	7161	7178	7194	7211	7228	2	3	5	7	8	10	12	13	15
·86	7244	7261	7278	7295	7311	7328	7345	7362	7379	7396	2	3	5	7	8	10	12	14	15
·87	7413	7430	7447	7464	7482	7499	7516	7534	7551	7568	2	3	5	7	9	10	12	14	16
·88	7586	7603	7621	7638	7656	7674	7691	7709	7727	7745	2	4	5	7	9	11	12	14	16
·89	7762	7780	7798	7816	7834	7852	7870	7889	7907	7925	2	4	5	7	9	11	13	14	16
·90	7943	7962	7980	7998	8017	8035	8054	8072	8091	8110	2	4	6	7	9	11	13	15	17
·91	8128	8147	8166	8185	8204	8222	8241	8260	8279	8299	2	4	6	8	10	11	13	15	17
·92	8318	8337	8356	8375	8395	8414	8433	8453	8472	8492	2	4	6	8	10	12	14	15	17
·93	8511	8531	8551	8570	8590	8610	8630	8650	8670	8690	2	4	6	8	10	12	14	16	18
·94	8710	8730	8750	8770	8790	8810	8831	8851	8872	8892	2	4	6	8	10	12	14	16	18
·95	8913	8933	8954	8974	8995	9016	9036	9057	9078	9099	2	4	6	8	10	12	14	17	19
·96	9120	9141	9162	9183	9204	9226	9247	9268	9290	9311	2	4	6	9	11	13	15	17	19
·97	9333	9354	9376	9397	9419	9441	9462	9484	9506	9528	2	4	7	9	11	13	15	17	20
·98	9550	9572	9594	9616	9638	9661	9683	9705	9727	9750	2	4	7	9	11	13	16	18	20
·99	9772	9795	9817	9840	9863	9886	9908	9931	9954	9977	2	5	7	9	11	14	16	18	21
	0	1	2	3	4	5	6	7	8	9	1	2	3	4	5	6	7	8	9

NATURAL SINES

	0′	6′	12′	18′	24′	30′	36′	42′	48′	54′	1′	2′	3′	4′	5′
											Proportional Parts				
0°	0·0000	·0017	·0035	·0052	·0070	·0087	·0105	·0122	·0140	·0157	3	6	9	12	15
1	0·0175	·0192	·0209	·0227	·0244	·0262	·0279	·0297	·0314	·0332	3	6	9	12	15
2	0·0349	·0366	·0384	·0401	·0419	·0436	·0454	·0471	·0489	·0506	3	6	9	12	15
3	0·0523	·0541	·0558	·0576	·0593	·0610	·0628	·0645	·0663	·0680	3	6	9	12	15
4	0·0698	·0715	·0732	·0750	·0767	·0785	·0802	·0819	·0837	·0854	3	6	9	12	14
5	0·0872	·0889	·0906	·0924	·0941	·0958	·0976	·0993	·1011	·1028	3	6	9	12	14
6	0·1045	·1063	·1080	·1097	·1115	·1132	·1149	·1167	·1184	·1201	3	6	9	12	14
7	0·1219	·1236	·1253	·1271	·1288	·1305	·1323	·1340	·1357	·1374	3	6	9	12	14
8	0·1392	·1409	·1426	·1444	·1461	·1478	·1495	·1513	·1530	·1547	3	6	9	11	14
9	0·1564	·1582	·1599	·1616	·1633	·1650	·1668	·1685	·1702	·1719	3	6	9	11	14
10	0·1736	·1754	·1771	·1788	·1805	·1822	·1840	·1857	·1874	·1891	3	6	9	11	14
11	0·1908	·1925	·1942	·1959	·1977	·1994	·2011	·2028	·2045	·2062	3	6	9	11	14
12	0·2079	·2096	·2113	·2130	·2147	·2164	·2181	·2198	·2215	·2232	3	6	9	11	14
13	0·2250	·2267	·2284	·2300	·2317	·2334	·2351	·2368	·2385	·2402	3	6	8	11	14
14	0·2419	·2436	·2453	·2470	·2487	·2504	·2521	·2538	·2554	·2571	3	6	8	11	14
15	0·2588	·2605	·2622	·2639	·2656	·2672	·2689	·2706	·2723	·2740	3	6	8	11	14
16	0·2756	·2773	·2790	·2807	·2823	·2840	·2857	·2874	·2890	·2907	3	6	8	11	14
17	0·2924	·2940	·2957	·2974	·2990	·3007	·3024	·3040	·3057	·3074	3	6	8	11	14
18	0·3090	·3107	·3123	·3140	·3156	·3173	·3190	·3206	·3223	·3239	3	6	8	11	14
19	0·3256	·3272	·3289	·3305	·3322	·3338	·3355	·3371	·3387	·3404	3	5	8	11	14
20	0·3420	·3437	·3453	·3469	·3486	·3502	·3518	·3535	·3551	·3567	3	5	8	11	14
21	0·3584	·3600	·3616	·3633	·3649	·3665	·3681	·3697	·3714	·3730	3	5	8	11	14
22	0·3746	·3762	·3778	·3795	·3811	·3827	·3843	·3859	·3875	·3891	3	5	8	11	13
23	0·3907	·3923	·3939	·3955	·3971	·3987	·4003	·4019	·4035	·4051	3	5	8	11	13
24	0·4067	·4083	·4099	·4115	·4131	·4147	·4163	·4179	·4195	·4210	3	5	8	11	13
25	0·4226	·4242	·4258	·4274	·4289	·4305	·4321	·4337	·4352	·4368	3	5	8	11	13
26	0·4384	·4399	·4415	·4431	·4446	·4462	·4478	·4493	·4509	·4524	3	5	8	10	13
27	0·4540	·4555	·4571	·4586	·4602	·4617	·4633	·4648	·4664	·4679	3	5	8	10	13
28	0·4695	·4710	·4726	·4741	·4756	·4772	·4787	·4802	·4818	·4833	3	5	8	10	13
29	0·4848	·4863	·4879	·4894	·4909	·4924	·4939	·4955	·4970	·4985	3	5	8	10	13
30	0·5000	·5015	·5030	·5045	·5060	·5075	·5090	·5105	·5120	·5135	2	5	8	10	12
31	0·5150	·5165	·5180	·5195	·5210	·5225	·5240	·5255	·5270	·5284	2	5	7	10	12
32	0·5299	·5314	·5329	·5344	·5358	·5373	·5388	·5402	·5417	·5432	2	5	7	10	12
33	0·5446	·5461	·5476	·5490	·5505	·5519	·5534	·5548	·5563	·5577	2	5	7	10	12
34	0·5592	·5606	·5621	·5635	·5650	·5664	·5678	·5693	·5707	·5721	2	5	7	10	12
35	0·5736	·5750	·5764	·5779	·5793	·5807	·5821	·5835	·5850	·5864	2	5	7	9	12
36	0·5878	·5892	·5906	·5920	·5934	·5948	·5962	·5976	·5990	·6004	2	5	7	9	12
37	0·6018	·6032	·6046	·6060	·6074	·6088	·6101	·6115	·6129	·6143	2	5	7	9	12
38	0·6157	·6170	·6184	·6198	·6211	·6225	·6239	·6252	·6266	·6280	2	5	7	9	11
39	0·6293	·6307	·6320	·6334	·6347	·6361	·6374	·6388	·6401	·6414	2	4	7	9	11
40	0·6428	·6441	·6455	·6468	·6481	·6494	·6508	·6521	·6534	·6547	2	4	7	9	11
41	0·6561	·6574	·6587	·6600	·6613	·6626	·6639	·6652	·6665	·6678	2	4	6	9	11
42	0·6691	·6704	·6717	·6730	·6743	·6756	·6769	·6782	·6794	·6807	2	4	6	9	11
43	0·6820	·6833	·6845	·6858	·6871	·6884	·6896	·6909	·6921	·6934	2	4	6	8	11
44	0·6947	·6959	·6972	·6984	·6997	·7009	·7022	·7034	·7046	·7059	2	4	6	8	10
	0′	6′	12′	18′	24′	30′	36′	42′	48′	54′	1′	2′	3′	4′	5′

NATURAL SINES

Proportional Parts

	0′	6′	12′	18′	24′	30′	36′	42′	48′	54′	1′	2′	3′	4′	5′
45°	0·7071	·7083	·7096	·7108	·7120	·7133	·7145	·7157	·7169	·7181	2	4	6	8	10
46	0·7193	·7206	·7218	·7230	·7242	·7254	·7266	·7278	·7290	·7302	2	4	6	8	10
47	0·7314	·7325	·7337	·7349	·7361	·7373	·7385	·7396	·7408	·7420	2	4	6	8	10
48	0·7431	·7443	·7455	·7466	·7478	·7490	·7501	·7513	·7524	·7536	2	4	6	8	10
49	0·7547	·7559	·7570	·7581	·7593	·7604	·7615	·7627	·7638	·7649	2	4	6	8	9
50	0·7660	·7672	·7683	·7694	·7705	·7716	·7727	·7738	·7749	·7760	2	4	6	7	9
51	0·7771	·7782	·7793	·7804	·7815	·7826	·7837	·7848	·7859	·7869	2	4	5	7	9
52	0·7880	·7891	·7902	·7912	·7923	·7934	·7944	·7955	·7965	·7976	2	4	5	7	9
53	0·7986	·7997	·8007	·8018	·8028	·8039	·8049	·8059	·8070	·8080	2	3	5	7	9
54	0·8090	·8100	·8111	·8121	·8131	·8141	·8151	·8161	·8171	·8181	2	3	5	7	8
55	0·8192	·8202	·8211	·8221	·8231	·8241	·8251	·8261	·8271	·8281	2	3	5	7	8
56	0·8290	·8300	·8310	·8320	·8329	·8339	·8348	·8358	·8368	·8377	2	3	5	6	8
57	0·8387	·8396	·8406	·8415	·8425	·8434	·8443	·8453	·8462	·8471	2	3	5	6	8
58	0·8480	·8490	·8499	·8508	·8517	·8526	·8536	·8545	·8554	·8563	2	3	5	6	8
59	0·8572	·8581	·8590	·8599	·8607	·8616	·8625	·8634	·8643	·8652	1	3	4	6	7
60	0·8660	·8669	·8678	·8686	·8695	·8704	·8712	·8721	·8729	·8738	1	3	4	6	7
61	0·8746	·8755	·8763	·8771	·8780	·8788	·8796	·8805	·8813	·8821	1	3	4	6	7
62	0·8829	·8838	·8846	·8854	·8862	·8870	·8878	·8886	·8894	·8902	1	3	4	5	7
63	0·8910	·8918	·8926	·8934	·8942	·8949	·8957	·8965	·8973	·8980	1	3	4	5	6
64	0·8988	·8996	·9003	·9011	·9018	·9026	·9033	·9041	·9048	·9056	1	2	4	5	6
65	0·9063	·9070	·9078	·9085	·9092	·9100	·9107	·9114	·9121	·9128	1	2	4	5	6
66	0·9135	·9143	·9150	·9157	·9164	·9171	·9178	·9184	·9191	·9198	1	2	3	5	6
67	0·9205	·9212	·9219	·9225	·9232	·9239	·9245	·9252	·9259	·9265	1	2	3	4	6
68	0·9272	·9278	·9285	·9291	·9298	·9304	·9311	·9317	·9323	·9330	1	2	3	4	5
69	0·9336	·9342	·9348	·9354	·9361	·9367	·9373	·9379	·9385	·9391	1	2	3	4	5
70	0·9397	·9403	·9409	·9415	·9421	·9426	·9432	·9438	·9444	·9449	1	2	3	4	5
71	0·9455	·9461	·9466	·9472	·9478	·9483	·9489	·9494	·9500	·9505	1	2	3	4	5
72	0·9511	·9516	·9521	·9527	·9532	·9537	·9542	·9548	·9553	·9558	1	2	3	3	4
73	0·9563	·9568	·9573	·9578	·9583	·9588	·9593	·9598	·9603	·9608	1	2	2	3	4
74	0·9613	·9617	·9622	·9627	·9632	·9636	·9641	·9646	·9650	·9655	1	2	2	3	4
75	0·9659	·9664	·9668	·9673	·9677	·9681	·9686	·9690	·9694	·9699	1	1	2	3	4
76	0·9703	·9707	·9711	·9715	·9720	·9724	·9728	·9732	·9736	·9740	1	1	2	3	3
77	0·9744	·9748	·9751	·9755	·9759	·9763	·9767	·9770	·9774	·9778	1	1	2	2	3
78	0·9781	·9785	·9789	·9792	·9796	·9799	·9803	·9806	·9810	·9813	1	1	2	2	3
79	0·9816	·9820	·9823	·9826	·9829	·9833	·9836	·9839	·9842	·9845	1	1	2	2	3
80	0·9848	·9851	·9854	·9857	·9860	·9863	·9866	·9869	·9871	·9874	0	1	1	2	2
81	0·9877	·9880	·9882	·9885	·9888	·9890	·9893	·9895	·9898	·9900	0	1	1	2	2
82	0·9903	·9905	·9907	·9910	·9912	·9914	·9917	·9919	·9921	·9923	0	1	1	1	2
83	0·9925	·9928	·9930	·9932	·9934	·9936	·9938	·9940	·9942	·9943	0	1	1	1	2
84	0·9945	·9947	·9949	·9951	·9952	·9954	·9956	·9957	·9959	·9960	0	1	1	1	1
85	0·9962	·9963	·9965	·9966	·9968	·9969	·9971	·9972	·9973	·9974	0	0	1	1	1
86	0·9976	·9977	·9978	·9979	·9980	·9981	·9982	·9983	·9984	·9985	0	0	0	1	1
87	0·9986	·9987	·9988	·9989	·9990	·9990	·9991	·9992	·9993	·9993	0	0	0	1	1
88	0·9994	·9995	·9995	·9996	·9996	·9997	·9997	·9997	·9998	·9998	0	0	0	0	0
89	0·9998	·9999	·9999	·9999	0·9999	1·0000	·0000	·0000	·0000	·0000	0	0	0	0	0
	0′	6′	12′	18′	24′	30′	36′	42′	48′	54′	1′	2′	3′	4′	5′

NATURAL COSINES

Proportional Parts **Subtract**

	0′	6′	12′	18′	24′	30′	36′	42′	48′	54′	1′	2′	3′	4′	5′
0°	1·0000	·0000	·0000	·0000	·0000	1·0000	0·9999	·9999	·9999	·9999	0	0	0	0	0
1	0·9998	·9998	·9998	·9997	·9997	·9997	·9996	·9996	·9995	·9995	0	0	0	0	0
2	0·9994	·9993	·9993	·9992	·9991	·9990	·9990	·9989	·9988	·9987	0	0	0	0	1
3	0·9986	·9985	·9984	·9983	·9982	·9981	·9980	·9979	·9978	·9977	0	0	0	1	1
4	0·9976	·9974	·9973	·9972	·9971	·9969	·9968	·9966	·9965	·9963	0	0	1	1	1
5	0·9962	·9960	·9959	·9957	·9956	·9954	·9952	·9951	·9949	·9947	0	1	1	1	1
6	0·9945	·9943	·9942	·9940	·9938	·9936	·9934	·9932	·9930	·9928	0	1	1	1	2
7	0·9925	·9923	·9921	·9919	·9917	·9914	·9912	·9910	·9907	·9905	0	1	1	1	2
8	0·9903	·9900	·9898	·9895	·9893	·9890	·9888	·9885	·9882	·9880	0	1	1	2	2
9	0·9877	·9874	·9871	·9869	·9866	·9863	·9860	·9857	·9854	·9851	0	1	1	2	2
10	0·9848	·9845	·9842	·9839	·9836	·9833	·9829	·9826	·9823	·9820	1	1	2	2	3
11	0·9816	·9813	·9810	·9806	·9803	·9799	·9796	·9792	·9789	·9785	1	1	2	2	3
12	0·9781	·9778	·9774	·9770	·9767	·9763	·9759	·9755	·9751	·9748	1	1	2	2	3
13	0·9744	·9740	·9736	·9732	·9728	·9724	·9720	·9715	·9711	·9707	1	1	2	3	3
14	0·9703	·9699	·9694	·9690	·9686	·9681	·9677	·9673	·9668	·9664	1	1	2	3	4
15	0·9659	·9655	·9650	·9646	·9641	·9636	·9632	·9627	·9622	·9617	1	2	2	3	4
16	0·9613	·9608	·9603	·9598	·9593	·9588	·9583	·9578	·9573	·9568	1	2	2	3	4
17	0·9563	·9558	·9553	·9548	·9542	·9537	·9532	·9527	·9521	·9516	1	2	3	3	4
18	0·9511	·9505	·9500	·9494	·9489	·9483	·9478	·9472	·9466	·9461	1	2	3	4	5
19	0·9455	·9449	·9444	·9438	·9432	·9426	·9421	·9415	·9409	·9403	1	2	3	4	5
20	0·9397	·9391	·9385	·9379	·9373	·9367	·9361	·9354	·9348	·9342	1	2	3	4	5
21	0·9336	·9330	·9323	·9317	·9311	·9304	·9298	·9291	·9285	·9278	1	2	3	4	5
22	0·9272	·9265	·9259	·9252	·9245	·9239	·9232	·9225	·9219	·9212	1	2	3	4	6
23	0·9205	·9198	·9191	·9184	·9178	·9171	·9164	·9157	·9150	·9143	1	2	3	5	6
24	0·9135	·9128	·9121	·9114	·9107	·9100	·9092	·9085	·9078	·9070	1	2	4	5	6
25	0·9063	·9056	·9048	·9041	·9033	·9026	·9018	·9011	·9003	·8996	1	2	4	5	6
26	0·8988	·8980	·8973	·8965	·8957	·8949	·8942	·8934	·8926	·8918	1	3	4	5	6
27	0·8910	·8902	·8894	·8886	·8878	·8870	·8862	·8854	·8846	·8838	1	3	4	5	7
28	0·8829	·8821	·8813	·8805	·8796	·8788	·8780	·8771	·8763	·8755	1	3	4	6	7
29	0·8746	·8738	·8729	·8721	·8712	·8704	·8695	·8686	·8678	·8669	1	3	4	6	7
30	0·8660	·8652	·8643	·8634	·8625	·8616	·8607	·8599	·8590	·8581	1	3	4	6	7
31	0·8572	·8563	·8554	·8545	·8536	·8526	·8517	·8508	·8499	·8490	2	3	5	6	8
32	0·8480	·8471	·8462	·8453	·8443	·8434	·8425	·8415	·8406	·8396	2	3	5	6	8
33	0·8387	·8377	·8368	·8358	·8348	·8339	·8329	·8320	·8310	·8300	2	3	5	6	8
34	0·8290	·8281	·8271	·8261	·8251	·8241	·8231	·8221	·8211	·8202	2	3	5	7	8
35	0·8192	·8181	·8171	·8161	·8151	·8141	·8131	·8121	·8111	·8100	2	3	5	7	8
36	0·8090	·8080	·8070	·8059	·8049	·8039	·8028	·8018	·8007	·7997	2	3	5	7	9
37	0·7986	·7976	·7965	·7955	·7944	·7934	·7923	·7912	·7902	·7891	2	4	5	7	9
38	0·7880	·7869	·7859	·7848	·7837	·7826	·7815	·7804	·7793	·7782	2	4	5	7	9
39	0·7771	·7760	·7749	·7738	·7727	·7716	·7705	·7694	·7683	·7672	2	4	6	7	9
40	0·7660	·7649	·7638	·7627	·7615	·7604	·7593	·7581	·7570	·7559	2	4	6	8	9
41	0·7547	·7536	·7524	·7513	·7501	·7490	·7478	·7466	·7455	·7443	2	4	6	8	10
42	0·7431	·7420	·7408	·7396	·7385	·7373	·7361	·7349	·7337	·7325	2	4	6	8	10
43	0·7314	·7302	·7290	·7278	·7266	·7254	·7242	·7230	·7218	·7206	2	4	6	8	10
44	0·7193	·7181	·7169	·7157	·7145	·7133	·7120	·7108	·7096	·7083	2	4	6	8	10
	0′	**6′**	**12′**	**18′**	**24′**	**30′**	**36′**	**42′**	**48′**	**54′**	**1′**	**2′**	**3′**	**4′**	**5′**

NATURAL COSINES

Proportional Parts
Subtract

	0′	6′	12′	18′	24′	30′	36′	42′	48′	54′	1′	2′	3′	4′	5′
45°	0·7071	·7059	·7046	·7034	·7022	·7009	·6997	·6984	·6972	·6959	2	4	6	8	10
46	0·6947	·6934	·6921	·6909	·6896	·6884	·6871	·6858	·6845	·6833	2	4	6	8	11
47	0·6820	·6807	·6794	·6782	·6769	·6756	·6743	·6730	·6717	·6704	2	4	6	9	11
48	0·6691	·6678	·6665	·6652	·6639	·6626	·6613	·6600	·6587	·6574	2	4	6	9	11
49	0·6561	·6547	·6534	·6521	·6508	·6494	·6481	·6468	·6455	·6441	2	4	7	9	11
50	0·6428	·6414	·6401	·6388	·6374	·6361	·6347	·6334	·6320	·6307	2	4	7	9	11
51	0·6293	·6280	·6266	·6252	·6239	·6225	·6211	·6198	·6184	·6170	2	5	7	9	11
52	0·6157	·6143	·6129	·6115	·6101	·6088	·6074	·6060	·6046	·6032	2	5	7	9	12
53	0·6018	·6004	·5990	·5976	·5962	·5948	·5934	·5920	·5906	·5892	2	5	7	9	12
54	0·5878	·5864	·5850	·5835	·5821	·5807	·5793	·5779	·5764	·5750	2	5	7	9	12
55	0·5736	·5721	·5707	·5693	·5678	·5664	·5650	·5635	·5621	·5606	2	5	7	10	12
56	0·5592	·5577	·5563	·5548	·5534	·5519	·5505	·5490	·5476	·5461	2	5	7	10	12
57	0·5446	·5432	·5417	·5402	·5388	·5373	·5358	·5344	·5329	·5314	2	5	7	10	12
58	0·5299	·5284	·5270	·5255	·5240	·5225	·5210	·5195	·5180	·5165	2	5	7	10	12
59	0·5150	·5135	·5120	·5105	·5090	·5075	·5060	·5045	·5030	·5015	2	5	8	10	12
60	0·5000	·4985	·4970	·4955	·4939	·4924	·4909	·4894	·4879	·4863	3	5	8	10	13
61	0·4848	·4833	·4818	·4802	·4787	·4772	·4756	·4741	·4726	·4710	3	5	8	10	13
62	0·4695	·4679	·4664	·4648	·4633	·4617	·4602	·4586	·4571	·4555	3	5	8	10	13
63	0·4540	·4524	·4509	·4493	·4478	·4462	·4446	·4431	·4415	·4399	3	5	8	10	13
64	0·4384	·4368	·4352	·4337	·4321	·4305	·4289	·4274	·4258	·4242	3	5	8	11	13
65	0·4226	·4210	·4195	·4179	·4163	·4147	·4131	·4115	·4099	·4083	3	5	8	11	13
66	0·4067	·4051	·4035	·4019	·4003	·3987	·3971	·3955	·3939	·3923	3	5	8	11	13
67	0·3907	·3891	·3875	·3859	·3843	·3827	·3811	·3795	·3778	·3762	3	5	8	11	13
68	0·3746	·3730	·3714	·3697	·3681	·3665	·3649	·3633	·3616	·3600	3	5	8	11	14
69	0·3584	·3567	·3551	·3535	·3518	·3502	·3486	·3469	·3453	·3437	3	5	8	11	14
70	0·3420	·3404	·3387	·3371	·3355	·3338	·3322	·3305	·3289	·3272	3	5	8	11	14
71	0·3256	·3239	·3223	·3206	·3190	·3173	·3156	·3140	·3123	·3107	3	6	8	11	14
72	0·3090	·3074	·3057	·3040	·3024	·3007	·2990	·2974	·2957	·2940	3	6	8	11	14
73	0·2924	·2907	·2890	·2874	·2857	·2840	·2823	·2807	·2790	·2773	3	6	8	11	14
74	0·2756	·2740	·2723	·2706	·2689	·2672	·2656	·2639	·2622	·2605	3	6	8	11	14
75	0·2588	·2571	·2554	·2538	·2521	·2504	·2487	·2470	·2453	·2436	3	6	8	11	14
76	0·2419	·2402	·2385	·2368	·2351	·2334	·2317	·2300	·2284	·2267	3	6	8	11	14
77	0·2250	·2232	·2215	·2198	·2181	·2164	·2147	·2130	·2113	·2096	3	6	9	11	14
78	0·2079	·2062	·2045	·2028	·2011	·1994	·1977	·1959	·1942	·1925	3	6	9	11	14
79	0·1908	·1891	·1874	·1857	·1840	·1822	·1805	·1788	·1771	·1754	3	6	9	11	14
80	0·1736	·1719	·1702	·1685	·1668	·1650	·1633	·1616	·1599	·1582	3	6	9	11	14
81	0·1564	·1547	·1530	·1513	·1495	·1478	·1461	·1444	·1426	·1409	3	6	9	11	14
82	0·1392	·1374	·1357	·1340	·1323	·1305	·1288	·1271	·1253	·1236	3	6	9	12	14
83	0·1219	·1201	·1184	·1167	·1149	·1132	·1115	·1097	·1080	·1063	3	6	9	12	14
84	0·1045	·1028	·1011	·0993	·0976	·0958	·0941	·0924	·0906	·0889	3	6	9	12	14
85	0·0872	·0854	·0837	·0819	·0802	·0785	·0767	·0750	·0732	·0715	3	6	9	12	14
86	0·0698	·0680	·0663	·0645	·0628	·0610	·0593	·0576	·0558	·0541	3	6	9	12	15
87	0·0523	·0506	·0489	·0471	·0454	·0436	·0419	·0401	·0384	·0366	3	6	9	12	15
88	0·0349	·0332	·0314	·0297	·0279	·0262	·0244	·0227	·0209	·0192	3	6	9	12	15
89	0·0175	·0157	·0140	·0122	·0105	·0087	·0070	·0052	·0035	·0017	3	6	9	12	15
	0′	**6′**	**12′**	**18′**	**24′**	**30′**	**36′**	**42′**	**48′**	**54′**	**1′**	**2′**	**3′**	**4′**	**5′**

NATURAL TANGENTS

	0′	6′	12′	18′	24′	30′	36′	42′	48′	54′	Proportional Parts 1′	2′	3′	4′	5′
0°	0·0000	·0017	·0035	·0052	·0070	·0087	·0105	·0122	·0140	·0157	3	6	9	12	15
1	0·0175	·0192	·0209	·0227	·0244	·0262	·0279	·0297	·0314	·0332	3	6	9	12	15
2	0·0349	·0367	·0384	·0402	·0419	·0437	·0454	·0472	·0489	·0507	3	6	9	12	15
3	0·0524	·0542	·0559	·0577	·0594	·0612	·0629	·0647	·0664	·0682	3	6	9	12	15
4	0·0699	·0717	·0734	·0752	·0769	·0787	·0805	·0822	·0840	·0857	3	6	9	12	15
5	0·0875	·0892	·0910	·0928	·0945	·0963	·0981	·0998	·1016	·1033	3	6	9	12	15
6	0·1051	·1069	·1086	·1104	·1122	·1139	·1157	·1175	·1192	·1210	3	6	9	12	15
7	0·1228	·1246	·1263	·1281	·1299	·1317	·1334	·1352	·1370	·1388	3	6	9	12	15
8	0·1405	·1423	·1441	·1459	·1477	·1495	·1512	·1530	·1548	·1566	3	6	9	12	15
9	0·1584	·1602	·1620	·1638	·1655	·1673	·1691	·1709	·1727	·1745	3	6	9	12	15
10	0·1763	·1781	·1799	·1817	·1835	·1853	·1871	·1890	·1908	·1926	3	6	9	12	15
11	0·1944	·1962	·1980	·1998	·2016	·2035	·2053	·2071	·2089	·2107	3	6	9	12	15
12	0·2126	·2144	·2162	·2180	·2199	·2217	·2235	·2254	·2272	·2290	3	6	9	12	15
13	0·2309	·2327	·2345	·2364	·2382	·2401	·2419	·2438	·2456	·2475	3	6	9	12	15
14	0·2493	·2512	·2530	·2549	·2568	·2586	·2605	·2623	·2642	·2661	3	6	9	12	16
15	0·2679	·2698	·2717	·2736	·2754	·2773	·2792	·2811	·2830	·2849	3	6	9	13	16
16	0·2867	·2886	·2905	·2924	·2943	·2962	·2981	·3000	·3019	·3038	3	6	9	13	16
17	0·3057	·3076	·3096	·3115	·3134	·3153	·3172	·3191	·3211	·3230	3	6	9	13	16
18	0·3249	·3269	·3288	·3307	·3327	·3346	·3365	·3385	·3404	·3424	3	6	10	13	16
19	0·3443	·3463	·3482	·3502	·3522	·3541	·3561	·3581	·3600	·3620	3	6	10	13	16
20	0·3640	·3659	·3679	·3699	·3719	·3739	·3759	·3779	·3799	·3819	3	6	10	13	17
21	0·3839	·3859	·3879	·3899	·3919	·3939	·3959	·3979	·4000	·4020	3	7	10	13	17
22	0·4040	·4061	·4081	·4101	·4122	·4142	·4163	·4183	·4204	·4224	3	7	10	14	17
23	0·4245	·4265	·4286	·4307	·4327	·4348	·4369	·4390	·4411	·4431	3	7	10	14	17
24	0·4452	·4473	·4494	·4515	·4536	·4557	·4578	·4599	·4621	·4642	4	7	11	14	18
25	0·4663	·4684	·4706	·4727	·4748	·4770	·4791	·4813	·4834	·4856	4	7	11	14	18
26	0·4877	·4899	·4921	·4942	·4964	·4986	·5008	·5029	·5051	·5073	4	7	11	15	18
27	0·5095	·5117	·5139	·5161	·5184	·5206	·5228	·5250	·5272	·5295	4	7	11	15	18
28	0·5317	·5339	·5362	·5384	·5407	·5430	·5452	·5475	·5498	·5520	4	8	11	15	19
29	0·5543	·5566	·5589	·5612	·5635	·5658	·5681	·5704	·5727	·5750	4	8	12	15	19
30	0·5774	·5797	·5820	·5844	·5867	·5891	·5914	·5938	·5961	·5985	4	8	12	16	20
31	0·6009	·6032	·6056	·6080	·6104	·6128	·6152	·6176	·6200	·6224	4	8	12	16	20
32	0·6249	·6273	·6297	·6322	·6346	·6371	·6395	·6420	·6445	·6469	4	8	12	16	20
33	0·6494	·6519	·6544	·6569	·6594	·6619	·6644	·6669	·6694	·6720	4	8	13	17	21
34	0·6745	·6771	·6796	·6822	·6847	·6873	·6899	·6924	·6950	·6976	4	9	13	17	21
35	0·7002	·7028	·7054	·7080	·7107	·7133	·7159	·7186	·7212	·7239	4	9	13	18	22
36	0·7265	·7292	·7319	·7346	·7373	·7400	·7427	·7454	·7481	·7508	5	9	14	18	23
37	0·7536	·7563	·7590	·7618	·7646	·7673	·7701	·7729	·7757	·7785	5	9	14	18	23
38	0·7813	·7841	·7869	·7898	·7926	·7954	·7983	·8012	·8040	·8069	5	10	14	19	24
39	0·8098	·8127	·8156	·8185	·8214	·8243	·8273	·8302	·8332	·8361	5	10	15	20	24
40	0·8391	·8421	·8451	·8481	·8511	·8541	·8571	·8601	·8632	·8662	5	10	15	20	25
41	0·8693	·8724	·8754	·8785	·8816	·8847	·8878	·8910	·8941	·8972	5	10	16	21	26
42	0·9004	·9036	·9067	·9099	·9131	·9163	·9195	·9228	·9260	·9293	5	11	16	21	26
43	0·9325	·9358	·9391	·9424	·9457	·9490	·9523	·9556	·9590	·9623	6	11	17	22	28
44	0·9657	·9691	·9725	·9759	·9793	·9827	·9861	·9896	·9930	·9965	6	11	17	23	29
	0′	6′	12′	18′	24′	30′	36′	42′	48′	54′	1′	2′	3′	4′	5′

NATURAL TANGENTS

	0′	6′	12′	18′	24′	30′	36′	42′	48′	54′	Proportional Parts 1′	2′	3′	4′	5′
45°	1·0000	·0035	·0070	·0105	·0141	·0176	·0212	·0247	·0283	·0319	6	12	18	24	30
46	1·0355	·0392	·0428	·0464	·0501	·0538	·0575	·0612	·0649	·0686	6	12	18	25	31
47	1·0724	·0761	·0799	·0837	·0875	·0913	·0951	·0990	·1028	·1067	6	13	19	25	32
48	1·1106	·1145	·1184	·1224	·1263	·1303	·1343	·1383	·1423	·1463	7	13	20	27	33
49	1·1504	·1544	·1585	·1626	·1667	·1708	·1750	·1792	·1833	·1875	7	14	21	28	34
50	1·1918	·1960	·2002	·2045	·2088	·2131	·2174	·2218	·2261	·2305	7	14	22	29	36
51	1·2349	·2393	·2437	·2482	·2527	·2572	·2617	·2662	·2708	·2753	8	15	23	30	38
52	1·2799	·2846	·2892	·2938	·2985	·3032	·3079	·3127	·3175	·3222	8	16	24	31	39
53	1·3270	·3319	·3367	·3416	·3465	·3514	·3564	·3613	·3663	·3713	8	16	25	33	41
54	1·3764	·3814	·3865	·3916	·3968	·4019	·4071	·4124	·4176	·4229	9	17	26	34	43
55	1·4281	·4335	·4388	·4442	·4496	·4550	·4605	·4659	·4715	·4770	9	18	27	36	45
56	1·4826	·4882	·4938	·4994	·5051	·5108	·5166	·5224	·5282	·5340	10	19	29	38	48
57	1·5399	·5458	·5517	·5577	·5637	·5697	·5757	·5818	·5880	·5941	10	20	30	40	50
58	1·6003	·6066	·6128	·6191	·6255	·6319	·6383	·6447	·6512	·6577	11	21	32	43	53
59	1·6643	·6709	·6775	·6842	·6909	·6977	·7045	·7113	·7182	·7251	11	23	34	45	57
60	1·7321	·7391	·7461	·7532	·7603	·7675	·7747	·7820	·7893	·7966	12	24	36	48	60
61	1·8040	·8115	·8190	·8265	·8341	·8418	·8495	·8572	·8650	·8728	13	26	38	51	64
62	1·8807	·8887	·8967	·9047	·9128	·9210	·9292	·9375	·9458	·9542	14	27	41	55	68
63	1·9626	·9711	·9797	·9883	1·9970	2·0057	·0145	·0233	·0323	·0413	15	29	44	58	73
64	2·0503	·0594	·0686	·0778	·0872	·0965	·1060	·1155	·1251	·1348	16	31	47	63	78
65	2·145	·154	·164	·174	·184	·194	·204	·215	·225	·236	2	3	5	7	8
66	2·246	·257	·267	·278	·289	·300	·311	·322	·333	·344	2	4	5	7	9
67	2·356	·367	·379	·391	·402	·414	·426	·438	·450	·463	2	4	6	8	10
68	2·475	·488	·500	·513	·526	·539	·552	·565	·578	·592	2	4	6	9	11
69	2·605	·619	·633	·646	·660	·675	·689	·703	·718	·733	2	5	7	9	12
70	2·747	·762	·778	·793	·808	·824	·840	·856	·872	·888	3	5	8	10	13
71	2·904	·921	·937	·954	·971	2·989	3·006	·024	·042	·060	3	6	9	12	14
72	3·078	·096	·115	·133	·152	·172	·191	·211	·230	·251	3	6	10	13	16
73	3·271	·291	·312	·333	·354	·376	·398	·420	·442	·465	4	7	11	14	18
74	3·487	·511	·534	·558	·582	·606	·630	·655	·681	·706	4	8	12	16	20
75	3·732	·758	·785	·812	·839	·867	·895	·923	·952	·981	5	9	14	19	23
76	4·011	·041	·071	·102	·134	·165	·198	·230	·264	·297	5	11	16	21	27
77	4·331	·366	·402	·437	·474	·511	·548	·586	·625	·665	6	12	19	25	31
78	4·705	·745	·787	·829	·872	·915	4·959	5·005	·050	·097	7	15	22	29	37
79	5·145	·193	·242	·292	·343	·396	·449	·503	·558	·614	9	18	26	35	44
80	5·671	·730	·789	·850	·912	5·976	6·041	·107	·174	·243	11	21	32	43	54
81	6·314	·386	·460	·535	·612	·691	·772	·855	6·940	7·026	13	27	40	54	67
82	7·115	·207	·300	·396	·495	·596	·700	·806	7·916	8·028	17	34	51	69	86
83	8·144	·264	·386	·513	·643	·777	8·915	9·058	·205	·357	23	46	68	91	114
84	9·514	9·677	9·845	10·019	10·199	10·385	10·579	10·780	10·988	11·205					
85	11·43	11·66	11·91	12·16	12·43	12·71	13·00	13·30	13·62	13·95	p.p. cease				
86	14·30	14·67	15·06	15·46	15·89	16·35	16·83	17·34	17·89	18·46	to be				
87	19·08	19·74	20·45	21·20	22·02	22·90	23·86	24·90	26·03	27·27	sufficiently				
88	28·64	30·14	31·82	33·69	35·80	38·19	40·92	44·07	47·74	52·08	accurate				
89	57·29	63·66	71·62	81·85	95·49	114·6	143·2	191·0	286·5	573·0					
	0′	6′	12′	18′	24′	30′	36′	42′	48′	54′	1′	2′	3′	4′	5′

INDEX

The numbers refer to pages

Alternating voltages, 267, 288
Angles, general, 179
  of polygons, 186
  of triangles, 184
Areas, circle, 8
  formulæ (various), 102
  irregular, 230
  rectangle and triangle, 6, 274*a*
  trapezium, 229
Average value, 232, 234

Beam (tendency to bend), 155
Bisector of an angle, 200
  of a line, 204
Block diagrams, 24

Characteristic (of a logarithm), 55
Charge on a capacitor, 102
Chords (table of), 252
Circles, angle properties, 214–217
  cyclic quadrilateral, 215
  symmetry, 211
Circular measure, 269
Completing the square, 92, 159
Compliance, 306
Components of a vector, 286
Compression, 299
Cone, 73
Continuous and discontinuous graphs, 29
Cosine ratio, 256
Cosine Rule, 274*b*
Cursor, use of, 63
Curved surfaces, 71, 73, 102
Cylinder, 12, 71

Decimal system of units, 15
Difference of two squares, 86
Dovetail joints, 245

Equations arising from formulæ, 114
  historical note, 112
  involving brackets, fractions, 119
  quadratic, algebraic solution, 159
    by factors, 161
    graphical solution, 147

Equations, rules for solving, 117
  simultaneous, 125–132
Experiments, 73, 77, 81, 283

Factor of safety, 309
Factors, arithmetical, 1
  common, 84
  difference of squares, 85
  perfect square, 87
Fatigue, 309
Forces, effect on materials, 297
  in equilibrium, 282, 293
Formulæ, changing the subject, 94
  mensuration, 7–12, 71
  sources, 70
  summary of useful, 102
  warning note, 101
Fractional powers, 46
Frequency (resonant), 17, 66, 103

Galileo, 77
Graphs, aids to design, 36
  continuous and discontinuous, 29
  families, 143, 150, 152
  inverse proportion, 156
  quadratic, 149
  scales, 41
  straight-line, 83, 135–144
  trigonometrical, 265
  to solve equations, 140, 147
Gunther scale, 58

H.C.F., 2
Helical spring, 81, 302
Hertzian waves, 103
Histogram, 26
Hooke's law, 83, 102, 302
Hyperbola (rectangular), 157

Index laws, 13
Interpolation, 31

Kirchhoff's laws, 124

L.C.M., 2
Levers, 288, 291
Loci, 177, 191, 201, 203, 209

Logarithms, any number, 54
graphically, 47–50
tabular methods, 96
use of log tables, 51–57

Mantissa, 55
Materials (strength data), 308
Mid-ordinate rule, 239
Moments, 290, 294

Negative indices, 14
numbers (multiplying), 121

Ohm's law, 17, 73, 102

Parabola, 150
Parallels, construction, 183
properties, 181
Pendulum, 77, 102
Percentage tolerance, 18
Perfect square, 87
Perpendicular lines, 180
Polygons, angles of, 186
Power (electrical), 102
Powers of ten, 1, 45
Prime factors, 1
Pythagoras' theorem, 173, 175, 213, 224
trigonometrical form, 259

Quadrant rules (trigonometry), 263

Radian, 269
Resistance in series, 84
Resonant frequency, 17, 66, 103
Resultant, 281
Reversals (of stress), 309

Scalar quantities, 275
Sector of a circle, 72, 269
Segment of a circle, 218, 270
Shear stress, 300
Significant figures, 18, 67
Similar figures, 221–228
Sine-bar, 243
ratio, 240
Sine Rule, 274*b*
Slide-rule, 58–67
Slope (gradient), 245, 248
Springs, flat and helical, 303
Square root of indices, 14
Standard form of a number, 45
Statistical graphs, 23, 26
Steelyard, 295
Stiffness, 304, 306
Straight-line variation, 83, 135–140
Strain, 298
Stress, 297, 299

Table (4-figure), the use of,
cosine, 256
logarithm, 50
sine, 242
tangent, 250
Tangents to a circle, 216–221, 223
Tension, 299
Test to destruction, 307
Trapezoidal rule, 231
Triangles, angle properties, 185
areas, 274*a*
congruence, 199
equilateral, 205
of forces, 282
isosceles, 204
right-angled, 206
specifications, 192
stability of structures, 192
Trigonometrical ratios, angles greater than 90°, 260
cosine, 256
relations between, 259
sine, 240
tangent, 245

Ultimate strength, 307, 308

Variations of stress, 309
Vector quantities in one dimension, 276
in two dimensions, 277
in three dimensions, 278
parallelogram of vectors, 280
Volumes, 10, 102

Wavelength, 103

Yield point, 307
Young's modulus, 306, 308